Photomicrography

WILEY SERIES ON PHOTOGRAPHIC SCIENCE AND TECHNOLOGY AND THE GRAPHIC ARTS

WALTER CLARK, EDITOR
Kodak Research Laboratories

Photomicrography

A COMPREHENSIVE TREATISE

VOLUME I

Roger P. Loveland

Former Research Associate
Kodak Research Laboratories

John Wiley & Sons, Inc.

NEW YORK | LONDON | SYDNEY | TORONTO

10 9 8 7 6 5 4 3 2 1

Library of Congress Catalog Card Number: 70-88315

SBN 471 54830 8

Printed in the United States of America

Preface

This book is intended to be a text as well as a reference book. The problem is how to make the book useful to the neophyte in the field, yet a convenience for the experienced photomicrographer who wishes to refer to certain data or to look up a matter in a field outside his own routine (and we all do). It is also hoped that the device of printing essential steps of technique and fundamental statements in boldface type and the explanations in lightface will provide easy reference to them. Tabulation of the data, with a list of tables, has been utilized for this purpose, and it is hoped some of these tables will be especially helpful in the planning and preparation of photomicrographs.

Specific subject matter must necessarily be included repeatedly in considering the various fields of photomicrography. This situation could require many cross references. It is more efficient to a reader's time to repeat material, perhaps with a different emphasis whenever a small amount of text is involved. Hence the repetition in this book is appreciable.

Ideally, a subtitle should lead directly to all that is written on a subject in the book except for cross references to the specific pages on which it also is discussed. Citing specific page references in the text for each such case is impractical. The same inclusiveness, with ease of reference, has been attempted by carefully inserting the reference into the subject index with the pertinent key words. When these are not obvious, a parenthetical word or phrase may be added. The inclusion of the chapter of the reference is often helpful; it may be sufficient. When such devices are inadequate, the specific page reference is given.

The bibliography is a selected one; it is not intended to be complete. The entries are arranged in categories to aid in locating as well as to make possible the inclusion of useful ones to which no specific reference has been made.

Since I have been a member of the Kodak Research Laboratories, it is natural that the photographic illustrations here deal with Kodak materials. The tables that list these materials and their characteristics are very complete because the data were made available through the cooperation of the Eastman Kodak Company. Up-to-date information and literature were obtained from representatives of other manufacturers of photographic materials, although no attempt was made to include those materials not produced in the United States, as the only practical limitation. Equivalent products of other manufacturers can usually be used.

I acknowledge gratefully the cooperation of all the manufacturers and sellers of microscopes. These companies were most helpful, but it is unnecessary to list them here; their contributions are obvious inside this book. Thanks are specifically due and gladly given for the photographs that were furnished to apparatus on the market and schematics of optical systems. Some illustrations were made from photographs of apparatus in our laboratory purchased from these companies.

Many individuals have been helpful, especially within the structure of the Eastman Kodak Company, Bausch and Lomb, Inc., and the American Optical Company. I am happy to be able to give my special thanks to some whose aid has been essential in some phases of the preparation of this book. This includes Dr. Walter Clark, editor of this series. Mrs. Joanne Weber, as an associate during the preparation of more than half of this book, gave competent assistance as an editor, especially of the illustrations, so that her influence affected all parts of the book. Another associate, Mr. B. M. Spinell, had much to do with the production of the data, selected for inclusion, a great part of which has already appeared in mutually authored papers, and made many of the photomicrographs that appear here. He also provided invaluable help in reviewing the manuscript as did Mr. Lynn Wall for whose cooperation and suggestions I am very grateful. I thank also Dr. Rudolph Kingslake, of Eastman Kodak Company and the University of Rochester, who read some of the manuscript and offered valuable suggestions on optics. I am also happy to thank Mr. James Benford, of Bausch and Lomb, Inc., who furnished the material for Figures 1–19 and 1–21, for his advice.

Finally, I thank my wife for her patience and help, especially in the proofreading.

Roger P. Loveland

Rochester, New York

Contents

Tables

x   *Tables*

Photomicrography

Chapter 1

General Survey: Up the Magnification Scale

Many of the textbooks that deal with microscopy and photomicrography devote the majority of their pages to discussions of the specimens under study. This is properly so. The preparation of the specimen for microscopical examination is not only a fundamental part of microscopy, often requiring extreme skill, but is usually the most time-consuming part of the project; moreover, the interpretation of the microscopical field or a photograph of it may call for considerable knowledge and experience. All of this just adds to the fascination that is inherent in the use of the microscope. This book does not, however, discuss the general subject of specimen preparation.

There are many specimens, fortunately, that can be examined under a microscope, especially at relatively low magnifications, without special preparation. It is wise for the beginner to confine himself to such speciments or to use professionally prepared slides.

Most photography, including photomicrography, consists in setting up a lens so that the image that is formed behind it includes the objects of interest. This image is real; it occupies space and can be examined and measured much as the original object. Then photographically sensitive material is placed in this same position, specifically in the plane of best representation of the object, which allows a permanent record to be made of the image. This book discusses the subject from this twofold aspect: setting up the image and reproducing it. Those readers who own an integrated type of photomicrographic apparatus that has been permanently aligned and adjusted in the factory from light source to film plane may wish to skip much of the first section. The book has been arranged to allow selection of those portions pertinent to one's own interest and equipment.

However, even a person who intends only to use rather than to design optical elements may find it practical to acquire some pertinent optical principles, and the learning may be interesting; for instance, does it make a difference if a sheet of glass (including a microscope slide or cover slip) is placed in an optical beam; if so, when, how much, and why?

GENERAL OPTICAL PRINCIPLES

Nature of Light

Photomicrography is one of the fields of application of **optics**, which is "a science that deals with light" (*Webster's Third New International Dictionary*). Therefore an understanding of the nature of light is fundamental. General knowledge of the theory of light is widespread, much more so than the knowledge of some of its implications that are of importance to us.

Energy can be transmitted (i.e., transferred) by **radiation** that has **transverse waves**. The characteristics of radiant energy can change from those of cosmic rays, gamma rays, x-rays, light, radiant heat, and radio waves, yet these forms differ only by their wavelengths and their correlated properties. The total range, or even a part of this range, is called a **spectrum**. The **spectrum of light** is that part of the total that is perceived by the human eye. A photographic film or plate can be sensitive to a much broader spectrum than the eye, a fact of much importance to us. A composite of the wavelengths of the entire light spectrum is perceived as **white light**.

Such radiation can be spread out in the order of the component wavelengths by inserting into a beam a glass prism or a fine grating with grooves whose separation is of the order of the wavelengths. This produces the familiar rainbow arrangement of colors in which the invisible **ultraviolet** lies beyond the violet end and the **infrared**, beyond the red. The wavelengths of visible light are usually expressed in small units of the metric system, either in **millimicrons** or angstroms. Millimicrons (also called **nanometers**) are used in this book and abbreviated as mμ. The word **wavelength** is represented by the symbol λ. Association of wavelengths, as numbers, and perceived colors is extremely worthwhile for the photomicrographer. Even for photomicrography in black-and-white, selection of the proper color filters is important and most naturally involves some numerical knowledge of the light spectrum, which is represented in Figure 1-1. The color names are spread at equal intervals according to color perception; obviously this does not correspond to equal distance intervals along the spectrum. Each name is located at the centroid of a range (step) of almost the same color. A light source can be imagined that emits light of only one wavelength (i.e., **monochromatic**), or a whole series of such lamps, each emitting a different wavelength. The color of the light of each would be that of the corresponding wavelength of the spectrum. Therefore the specification of the equivalent or **dominant wavelength** is one method of specifying hue. When a hue is diluted with white or black, its hue may shift, especially if yellow is a component. Yellow-greens become olive and yellow-reds become brown as they darken.

Mixtures of wavelengths that correspond to names three or five places apart in Figure 1-1 will produce hues corresponding to the middle name. On the other hand, mixtures of light of wavelengths in the two end portions of the spectrum (i.e., λ 400–500 mμ and λ 600–700 mμ) produce new "nonspectral colors," that is purples, magentas, and pinks.

Color Produced by Absorption Bands

The color of nonluminous objects, including specimens in the field of a microscope, is usually due to the light that is reflected or transmitted after some wavelengths have been absorbed. This is still most easily thought about by imagining the light that would meet the eye spread out into its spectrum, now a residual, with one or more gaps due to **absorption bands.**

If, while a beam of white light is expanded into a spectrum, a portion of it is blocked out, as with a piece of cardboard, and the spectrum recombined with a second inverted prism (as Newton once did), the light beam will be seen as a color by the eye and will produce a colored spot on a white screen.

First consider a wide piece of cardboard to be inserted from the short-wavelength (violet) end. An unchanged beam of white light should be present for comparison. As the violet end of the spectrum is blocked out,

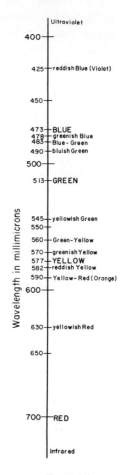

Figure 1-1

the final assembled beam becomes very light yellow, passing through lemon yellow to orange as more of the spectrum is removed; finally only the red color is left. If the wide board is advanced from the far red end, the first noticeable effect on the integrated beam and spot on the screen is to make it appear as a light sky blue, but this soon changes to a blue-green that becomes most intense (saturated) when the blocking board reaches wavelengths of about 600 mμ and all of the red is removed. The resulting color is called cyan, or minus red. As the board advances, the cyan gradually becomes blue and then violet.

In practice the absorption bands may not extend to the end of the spectrum; indeed, they may become very narrow and block only a small range of wavelengths; this is the case with quite a number of important biological stains. We can still use our imagined double-prism experiment by considering the interposition of a narrow strip of cardboard into the dispersed beam. Light coming from a colored object or dye solution with such narrow absorption band would have the same narrow absorbed gap in the spectrum produced by the cardboard, although not as a precise rectangular area. The effect of moving a narrow absorption band, by steps, through the spectrum is illustrated by Figure 1-2. In this case actual dye solutions were used; the hues that resulted are called by the names suggested by the ISCC-NBS Color Dictionary (8, U.S. Nat. Bur. Stand. No. 553, 1955). These colors form a somewhat more complicated system than the simple hues along the spectrum, for they include the "nonspectral hues," that is, the purples. The term magenta has been applied to a specific shade, but it is now frequently considered as a broader term that covers the color resulting from a single absorption band in the green, including all of the green. Cyan is a similar broader term. The observed color of the residual light from a narrow absorption band is changed by much smaller shifts of the absorption band near the middle of the spectrum than near the ends.

An important consideration from the viewpoint of the photomicrographer, and one that becomes evident later, is that the significant factor is the wavelengths of the light that are removed from the spectrum, although the observed color is that of the unaffected light. The color of the light removed would, of course, be complementary (by definition) to the observed color.

Now, more normally, light becomes colored because it is transmitted or reflected from an object that *absorbs* a portion of the spectrum. Moreover, these **absorption bands** are not so sharply limited at a wavelength edge as the blocking strips in the spectrum The same principles and rules hold, however. Actually the stains and also the color filters used in microscopy have quite sharp absorption bands.

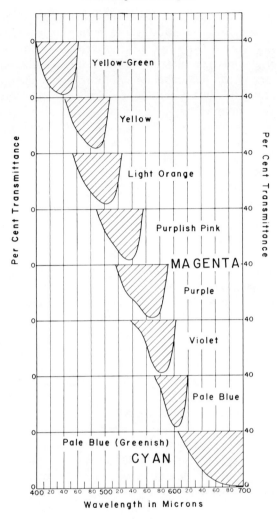

Figure 1-2

Color Attributes and Nomenclature

A complete specification of the **spectral quality of light** from a body obviously requires a wavelength-by-wavelength specification of the relative intensity of the light from it, which amounts to a description of the absorption bands. This is most conveniently presented as a spectral absorption, transmission, or density curve on graph paper. Examples of this occur thoughout this book.

Color, on the other hand, **is a visual sensation**. A given spectral quality of light will have a specific color, it is true, but this is more than necessary, it has been discovered. **Any color** (visual sensation) **can be matched by some appropriate mixture of light of three wavelengths, or three hues, called primaries.** A great number of different wavelengths and hues can be used as primaries, but a certain set of blue, green, and red spectral colors have been adopted as standard for reasons that can be studied in any good book on color, (8, R. M. Evans, 1948). The problems in the photographic reproduction or use of color are discussed by Hunt (8, 1967).

Color is the generic term for the common visual sensation and includes black, gray, and white. Color has **three attributes**, hue, saturation, and brightness. **Hue is that attribute of a color that causes it to differ from black, gray, and white.** A pure hue is equivalent to the effect of a single wavelength of the visual spectrum, as in its blue, green, or red portions, but two wavelengths near opposite ends of the spectrum are required to match a pure purple or magenta hue. **Saturation of a color expresses the degree with which it differs from a gray** of the same lightness. Conversely stated, **saturation is a measure of the degree with which black and white** (or the equivalent gray) **are mixed with a pure hue.** In the famous Munsell color system this attribute is termed chroma. Finally, the term **brightness expresses the luminous intensity of the color** as it could be most nearly matched, visually, to some gray. Although this technical term is used in the same sense as in ordinary English speech, it is not so commonly recognized how the degree of a neutral gray mixed with a pure hue markedly affects its visual appearance. Pure bright yellows become olive; as they darken, oranges and reds become brown.

Although any color can be evoked by adding blue, green, or red light, **the three additive primaries**, in suitable mixture, a much more frequent concern is how to obtain particular colors by mixing colored materials such as dyes and pigments. This is, of course, the problem of most color photography in which the final color picture is viewed by looking through three superposed dyed or pigmented photographs. In this case the light from one colored layer may be particularly absorbed by the next. However, **the effect of the absorption bands of the color components is additive** and, from the incident white light, controls the spectral quality of the light that finally reaches the eye. In dealing with colored materials, therefore, it is more convenient to consider as primaries the colors produced by the bands that absorb the three additive primaries, called **the substractive primaries**: yellow (minus blue), magenta (minus green), and cyan (minus red). These subtractive primaries can also be designated by the additive primary light that each permits to be sent to the eye: yellow (green and red), magenta (blue and red), and cyan (blue and green). Thus, since absorption bands are additive, any mixture of the additive primaries may be made to reach the eye by suitable mixture of subtractive primary colorants; but, since absorption, only, is used to produce each color, both the vividness of hue and the available brightness are more limited than when colors are produced with lights of the additive primaries.

Several practical aspects of color experience are especially important for the use of color photography. One involves the term **color constancy**, which expresses the human tendency to ascribe a color to an object or surface and to maintain the color irrespective of the quality of the illumination. The appearance of the colors in one's living room does not seem to alter so much from daylight to the artificial illumination of evening as does the quality of the illumination. It is only a tendency; with certain colors, such as magentas and purples, the change is too great and the color constancy effect breaks down. Objects in color slides that are projected in a dark room are usually seen with their daylight appearance, with wide tolerance for changes of quality of the projection light and even the color balance of the photograph. On the other hand, when a photograph is viewed in a lighted room, either by transmission on a viewer or as a color reflection print, the presence of the familiar surroundings makes us critical of the color balance of the photograph and the tolerance for the viewing illumination becomes much less.

Another matter is involved also. If the photograph reproduced the original spectral quality of the scene with all of the absorption bands well reproduced, the tolerance for the viewing illumination would be great; but all scenes reproduced by a given photographic color process are represented by three standard absorption bands that usually fool the eye quite well. Just as in motion pictures the eye is deceived by apparent motion on the screen, situations will occur in which the deceit breaks down; for example, when carriage wheels are seen to revolve backward. Some absorption bands can cause trouble as well in color photography.

Polarization of Light

It is obvious that if light vibrates transversely during propagation through space, since the latter has three dimensions, only one of which is specified by the direction of propagation, the vibrations can be oriented in all of the directions mutual to the other two direction axes. which might be called x and y. By analogy it is as if a group of parallel strings were all vibrating transversely with waves whose planes were oriented randomly with respect to one another. Now, if a grill, similar to a picket fence, with parallel slots is placed as a barrier in the path of the vibrating strings, only those vibrations parallel to the slots will get through; these would be components of planes that had slanted with respect to the slots. All the vibrations beyond the grill, being parallel to it, are said to be **plane-polarized**. Aligned groups of atoms and molecules of the correct spacing can act as a grill to light waves; such materials are called **polarizers**. The phenomenon of **polarization** is an important one for microscopy. This book can supply only the definitions that are needed in general in microscopy and photomicrography. A much more extensive discussion is required and can be found, along with exceedingly helpful diagrams, in any good text on petrography, which is a branch of microscopy based on an application of this phenomenon. Some references are given in the bibliography.

When a beam of light is passed through slabs, or plates, of some materials, it is broken into two beams, each of which is now plane-polarized and vibrates at right angles to the other. For this reason, if a plate of transparent calcite or Iceland spar is placed over a dot on a piece of paper, two dots will be seen. The material is said to **anisotropic**; it has different properties in different directions. Optically anisotropic materials, such as calcite, also are **birefringent** because they break an entering beam of light into two beams that travel at different velocities within the crystal. One is called the **ordinary** ray; the other, the **extraordinary** ray. In an **isotropic** material, such as rock salt or well annealed glass, light travels with equal velocity in all directions and is not split into a double beam. **A polarizer is a material in which one of the two rays is suppressed.** In tourmaline this is done internally and naturally; in a manufactured polarizing prism, such as a Nicol prism, it is done by correct cutting of the crystal plus absorption of the unwanted beam. Note therefore that a polarized beam can have only half the intensity of the original incident light as a maximum.

Reverting to the analogy of the vibrating strings, with a second grill inserted behind the first and parallell to it, we can see that if the directions of the slots of the two grills are parallel the vibrations will be unaffected, except for frictional losses. If, however, the second grill is rotated in its plane, by the time it has been turned 90° all components of the originally transmitted vibrations will have been damped out. This second grill is called an **analyzer**; except for its use, it may be identical with the polarizer. A microscope may be equipped with a polarizer in the illumination beam below the substage condenser and an analyzer above the specimen, either in the body tube or above the eyepice. If the specimen is anisotropic, it will act as an analyzer; the analyzer of the microscope can then be used to diagnose the directions of the different optical axes of the specimen.

Originally all polarizers and analyzers were relatively long prisms of natural crystals. This greatly limited the microscopical applications and the quality of the images with economically practical prisms. With the advent of Polariod sheet polarizer and analyzer, this limitation was removed, except for the new one that these materials fail at the two extremes of the spectrum. They represent, however, an enormous practical advance. The first type consisted of a colloidal dispersion of microscopical anisotropic crystals all aligned within the sheet. This type has been almost displaced by sheets of plastic made anisotropic with aligned molecules.

Interference of Light

Another result of the wave structure that is fairly commonly experienced is the phenomenon of Newton's rings, which are observed as a nuisance when two glass plates come into contact at one point. The iridescent colors of an oil film on water are caused by the same phenomenon. Newton placed a weak convex lens on a glass plate and observed it by reflected light. The center was dark surrounded by rings of color. (By transmitted light the center is bright but is still surrounded by Newton's rings.) With light of one color, the central spot was surrounded by bright and dark rings of increasing size. Newton was unable to explain the phenomenon; it remained for the wave theory adopted after Young's experiments to provide this explanation. If two parallel slits are illuminated from a common point light source (or preferably by a parallel slit) emitting or filtered to one wavelength, most of the transmitted light will radiate in a broad fan, and a series of dark and bright bands will be seen. (This may need magnification to be visible.) It will be found that a line lying at an equal distance from both slits, as at C in Figure 1-3, is bright. At a point at which the distance *from one slit* is $\frac{1}{2}$ wavelength longer than from the other, a dark line results. Such a line will exist on each side of the center C; and so in succession, but with decreasing brightness with distances from the two slits (which act as two sources), there will be a series of bright bands in which the distances *from the two slits* differ by integral multiples of the wavelength and dark bands in which the distances differ by $\frac{1}{2}$ wavelength. For equal distances from the source and distances differing by a whole wavelength, waves from the two slits will arrive **in phase**, that is with troughs and crests matching, and so will reinforce each other. With a $\frac{1}{2}$-wavelength difference in distance, the two beams are exactly **out of phase**, with trough matching crest, so that the two light beams cancel each other and a dark band results. It will be noted, however, that this phenomenon of **interference**, originally considered startling,

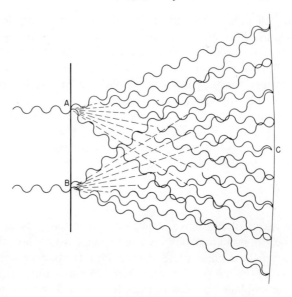

Figure 1-3

does not contravene the law of conservation of energy, since the energy in the final pattern is merely displaced. If white light, consisting of a mixture of wavelengths, is used, the bands will obviously be displaced according to wavelength (again to make equal distances from the two slits) and now become a series of colored bands. It is this appearance, sometimes as a series of colored rings, that is quite common, since interference is a fundamental and common characteristic of wave propagation that can be produced in a variety of ways.

For two light beams to interfere they must be **coherent**. This means that they must be identical in wavelength and so oriented in their vibrations in space and time that if part of a wave train coincides the whole wave train will coincide, and so for all the wave trains of the two beams. Obviously, if the angles of vibration were even slightly different in plane or pattern, this could not happen. Even the spacing of the wave trains, or **quanta**, could be different. Hence about the only practical way to ensure coherence between two beams is to split a single beam, as Young did. A common way is to use a partially reflecting mirror for this purpose. The two beams will then interefer when recombined **if the independent paths of the two beams have been optically equal**. A path can be changed by being a bit longer or by something being introduced into the beam to slow it down temporarily. Figure 1-4*a* illustrates the effect of introducing a cover glass into one beam of an interfering pair of beams. Thus in **interferometry** the thickness of a material can be measured to a small fraction of a wavelength—in fact, to about $\frac{1}{200}$ of a wavelength with proper optical aids. This is applied in microscopical interferometry, usually called **interference microscopy**.

There are many types of interferometer but all depend on splitting a beam of light from a single source to obtain two coherent beams with the same phase or which can be brought into phase if the act of splitting the beam has had that effect. The method of splitting the beam is important; a mirror that transmits part and reflects part (the internal face of a compound prism may do the same thing) or a birefringent crystal may be used.

The first and very famous form of interferometer was that of Michelson which has been much in demand, with some variation, for example, by Twyman, who used a lens to collimate the light from a point source. In this form light from the source S is sent to both complete mirrors M_1 and M_2 by the partially reflecting mirror M_3. Now, if mirrors M_1 and M_2 were both fantastically flat and at exactly the same distance from the junction at M_3 *and* absolutely perpendicular to the light axis, the surfaces of the light wave train would come back in phase to give **constructive interference**, as the two intensities of the two beams added to each other over the whole field of view. If one mirror is advanced or drawn back by the distance of one-half wavelength, **destructive interference** takes place over the entire field and the two beams cancel one another. If either mirror M_1 or M_2 were tilted *slightly*, a series of light and dark bands would instantly appear, as in Young's experiment. With Michelson's instrument these bands can be projected on a screen or viewed through a telescope; in the Twyman form a telescope must be used. Now, with the tilted mirror, as one mirror is slowly advanced, the interference bands move across the field and the distance of movement can be measured to a fractional wavelength, since the movement of one band past a point in the field corresponds to a distance of $\frac{1}{2}$ wavelength. The width of the bands corresponds to the angle of tilt.

Because the Michelson form of interferometer caused the two beams to travel back and forth in the same path, it was not easy to use in measurements, by transmission, of materials inserted in one beam to change the optical path. So first Jamin in 1860 and later Mach and Zender, independently in 1891, devised forms in which the light beams formed a rectangle with single paths. In the Jamin form (see Figure 1-4c) the beam is split by thick plate glass blocks, which act as mirrors. Two independent blocks of glass, *Jamin compensators*, which are much used, are in both beams; by slight geared rotation they can vary the path and bring the phase into coincidence. In the Mach-Zender form (see Figure 1-4d) the two beams can be widely separated, and with more independence, at the cost of two

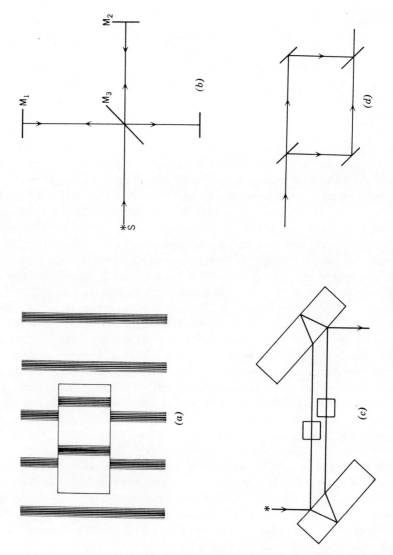

Figure 1-4

12

more mirrors to perfect and adjust critically. All three forms have been applied to microscopy. In addition, F. H. Smith used a birefringent crystal to split the beam for a microscopical interferometer.

Diffraction of Light

Another fundamental characteristic that is important to us is due to the wave structure of light. When an advancing wavefront is cut by passing perpendicularly to an edge, we should not expect a clean sharp edge of light waves to be propagated past the edge by the waves, but rather that a wave would tend to curl around the edge, as do water waves, radiating in all directions. Grimaldi discovered that this was so and named the phenomenon **diffraction**. Fresnel showed that the phenomenon was best explained by utilizing Huygens' proposal that in an advancing wave of light, from each point of the wavefront, the light itself would expand equally in all forward directions. By mutual interference effects the result may be spherical waves advancing from a source that become plane waves as the radius of travel gets very large. In Figure 1-5*a* consider *P* a point source of monochromatic light illuminating a white screen *ABC* from a considerable distance. The latter should be uniformly illuminated by the almost plane waves. Now a sharp edge *E*, such as a razor blade, is interposed. There will be a shadow on the screen. However, no matter how sharp the edge, if the shadow is magnified to correspond to the wavelength of the light, it will be found that not only is the edge of the shadow wedged but some light is inside the straight line that lies between *P* and *B*; that is, there is light toward *A*. When the edge *E* is observed from any point behind the illuminated portion *CB*, it will gleam as if it were a line source of light sending out a cylindrical wave in all directions; and it

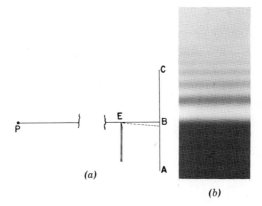

Figure 1-5 Diffraction at a single edge.

can be considered to act in that way. Consider the wavefront as it reaches the plane of the razor blade and its further behavior. With each point of light expanding hemispherically, the edge E will, in effect, generate a cylindrical wave that expands in a forward direction. Thus it sends light into the shadow area but with a rapidly decreasing intensity on the white screen as the inclination of the radii increases. Therefore the edge of the shadow is wedged rather than sharp. The unshadowed area of the screen, from B decreasingly toward C, has not received the light from the expanding points of a broad wave front that maintained the uniform wavefront. From another but consistent viewpoint the light of the cylindrical wavefront from E interferes with the unimpeded wavefront along $B-C$, with waves reinforcing or nullifying each other at half wavelengths. (This is the reason monochromatic light shows the pattern best.) Either of the two viewpoints accounts for the banded structure found near the edge of the illuminated field and illustrated in Figure 1–5b. This structure was made with monochromatic blue light impinging directly from a concentrated source onto a razor blade. The screen was 6 ft from the edge E, and the image of Figure 1–5b magnified about 6X. Characteristically the first light band is brightest, even brighter than the unshadowed area beyond this pattern.

With a single edge the fraction of the total light beam involved is small and the visibility of the phenomenon is low. However, if a second parallel edge is brought up to the first to form a slit so narrow that all or most of the light is involved in the diffraction band, it becomes much more easily visible; this effect is much enhanced in a series of slits, placed to reinforce one another. Moreover, this diffraction phenomenon caused by all edges becomes exceedingly important to explain the nature of optical images at high magnification.

Behavior of Light at Surfaces

From the standpoint of the design and much of the use of optical systems light can be considered as made up of rays and beams. A ray of light is exemplified by the appearance of the light through a small hole in the wall of a smoke-filled space. That such a ray is exceedingly straight in homogeneous media is one of the fundamental characteristics of light. This characteristic is utilized in the basic first step in aligning an optical system, which consists in removing all optical elements and leaving only mechanical elements in place. However, if the light ray meets a surface, it will be absorbed, reflected, or transmitted or undergo some combination of these effects. Absorption as the principal occurrence, is discussed later. If reflected, the ray is bent back at an angle equal to the original

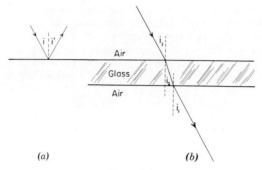

(a) (b)

Figure 1-6

incident angle. Optical angles are measured from a line perpendicular to the surface (see Figure 1-6). If the light is transmitted, the ray will be bent or **refracted**, as shown in Figure 1-6b and the ray will be steeper in the denser medium. This behavior follows from the fact that the velocity of light decreases with the density of the medium. A simple analogy for use in remembering the nature of this behavior is the famous one of a regiment of soldiers marching at an angle across and beyond a parade ground to the edge of a plowed field. The soldiers reaching the field will be slowed down and the line of march will be turned as illustrated in Figure 1-7. The relation here is known as Snell's law: the ratio of the sines of the two angles is a constant for any angle of incidence; that is,

$$\frac{\text{sine } i_1}{\text{sine } i_2} = \text{a constant} = n. \tag{1-1}$$

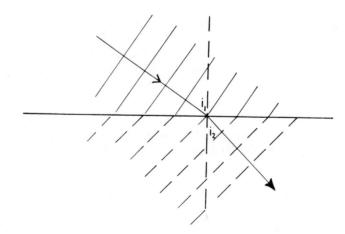

Figure 1-7

If the term **sine** is unfamiliar, it is worth learning, since it is merely a ratio of a side of a right triangle, opposite an angle, divided by the hypotenuse; that is, sine $i = a/c$ (see Figure 1-8). Since this ratio is the same, regardless of the size of the right triangle, it is a useful measure of an angle that will be met again in using a microscope intelligently. The cosine (often written as cos i) is merely the other ratio; that, cos $i = b/c$. This also is an important term in optics. It is obvious that these terms can be used as a measure of the angle.

Snell's law can also be written

Figure 1-8

$$n_1 \text{ sine } i_1 = n_2 \text{ sine } i_2 = n_3 \text{ sine } i_3, \text{ etc.,} \qquad (1\text{-}1a)$$

for a series of materials. If the space beyond the boundary of a material is a vacuum and we define $n_1 = 1.0$ for vacuum, the constant n of (1-1) will be property of the single material called its **refractive index.** On this basis the refractive index of air is 1.000293 at 0°C and 760 mm. In practice workers in optics usually measure and keep the refractive index of materials relative to air at unity, therefore the refractive indices of glasses are usually reported in this manner. Note in Figure 1-6b that if the ray entered another material than air after passing through the first sheet of material it would be bent more steeply, or less so, according to whether the refractive index of this next material were greater or less than that of the first sheet. Moreover, it would be possible to have a sheet of a different material, chemically, but with the same refractive index, so that the direction of the ray would remain unchanged.

A rather important phenomenon for microscopists may occur when the light is traveling from a denser medium to one less dense – specifically through a surface to a medium of lower refractive index. As the ray emerges into the medium of lower index, for example, when a ray comes from water or glass into air, it will be bent more steeply from the normal (or perpendicular); this is noticeable in Figure 1-6b. As the beam is inclined more steeply from the normal in the glass, the emergent beam will be even more so (from the normal) until it grazes the surface. This must be some sort of limit! It can be foretold from Snell's law, since the sine of an angle cannot exceed 1.0. If the ray in the glass is inclined still more steeply from the normal, it is internally reflected back by the surface, that is, for all angles greater than the **critical angle I**. Therefore the critical angle of incidence is that angle for which the sine of the angle of refraction is 1.0.

$$\frac{\text{sine I}}{\text{sine } i_2} = \frac{n_2}{n_1} \qquad (1\text{-}1b)$$

where sine $i_2 = 1.0$.

When the ray penetrates from glass to air, $n_1 = 1.52$ (about) and $n_2 = 1.0$. Therefore for rays emerging into air the critical angle is that angle whose sine is the reciprocal of the refractive index of the denser medium. This is 48.7° for water and about 41° for glass. Any one who has looked up from below a water surface, slantingly, even at a glass of water, is familiar with this phenomenon.

Note that the direction of the ray after emerging from the sheet back into the air parallels that of the incident ray, but *the line of the ray is displaced*. Obviously the thicker the sheet, the greater the displacement. Now, if a bundle of rays emanates from a point, each ray will be incident on the sheet, say of glass, at a different angle. Each will be parallel to its original direction after traversing the sheet but will be displaced. If the reader will draw this figure, he will find that apparently the rays no longer originate at a point and the discrepancy is proportional to the thickness of the glass! The converse of this statement is just as true if a bundle of rays were converging to a point before the introduction of a transparent sheet. Just how sensitive any given optical image is to the thickness variation of a sheet of glass introduced into the beam will vary with the maximum slope (from the normal) of the extreme rays of the bundle and also on the tolerance allowed in the size of the point from which they are apparently leaving or to which they are converging. We have often looked through panes of glass or seen pictures taken through panes of glass. Yet we shall meet cases in which a variation of cover-glass thickness of more than 0.01 mm is unacceptable!

Referring again to Figures 1-6a and 1-6b, we note that for homogeneous materials the light rays are straight except at the surfaces. This greatly simplifies optical considerations.* Obviously by bending the surfaces of a reflecting or a refracting material appropriately we should be able to bend all of the rays emanating from a point back again to a point which would constitute a perfect image of that point. If we can do this for a group of points in an area, we shall have a picture image. Such a surface is a lens, which can be a mirror (**catoptric**) lens or a transmitting (**dioptric**) lens. To make lenses that would reform points with light rays perfectly, it would be necessary to bend the surfaces into complex curves and with accuracies precluding deviations of less than fractions of a wavelength. In practice such **aspheric** surfaces are so difficult to make that only a few types are available, mostly for use as condensers. However, another simple device can be utilized; that is, the reduction of the maximum curvature of the lens surfaces the light rays may encounter. Since the imperfection of the image (i.e., its "errors") will increase with the curvature of the spherical surfaces, it is advantageous to substitute a group of lenses, each having surfaces of less curvature but all adding up to the original power. With

---

*This is not true in some other optics, such as electron optics.

modern coated lenses this method is especially valuable in making a lamp condenser from a series of spectacle lenses.

Reflection optics have one great advantage over optics based on refraction. The angle of reflection, as indicated in Figure 1-6a, is completely unaffected by the wavelength of the light so that white light can be used as simply as that of one wavelength. This is not true of refraction optics.

From the statement that the refractive index of material normally decreases with increasing wavelength it can be seen that in the situation depicted in Figure 1-6b, after incidence on a surface at a slope, a ray of white light would break up into a diverging fan of rays of different wavelengths, with the rays of shortest wavelength bending the most (see Figure 1-9). (In a sheet of glass each ray, on emerging, would again become parallel to the original ray.) This phenomenon is called **dispersion** and its extent depends on the material. This important defect in lenses is discussed later.

It would seem desirable to make all lenses of catoptric type, but in these lenses the light beam is sent back toward the object by the mirror. At least one more mirror is usually needed to send it forward; a hole in one of the mirrors lets the beam through. This obscuration can be serious. Even then catoptric lenses are important in telescopes and in ultraviolet

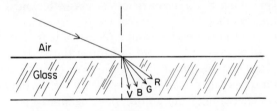

Figure 1-9

and infrared photomicrography. Obviously an image can be focused visually and the ultraviolet and infrared components will also be in focus.

The Nature of Lenses

A familiar phenomenon is that of a burning glass. If a lens can form a brilliant spot when held perpendicular to a sun beam, by definition it is a **positive lens**; at least one surface will be convex. If we consider that a sun beam is composed of exactly parallel rays, we have at once a demonstration of the two most important characteristics of a lens. By definition again the bright spot (Figure 1-10F) is the **focal point** or **focus** of the lens, that is, the image point of parallel rays, and its distance from the lens is the **focal length** of the lens. It is a fundamental principle of optics, and one useful in photomicrography, that **the reversed pattern of light rays would be the same, given the same optics**. Therefore, if a point source of light is placed at the focus of a lens, as at *F* in Figure 1-10, a parallel beam will be formed on the other side. When used to produce parallel light, a lens is called a **collimator**, a term frequently employed in discussing illumination for photomicrography. The light beam of Figure 1-10 could have come from either side of the lens; that is, a lens has two focal points, one on each side. However, corrected commercial lenses are often not symmetrical; that is, the two focal lengths may not be equal. These lenses cannot always be reversed in use indiscriminately, a point that should be remembered when setting up an optical system. The term focal length is discussed further later in this chapter under the subject heading of "thick lenses."

As we shall see later, a lens may not always be used with air as the medium of object or even image space. A useful rule here is

$$\frac{f}{f'} = \frac{n}{n'}, \tag{1-2}$$

where *n* and *n'* are the refractive indices of the object and image space, respectively, for a given lens.

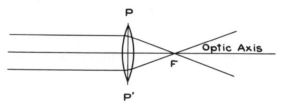

Figure 1-10

From this it is evident that if the refractive index of object or image space is changed from that of air ($n = n' = 1.000$) the corresponding focal length of the lens will be changed proportionally to the refractive index of the new medium.

Since many light sources are too large even to approximate a point, what is the result when such an extended source is placed in the focal plane of a lens, that is, at the focal point and perpendicular to the axis? This is most easily answered by considering the point at a focal distance from the lens but not on the optical axis. A lens will produce a parallel beam of rays from such a point, as might be expected, but now it will be tilted with respect to the axis of the lens (see Figure 1-11). Other points on the light source (the latter may be considered to be made up of points) will also cause the formation of parallel bundles of rays, but all rays except

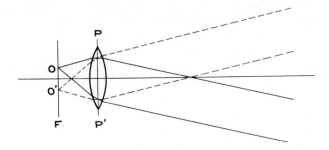

Figure 1-11

the bundle from the point on the axis will also be inclined to it. Thus the whole beam of light will be divergent but composed of parallel bundles. Another lens behind the collimator would, of course, again bring such a beam to a collection of points; that is, it would reconstitute the image of the source (see Figure 1-12*a*). If the other lens had a different focal length, the image would not be the same size as the object; it would be larger if the focal length of the second lens were longer (Figure 1-12*b*). The rule is **with such a pair of lenses, and with the object at the focus of the first component, the relative size of the object and the image is proportional to the focal lengths of the lenses.**

Of course, a single lens can form an image. To describe this the formula often used is the simple one directly applicable to simple lenses of negligible thickness;

$$\frac{1}{f} = \frac{1}{l'} + \frac{1}{l},$$ (1-3)

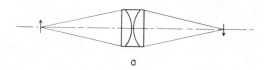

a

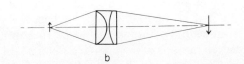

b

Figure 1-12

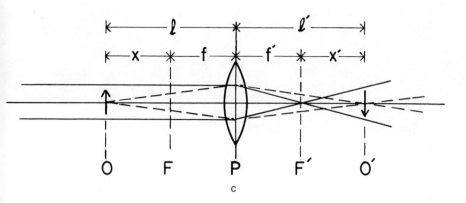

Figure 1-12c

where f = focal length and l and l' are the object and the image distances, respectively, measured from the lens. Moreover,

$$\text{magnification} = m = \frac{l'}{l}. \tag{1-4}$$

These two equations can be combined to give two more that are especially useful for determining optical distances with thin lenses. For reasons discussed later they are all also useful for most practical lenses. To get the following equations merely multiply (1-3) first by l and then by l' and substitute m for l'/l or inversely:

$$\text{object distance} = l = f\left(\frac{1}{m}+1\right) = f\left(\frac{m+1}{m}\right), \tag{1-5}$$

$$\text{image distance} = l' = f(m+1). \tag{1-6}$$

The magnification m can be a small fraction or a large number or any intermediate value, of course. In any case, the image distance is greater (or less) than the object distance by the factor m; that is, $l' = m \cdot l$. Constantly bearing this very simple fact in mind helps optical thinking. A practical device for calculating optical distances, based on (1-5) and (1-6), is described on p. 33.

If the lenses can still be considered to be of negligible thickness, there is another useful rule: two or more lenses can be combined by placing them together coaxially, and the focal length of the combination can be obtained by adding the reciprocals. Thus

$$\frac{1}{F} = \frac{1}{f_1} + \frac{1}{f_2} + \frac{1}{f_3}, \text{etc.} \tag{1-7}$$

Sometimes the following relationship is useful. Let $l = f + x$, $l' = f' + x'$:

$$f^2 = xx', \tag{1-8}$$

$$m = \frac{x'}{f'}. \tag{1-9}$$

In most "thin" lenses $f = f'$; then $m = f/x$. (N.B. The writer has neglected the sign convention for this purpose.) The rule illustrated by Figures 1-12a and 1-12b is consistent with this one and can be obtained by simple calculation. It is so useful that it is constantly being applied even when the component lenses are not of negligible thickness; in this case, of course, the rule is only an approximation. Usually, the actual F is somewhat longer than that calculated from (1-7).

Either because of their thickness or for other reasons two thin lenses used in combination may not lie in the same plane, even approximately, but at some distance d apart. This can be taken into account in the calculation of the focal length of the unit lens formed by the combination of two lenses of focal length, f_a and f_b:

$$F = \frac{f_a f_b}{f_a + f_b - d}. \tag{1-10}$$

In any discussion of microscopy we must consider the visual use of the **simple microscope**, which is any single unit lens used to produce a magnified image. The lens itself may be, and often is, a complex one, such as a commercial magnifier or an enlarger lens. Since the apparent size of an object increases as it approaches the eye until it reaches the minimum distance of clear perception, the simple microscope can be considered as a device to bring the object optically to its focal distance from the eye. Therefore*

$$\text{magnification} = \frac{\text{normal viewing distance}}{\text{focal length}}.$$

---

*This formula does not apply to a weak lens used for low magnification such as a reading glass where eye accommodation is important. Here the formula is

$$m = \frac{\text{normal viewing distance}}{f} + 1$$

This "normal viewing distance" has been arbitrarily standardized as 10 in., or 250 mm (although in practice it is usually greater); that is,

$$m = \frac{10 \text{ in.}}{f} = \frac{250 \text{ mm}}{f}.$$

Simple lenses are extensively employed in photomicrography. A knowledge of their limitations and of the general characteristics of "corrected" lenses is fundamental to the proper use of both types and the wise selection of corrected lenses. All lenses usually have spherical or plane surfaces (Figure 1-13). Referring again to Figure 1-11, we can see that the focal plane, *F-F* cannot really be extensively flat, since a focal length's distance from the lens would trace out a curved surface, not a plane. This is related to the well-known defect of most lenses and all simple ones, that is, **curvature of the field.**

Worse than that, with a simple lens only the rays very close to the optical axis pass through a point if they arrive parallel to the axis. The further they are from the axis when parallel, the closer to the lens they will be when, on the other side of the lens, they cross the axis and the other rays that passed the lens nearer the axis. This defect is called **spherical aberration.** Such a lens would exhibit the same defect when a point on the axis is to be imaged; that is, the rays furthest from the central ray would cross the latter closer to the lens in the image space (Figure 1-14). Obviously such a defect will be worse the greater the diameter or aperture of a lens that is used. This defect in lenses cannot only be corrected but may

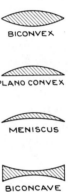

BICONVEX

PLANO CONVEX

MENISCUS

BICONCAVE

Figure 1-13

sometimes be overcorrected; that is, the rays closer to the axis may cross it nearer the lens than the others after they have passed through the lens. This is an important defect in the optical elements used in photomicrography; overcorrection is often met in practice in microscope objectives.

If the beam of parallel rays is not also parallel to the axis, or if the object and the image points do not lie on the optical axis, the same defect is exhibited by simple lenses. However, now the pattern is unsymmetrical (see Figure 1-15) and the defect is called **coma.** It will be noted that there is no position at which the rays from an object point again meet in a point to form an excellent image. A lens corrected to eliminate both spherical aberration and coma is called an **aplanat** or an **aplanatic lens.** Lenses with no further correction than this are extensively used as condensers for illumination, especially in microscope substages.

Figure 1-14   Figure 1-15

Because coma involves the rays from points off the axis, its presence is a limitation to the size of the acceptable field in the image plane and its elimination is necessary for an extended field. In any transparent sphere, such as the one represented in Figure 1-16 with radius r, there exists a hypothetical inner sphere of radius $r' = rn/n'$, (where n and n' are the refractive indices outside and inside the sphere) such that all rays originating at this inner surface are aplanatic; P in Figure 1-16 illustrates such a point. Although the surface is within a glass sphere, it is obvious that if the sphere were cut in two (e.g., along the diameter d–d) and the intervening space were filled with a liquid having optical properties identical to those of the glass, the rays from P would be unaffected and a practical lens element would be obtained. Moreover, it is possible to put other lens elements behind the hemisphere that will bring the rays back to a point, that is, form an image of P, without introducing spherical aberration or coma. This is the basic principle of the immersion microscope objective.

It has also been shown that a lens will be free from coma for each

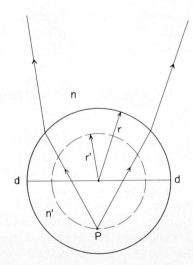

Figure 1-16   Aplanatic surface of a sphere.

successive aperture zone from the center whenever the following relationship applies:

$$nh \sin \alpha = n'h' \sin \alpha', \qquad (1\text{-}11)$$

where n and n' are refractive indices in object and image space, h and h' are the heights of object and image points from the axis, and α is the angle subtended by the lens aperture from the axial points corresponding to h and h' in the object and image spaces. This is the well-known **sine theorem** which is often referred to and which we shall use again. Since h'/h is the magnification m, the required condition for freedom from coma is that the ratio $(n \sin \alpha)/(n' \sin \alpha')$ must be constant. This sine condition is not always fulfilled for lenses in practical use.

Another aberration, normally of importance to lens designers rather than to users especially of microscope objectives, is somewhat sophisticated, and, like coma, applies to rays lying outside the paraxial field. To explain it, first consider a perfect lens that gives a perfect point image P' of an object point P. Now consider a thin planocylindrical lens placed against it so that the axis of the cylinder is in the plane of the paper (Figure 1-17a). This plane is called the **meridional plane** and the rays in it, the **meridional rays**. In this and parallel planes the cylindrical lens has no power and therefore the focus at P' will not be altered, although the image of the point will be stretched to a line perpendicular to the meridional plane. The plane perpendicular to the paper is called the **sagittal plane** and the **sagittal rays** are affected by the full power of the cylindrical lens (Figure 1-17b). This will, of course, draw the image point P' closer to the lens, as to P'', the point representing a line lying in the meridional plane. A lens that shows this effect without the added cylinder is said to exhibit **astigmatism.** The analogy just explained falls down in that astigmatism is

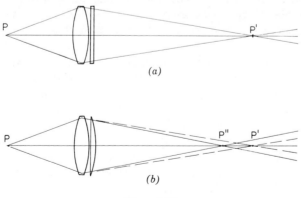

(a)

(b)

Figure 1-17

normally exhibited from points lying outside the paraxial region, most frequently being pronounced only in the outermost zones of an object-image field.

The nearer image lines from points lying in the meridional plane are parallel to radii from the center of the object field, whereas the lines lying in the sagittal plane are parallel to tangents to a central circle in the object field and are often designated in these terms.

A simple aplanat fulfilling the sine condition, however, does so for only one wavelength, and so would not usually be satisfactory as a camera or microscope lens. This is because the focal length of simple lenses is different for different wavelengths of light, a variation called **chromatic aberration**. This defect causes a series of images to be formed when a mixture of wavelengths is used, as with white light, those of shorter wavelength being focused closer to the lens (see Figure 1-18). Points in any one image, therefore, are surrounded by colored halos from the rays

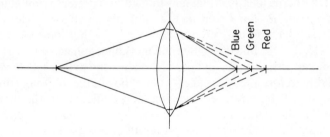

Figure 1-18

differently focused. This can easily be observed by focusing a simple commercial microscope lamp with a focusable lens to give an image of the filament and then moving a sheet of white paper along the optical axis and noting the color fringes. An even better method is to interpose a series of color filters and note the positions of the filament image strung along the axis.

It has been found possible, by combining simple lenses made from more than one kind of glass, to bring the images formed by two colors to the same focus, usually with appreciable improvement for images made from those colors that lie between them in the spectrum. Such lenses are called **achromatic lenses** or **achromats**. They are usually strictly aplanatic for only one color. It has even been found possible to bring three colors to the same focus. The image is improved for all intermediate colors and usually even for colors of somewhat longer wavelengths than those that are exactly corrected. Such lenses are called **apochromats**. They are usually aplanatic for two colors.

The actual behavior of the practical objectives used by the critical microscopist with respect to spherical and chromatic aberrations is important to him. Therefore a brief discussion of **the interaction between these important defects from perfection** will illustrate the compromises that must be made because of the lack of perfect lenses when we attempt to show unknown detail. Both spherical and chromatic aberrations are corrected by combining positive and negative lenses that have defects of opposite sign but that leave some positive power for use as an image-forming lens. For good economic reasons achromatic objectives are most in demand; they were the first to be developed and are usually made of only two different kinds of glass, for example, a crown and a flint. A complex lens, such as a 4 mm microscope objective, which has several achromatic component lenses, may therefore contain more than two different kinds. Achromatic lenses are designed so that the image rays from two **zones** (annular areas including the center) of the lens focus together in the

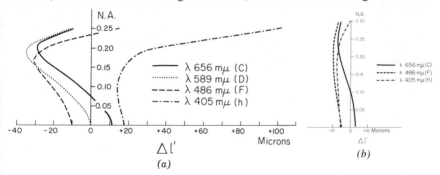

Figure 1-19 Zonal aberrations of 16 mm microscope objectives: (*a*) achromat; (*b*) apochromat.

same plane, that is, at a distance l', which is measured along the axis. Usually the selected zones are the center (paraxial) and the outermost (marginal) zones. These zones are illustrated in Figure 1-19*a*, which shows the **variation in the distance along the axis at which the rays from the various zones meet it**, as measured by $\Delta l'$, with a 16 mm achromatic microscope objective and for several wavelengths of light, as denoted by their Fraunhofer letters. Note that the focus is *not* the same for zones at intermediate apertures. This characteristic is called **zonal aberration.** It must be noted that this figure shows the behavior of the zones; the behavior of the whole lens at any aperture is an integral function of the curve for one wavelength up to that aperture. **As the wavelength of the light used through the lens is changed, the zonal aberration curve shifts.** Note how badly off the far blue is from the others in an achromat, yet this is a photographically important wavelength. Although the values on the abscissa of Figure

1-19 state the *difference* in distance from the lens at which the rays from the various zones cross the optical axis, that is, $\Delta l'$, reference to Figure 1-14 will show that this is a good measure of the departure from a point in any one vertical plane.

This wavelength effect is more directly illustrated by Curve *A* of Figure 1-20, in which the axial differences are plotted against the wavelength for the paraxial zone. Actually this curve was determined by the author by measuring the *shift* in the position of a 10X achromatic objective as the wavelength from a monochromater was changed and a pinhole test spot on the axis was kept in focus. This really measured Δl, but $\Delta l'$ in the image space differs only by a magnification factor. It will be seen from this curve that **with an achromat there are only two wavelengths that focus at the same axial distance**, each member of the sets lying on the opposite sides of a minimum axial image distance. The latter is usually assumed to represent **the wavelength of best performance.**

We can also plot the spherical aberration for the wavelengths of the spectrum for the same 16 mm achromatic objective used for Figure 1-19*a*; the curves in Figure 1-21*a* are obtained. Remember that negative spherical aberration means overcorrection; that is, the innermost rays focus closer to the lens. This corroborates the old rule, which is discussed in the chapter dealing with color filters. It is obvious that with **achromats a great**

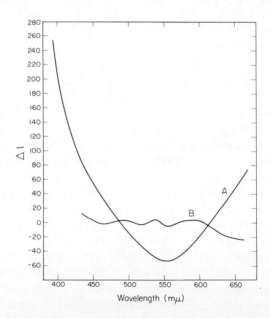

Figure 1-20 Paraxial shift of image with wavelength. A. Achromat. B. Apochromat.

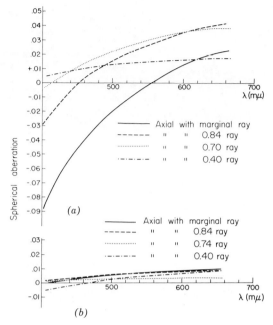

Figure 1-21 Spherical aberration versus wavelength from various zonal apertures: (*a*) for a 16 mm achromatic objective, 0.25 NA; (*b*) for a 16 mm apochromatic objective, 0.30 NA.

gain in quality should ensue if the light of a single wavelength were used instead of white light.

After E. Abbe showed the way, **apochromatic objectives** were made by using three component materials, usually two glasses and fluorite, in which the image distance *l'* is designed to be the same for three zones and three wavelengths. As illustrated by Figure 1-19*b*, **the aberrations are distinctly less for the zones of higher aperture**; this figure was made from design data for a 16 mm apochromatic objective. The data for Curve *B* in Figure 1-20 were obtained experimentally by using a 16 mm (10X) apochromatic objective; both curves apply to the central field. It is difficult to state the significance of the various waves of the curve, for the differential wavelength behavior of fluorite and quartz can be complex. The significant factor is the great improvement of performance. Since the dispersion of wavelength by all glass is smaller for longer wavelengths, the improvement in the far red and even infrared is marked.

One characteristic difference between achromats and apochromats may be confusing. In Figure 1-22 the equivalent focal lengths are given for an achromat and an apochromatic objective. **The variation of equivalent focal length with wavelength is much worse for the apochromat** in spite of

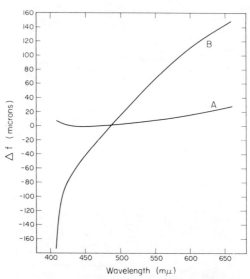

Figure 1-22 Equivalent focal-length variation with wavelength. (A. Achromat; B. Apochromat)

the fact that the variation of the image position is less with wavelength variation. This is because the separation of the principal plane of the very extended lens system varies with wavelength. However, magnification is determined by the equivalent focal length, as we have seen, so that although the various **colored images** of different wavelengths formed by an apochromat will be brought to focus more nearly in the same plane **they will vary more in size than with the achromat.** Since the focal length of the lens increases with wavelength (i.e., the power $P = 1/f$ decreases), the images near the blue end of the spectrum will be slightly larger and blue-fringed. This defect is overcome by using **compensating eye pieces** with apochromats for which the variation of equivalent focus with wavelength has been made just the opposite. Moreover, Abbe succeeded in making the $(\Delta f - \lambda)$ variation curve of a series of apochromats so nearly the same that all could be used with the same compensating eye pieces. This important application is discussed in Chapter 6. It is obvious that **when white light must be used**, as in color photomicrography, **apochromats must be used for superior definition.** This rule becomes the more exacting, the greater the aperture of the objective, so that it is dangerous to disregard it **when using objectives with numerical apertures greater than 0.5.** Moreover, with both achromats and apochromats the higher-power objectives of shorter focal length cannot be corrected quite so well as can those of longer focal length.

Recently M. Herzberger showed the possibility of making **superachromats** with four selected glasses (or minerals) in which the curve bends again and may be so flattened that a lens may be said to be excellent from the ultraviolet region into the infrared. At this writing this devopment has not been applied to microscope objectives in which the high apertures are such a problem.

Lens designers list at least five other abberations, but the knowledge of only two more is of much practical importance to the user. One is **curvature of field**, which has already been defined. Photographers refer to the same phenomenon when they discuss the "size of the field" of a lens. This is so because **field curvature is usually the limiting factor of field size**, and almost invariably it is so in photomicrography, in which the ground glass or photograph records only the appearance of a single plane, whereas the best focus of an image of the plane of the specimen lies on a curved surface. Moreover, the shorter the focal length of the lens, the greater the curvature of the focused image, so that for microscope objectives of very high magnification the curvature may be high indeed. In actual practice, because of the phenomenon of depth of field, a zone of best focus rather than a surface better describes the location of the image as it is observed by anyone. This zone, in which the definition is indistinguishably best, is represented in longitudinal section in Figure 1-23 with the lens on the left side of the figure. If the vertical lines represent various possible positions of the ground glass or photographic film, an important relation between field size and point of image focus becomes evident. In both positions *A* and *D* the central area of the image is definitely out of focus, but a ring of better focus on the negative, surrounding the center is a clue when the picture was made inside of focus, as in Case *A*. Pictures of good focus will be obtained from positions *B* and *C* and anywhere in between, but the acceptable field will be largest when the picture is taken at position *B*. The best technique for focusing an image is discussed in Chapter 6.

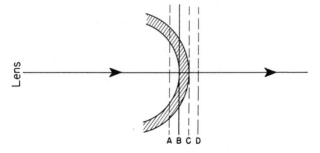

Figure 1-23

Although curvature of field can be decreased in the lens design, this decrease can generally be effected only by sacrificing some central definition that might otherwise be obtained. Therefore we can divide "objective" lenses into two classes:

1. A microscope objective (as normally used with a compound microscope) and a telescope objective are lenses in which best definition is so important that no effort has been made to decrease field curvature and the fields may be small. (But see Chapter 3).

2. A photographic objective, on the other hand, is a lens in which a correction has been made by the designer to decrease curvature of field. For certain ranges of magnification, normally up to about 50X, both types are available, and the photomicrographer must decide which to use. Frequently, obtaining a picture of a whole object, such as an insect, is more important than obtaining the best possible definition in a smaller area.

The other aberration that may be of importance to the photomicrographer is that of **distortion. Distortion results when magnification varies over the image field.** If it increases with distance from the center, the corners of a square object would apparently be pulled out in the image to give pincushion distortion. For the converse situation, in which the sides of the square would bulge, barrel distortion is observed. **Symmetrical lenses do not exhibit this aberration**. The compound microscope however, is usually extremely unsymmetrical and may exhibit pronounced distortion. The fact that the magnification becomes different as the edge of the field is approached may be important in making microscopical measurements.

Focal Scale

The following device is valuable to any user of lenses, especially when making decisions concerning the selection or placement of a lens. It makes use of the common artifice of utilizing units that will simplify an equation and of displacing a zero point to create simple proportionality.

Figure 1-24 represents the optical axis of any lens that is assumed to be in a plane perpendicular to the paper. The scale on each side is laid out with the focal length of that side of the lens as the unit. Since no object imaged by the lens and no image formed by it can lie within a focal length of the lens, the existence of this space is temporarily ignored and the zero point of the scale is split to exist on each side of the lens and a focal length away. Use of this scale is simple; it can be quickly sketched on any scrap of paper for use as a mental aid. The aim is to get the optical calculations needed out of the particular units involved (inches or millimeters), into

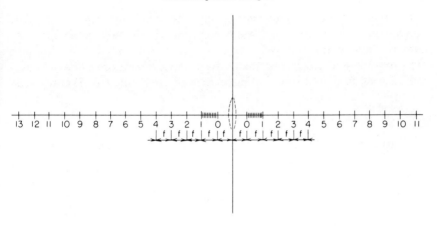

Figure 1-24 Focal scale.

focal units. If m is the magnification, the object will lie on one side at $1/m$ units away, whereas the image is at m on the other side.

Let us use this simple scale. If the object is at a distance of 25 units on the left, the image is at $\frac{1}{25}$ unit on the right and is $\frac{1}{25}$ the size of the object. If the object is at 5 units, the image is at $\frac{1}{5}$ and is $\frac{1}{5}$ as high as the object. When the object is at 1.0, the image is at 1.0 and of equal size. If the object is at $\frac{1}{10}$, the image is at 10 and now is 10 times larger.

We can usually make the calculations easily, often mentally. Then we can convert the system back to the real one by multiplying the resulting number by the focal length of the lens involved, but we must remember to shift back the zero point *by adding one focal length to each side of the lens that is involved.*

Example 1. A 2-in. (50 mm) lens is available for making a picture at 10X magnification. What bellows length is required and what is the total distance from object to image?

We find that in the focal scale units the object will be at $\frac{1}{10}$ and the image at 10. On converting back to actual units, where $f = 2$ in., the image will be at $2X (10 + 1) = 22$ in. from the lens. In the total distance there are two focal lengths to be inserted to reset the zero. Therefore the total distance D, which is obtained by counting along the scale, is $2X(\frac{1}{10} + 1 + 1 + 10) = 24.2$ in.

Example 2. What lens should be chosen to magnify the filament of a lamp $\frac{1}{5}$ of its original size when a distance of only 18 in. is available?

The filament will be at 5 focal units, its image at $\frac{1}{5}$ on the scale. Converting back, with f representing the desired focal length, we find that $(5 + 1) f = 18$. The focal length of the lens chosen must be 3 in. or less.

In optics, particularly in microscopy, the metric system is especially convenient for expressing dimensions, and the focal lengths of many camera and microscope objectives are quoted in millimeters. Six-inch (15-cm) celluloid rules, obtainable from most laboratory supply houses and bearing both inch and millimeter scales, are handy (see Figure 1-25).

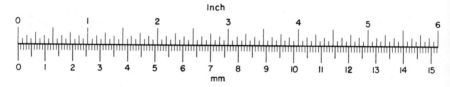

Figure 1-25

The following relationships hold:

1. 1 inch = 25.40 millimeters (mm) = 2.540 centimeters (cm).
2. 1 mil = 25.40 microns (μ) [dividing (1) by 1000], where 1 mil = 0.001 in. and 1 μ = 0.001 mm.
3. 1 microinch (μ in.) = 25.40 millimicrons (m μ); 1 μ in. = 0.001 mil = 0.000001 in.
4. 1 millimicron = 0.001 micron = 0.03937 microinch.
5. 1 nanometer (nm) = 1 millimicron (m μ).
6. 1 Ångstrom unit = 0.1 millimicron = 0.003937 microinch.

Thick Lenses

So far we have assumed that the thickness of a lens is negligible, but actually the thickness or length of a modern corrected lens can be great. Lens designers, and users also, have long applied a simplifying principle similar to that of the focal scale. A thick corrected lens can be treated as if there were two surfaces within it at which all the bending of the light occurs and the space between them can be assumed to be optically nonexistent, so that a ray of light at one surface is automatically at the same height from the axis at the other. These two surfaces are usually called the **principal planes** of the lens, in spite of the fact that they are not planes at all but surfaces that usually more nearly approximate portions of a sphere, especially outside the central area (Figure 1-26). These principal planes cut the axis at the **principal points** of a lens. As an approximate but useful rule, many lenses can be divided into three equal parts with the principal points at the divisions. Therefore in Example 1 the distance between the principal points should have been added to the total object-image distance in converting back to the real distance units. When the object and image space of a lens is air ($n = 1.0$), as it almost always is in a photographic objective, the principal points of the lens coincide with its **nodal points.** The latter are defined as the two points (lying on the axis) around which the lens can be rotated without affecting

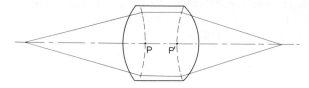

Figure 1-26

the position of the image. A professional lens bench is therefore equipped with a **nodal slide**, which is a vertical axis around which the lens can be swiveled to locate the nodal points – hence the principal points.

These considerations, plus the fact that thick lenses may be un-symmetrical (sometimes extremely so!), complicate the expression of their focal length. As discussed before, focal length is best defined as the distance from the lens of a point light source when the emergent beam is collimated or, conversely, the distance of the image point when a parallel beam is incident on one face of the lens. In an unsymmetrical lens these distances are different when measured from the surface of the lens. They constitute the **front vertex focus** and the **back vertex focus**; the latter is often important and may merely be called the **back focus**. Note that in an unsymmetrical lens there is a front and a back; the lens designer has decided which sides should have the greater and the lesser converging angle of the light beam.

The term in most frequent use by lens users, however, is the **equivalent focal length** or just e.f. This is **the focal length of a thin lens that has the same magnifying power as the thick lens.** Like all thin lenses (in air), the two equivalent focal lengths in object and image space are equal. The focal points are the same as those for the vertex foci but **the focal lengths are measured from the principal points of the thick lens.** This is a greatly simplifying concept and, in effect, has already been discussed in part.

We have seen that the equivalent focal length of a thick lens cannot be readily found, as can that of a thin lens, by measuring the distance from the lens at which the sun or some distant object is focused; that furnishes the back focal length. A nodal slide is required. Therefore the value of the equivalent focal length is usually furnished with a corrected lens. When it must be determined, however, a good approximation can most readily be obtained by measuring the distance for a given magnifying power according to the fundamental definition. It is simplest to set up an image of a plane at 1X magnification. This is easily done by illuminating a piece of ground glass or matte film that bears a scale or marks that are imaged on another piece of matte glass or film or even paper. The total distance from object to image planes is then measured. The focal scale tells us that the two planes are separated by four focal lengths, plus the distance between the two principal points. This distance must be estimated and subtracted from the total before dividing by four to find the focal length.* Any error in the estimated distance t, that affects the value of the focal length found, will also be divided by four.

Sometimes the photomicrographer may wish to combine some simple lenses, such as spectacle lenses, into a unit. This is most frequently done to make a condenser. If the total is still of negligible thickness, the composite is merely an addition of powers, as given in (1-7). When there is space between the components, (1-10) can be used, but the unit can also be treated as a thick lens, as just discussed, with its peculiar focal lengths.

Use of very thick lenses, especially highly unsymmetrical ones, does involve more complex considerations than use of thin lenses, especially if one is to be used under other than the exact conditions assumed by the designer of the lens. One of the problems is that of **vignetting**. If we look at the back of a lens, set up to form an image, through a small hole held in the image plane, the aperture of the lens will look round from a point on the optical axis. When the pinhole and the eye are moved from the axis, however, the aperture will appear elliptical. The eccentricity of this ellipse will depend on the thickness of the lens and design factors. The brightness of points in the image plane, that might be located by the position of the observer and the pinhole, will fall off from the central brightness according to the narrowness of the apparent aperture as seen from the image point.

Other thickness effects are again more noticeable in unsymmetrical lenses. **If a lens is used at a magnification decidedly different from that assumed by the designer and manufacturer** of the lens, a likely possibility is that the bundle of light rays will be limited by some other part of the lens barrel than the diaphragm provided for this purpose. This will not only make the brightness different from what it should be but the brightness and image illuminance relations may not even be normal. Moreover, certain quality degradations may occur. If especially looked for, however, this transfer of the position of the effective light diaphragm can be observed and the troubles avoided or expected.

---

*A more systematic yet practical discussion of the determination of the focal length of a lens, including a camera lens, is given in Appendix 2.

Aperture

Now let us consider an important characteristic of a lens and a term we shall need to use frequently, that is, the aperture. An aperture is a hole, of course, and a simple aperture in a thin wall, or diaphragm, as it would be called, **has certain properties in itself.** When the aperture is filled by a lens, the size of the aperture becomes the needed measure of the lens and thus designates its "speed." The diameter of the aperture is no exception to the rule, noted with the focal scale, that optical distances become simpler if focal units are used, that is, if they are used in a ratio to the focal length involved. This is the basis of the familiar *f*-system of specifying the size of the lens aperture on a camera. Specifically, **the *f*-number α of any lens aperture is the focal length divided by the diameter of the aperture.** This reciprocal of the more consistent unit formed by dividing by the focal length, usually keeps the value of the *f*-number above 1.0, for example, *f*/8 instead of $\frac{1}{8}$, at the price of having the **aperture grow smaller as the *f*-number becomes larger.** Since it is the *area* of a lens that governs the amount of light admitted by a lens, hence governs the exposure, the latter is proportional to the square of the *f*-number. That is why the apertures of most lenses have been marked in a series with a successive ratio of $\sqrt{2}$ (*f*/4, *f*/5.6, *f*/8, *f*/11.3, *f*/16, etc.) to allow the exposure to be doubled or halved easily.

Another way of measuring the aperture, used for the very near objects of microscopy is to **specify the angle subtended** by the diameter of the aperture (or its radius) **from the axial object point.** In practice we use the half-angle subtended by the radius of the aperture, shown as angle *AOB* in Figure 1-27. Specifically, the ratio of the lengths *AB/AO* is the measure of the aperture, since this ratio is a common measure of the angle. Some readers will immediately recognize this ratio as the sine of $\angle AOB$, usually called "sine *u*." (See p. 16 for the definition of sine.)

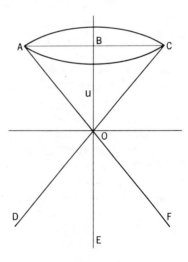

Figure 1-27

Although in photographic objectives the optical medium between the lens and the object is almost always air, with a refractive index of 1.0, this is not true of all microscope objectives and represents a more general case. The object space may be filled with oil or water and so have a different refractive index. The measure of the aperture can still be obtained simply by multiplying its value in air by the refractive index of the object space. It should be a constant for any objective. Thus we now have as a **measure of aperture** the expression

$$\text{refractive index (object space)} \times \frac{AB}{AO} = n \text{ sine } u = \text{NA}. \quad (1\text{-}12)$$

The last two letters, NA, stand for numerical aperture, which is the term arbitrarily given to this expression of the size of the aperture. Its value is engraved on most microscope objectives, and it is used frequently in this book and in other books on microscopy and photomicrography.

As pointed out later, when used properly, either system of specification of the aperture, the f-number or NA, can be used, whatever the object distance or magnification involved. The f-number is more convenient with camera lenses and the simple microscope; the NA is more convenient with a compound microscope objective, which is normally used at a fixed distance and (primary) magnification.

Other fundamental characteristics of the aperture are discussed later in this chapter.

Exposure Factors

Because aperture is one of the chief factors needed for determining a photographic exposure, we can now discuss how this exposure value can be determined. The brightness of the image on the photographic film will fix the exposure time needed for any film, of course, and this, in turn, is limited by the size of the aperture. Let

  I' = the illumination of the image,
  a = the diameter of the aperture,
  f = the focal length of the lens,
  α = the f-number,   aperture/focal length = $1/\alpha$),
  m = magnification,
  n = refractive index of the object space,
  b = an empirical constant, which has been considered (Gleichen, 1921) equal to the ratio of the entrance to exit pupil, a relation that is not always valid; however, $b = 1.0$ (or nearly so) for most symmetrical lenses,
  τ = transmission of lens; $\tau = 0.9$ (about) for most coated lenses, but may be considerably lower in the blue; the transmission of uncoated lenses can become quite low. For one lens of four uncemented elements $\tau = 0.65$.

$$I' = K \times \frac{a^2}{f^2} \times \frac{1}{(bm+1)^2}, \tag{1-13a}$$

$$I' = \frac{K}{\alpha^2(bm+1)^2}, \tag{1-13b}$$

$$I' = \frac{K \ (NA)^2}{m^2}, \tag{1-13c}$$

$$4K = n^2 K', \tag{1-13d}$$

$K = \frac{\pi}{4} \times \tau \times B = $ constant dependent on brightness B of the object and the transmission τ of the lens. $\tag{1-13e}$

Almost never do we determine explicitly the value of the constant K, although we do use it implicitly whenever an exposure test is made or an exposure table is consulted.

The photographic exposure can be defined as

$$E = I' \times \text{time of exposure} = I' \times t \tag{1-14}$$

Since there is *some* existent exposure value E_s that is optimum for each film, the brighter the illumination I' on the film, the shorter the required exposure time. Equations 1-13b and 1-13c are most useful, but for the moment we need note only the first one and that the *illumination* on the film is inversely proportional to the square of the f-number, whereas the required exposure *time* is directly proportional to it; that is, four times as much exposure is required at $f/16$ as at $f/8$. Note also that the exposure is also regulated by the magnification but by no other factors except as they affect these two.

Limitations of Corrected Lenses

There is another principle that we, as users of lenses, should know. We have noted some of the more pertinent defects of simple lenses; their possible image defects have been categorized into a still longer list. **There is no such thing as a perfectly corrected lens of universal applicability.** Corrected lenses are designed, usually by point-to-point ray-tracing but always by compromise, in which some perfection of qualities not needed for the specific purpose of the lens is given up to gain more of the qualities that will be needed. Among the assumptions of a specific design is the distance between the object and the lens, hence, with the specified focal length, the image distance also. Because of the reversibility principle of light the object and the image can be interchanged with no loss of efficiency. However, **if the optical assumptions made during the designing of a lens,** particularly the object and image distances, **are not adhered to, there will be some loss of efficiency,** that is, of the corrections to give good quality. Yet everyone who has a camera knows that the use of a corrected lens is not confined to a single object distance. It is still true that image degradation commences when departure from the assumptions of the lens designer is begun. If the departure is too great, the image degradation may become appreciable and finally unacceptable. It should be remembered that for this purpose also, object and image distances should be considered in focal-length units, although there are some corrections that in practice are more difficult to make on lenses with larger surfaces, and that these lenses as well will normally be of longer focal length.

We now have enough background of principles to enable us to use lenses with some ability, so let us consider this matter. Inevitably, consideration of illumination and image brightness, or photographic exposure, will also be involved, and before we are through so will image definition.

USE OF LENSES: UP THE MAGNIFICATION SCALE

Let us start with a familiar yet pertinent situation; let us take a camera on a sunny day into a field that offers a lovely landscape, perhaps with a few people in the background. The object distance is very great, especially when expressed in focal lengths, but the focal rule (Figure 1-24) still holds. Moreover, the assumptions made by the lens designer are holding rigorously. This is because the corrections incorporated in a camera lens design are almost always made with the assumption that the object is at infinity, which merely means at a very great distance when measured in focal lengths. As already noted, corrections for field curvature have also been made to allow the image to "cover" the prescribed negative size at the expense of some central definition.

We shall undoubtedly point and focus the camera with the sun directly behind us or somewhat to one side. The latter choice will affect the contrast within the camera image from flat front lighting to the extreme of pure side lighting.

How shall we determine the photographic exposure? Equations 1-9*b* and 1-12 prescribe the factors. If the subject is assumed to be of average reflectance, as is usually the case, the brightness of sunlight is the variable to be considered. This variable, however, is sufficiently constant for us to look up its value in a table. (The specific brightness value of sunlight will probably be only implicitly involved in the exposure factor.) The speed of the photographic film can also be found in a table. The transmission τ of the lens is usually sufficiently close to unity that it can be neglected. The magnification m is determined by the object distance in focal units. In this case it is so small with respect to the value 1.0 in the expression $(m+1)$ that it too, can be neglected. We normally state that the photographic exposure is solely dependent on the aperture setting of the camera lens, that is, its f-setting, with a given film on a given day, and we adjust the iris diaphragm on this assumption. Since the aperture scale is already marked in a $\sqrt{2}$ series, we do not even have to consider the fact that it enters our exposure (1-13b) rule as a squared factor ($1/\alpha^2$): for instance, we might decide to let $\alpha = 11.3$ by setting the diaphragm at $f/11$ on the scale, or, if we increased the exposure time by twice, from $\frac{1}{50}$ to $\frac{1}{25}$ sec, we would decrease the aperture by one-half the area, that is, to $f/16$ ($\alpha = 16$).

It is true that the size of the aperture has other effects than governing the brightness of the image. We know, for instance, that the depth of the field in focus decreases as the aperture is enlarged. On the whole, at these long object distances only considerations of exposure or depth of field determine how the diaphragm is set. Note that we need not deliberately consider centration of optics or filling the aperture of the lens with the illumination from the objects. This is true, in general, of reflected oblique illumination at all magnifications, since we can depend on the scattered light.

Now let us come close to the subject, taking pictures as we go. The illumination principles remain the same as we approach the object and the image grows larger up to the point at which photography of the face is important. We know that direct light coming down from about 45°, and the same angle from the side, brings out the contours of the face, but detail under the eyes and even beside the nose may be lost. To avoid this we may set up a reflector on the other side to control the brightness range to stay within the capacity of the reproduction scale of the paper print or reversal color film, which is the usual limitation. We may, instead, use an auxiliary light source to balance the lighting. If it is a cloudy day or if we are working in a studio and we put broad diffusers on the light sources, we shall have "flat" lighting and lowered tone contrast, with little difference between highlights and shadows. When photographing familiar faces, we may prefer a "Rembrandt" type lighting, but we must remember that when detail is unknown, as it may be at higher magnifications, its loss in the shadows is normally unacceptable.

How closely can we approach the subject in this simple manner by merely refocusing the camera lens and neglecting the fact that we are departing from the lens designer's assumption? The answer is dependent on a number of considerations but principally on our own standards of quality for the situation. It certainly depends on the subject; there may be no important fine details to be considered. The answer also depends on the aperture, since we might be able to ignore the assumptions of lens design at $f/11$ when we could not at $f/4.5$. Finally, it is dependent on the nature of the lens. An excellent wide-aperture lens, designed to be used at an object distance of infinity, is also most suitably quite unsymmetrical. Usually such a lens can be used down to 20 or even 10 focal lengths distance before the image quality becomes unacceptable. Therefore the focusing scales of most cameras of moderate size are discontinued at this point.

Close-ups

Somewhere in this region of object distance we may decide that it is impractical to continue to disregard the magnification factor m in (1-13b) for determining exposure. This can always be decided at any time by including the whole factor* $(m+1)^2$ by which the exposure time should be multiplied, instead of considering its value to be 1.0; for instance, when $m = \frac{1}{10}$ (11 focal lengths), the correct exposure time is $(1.1)^2 = 1.2$, or 20% longer than when the factor is neglected. By the time $m = \frac{1}{2}$ the correct exposure will be $(1.5)^2$ or $2\frac{1}{4}$ times longer than if the factor were neglected, as in work at a distance.

When taking photographs closer than the distance at which the lens design assumptions hold, **the simplest remedy** is to restore the former condition purely optically, that is, with the object far away in focal units, by *inserting in front* of the camera lens an auxiliary lens whose focal length is equal to the subject distance. This auxiliary lens acts as a collimator, of course, and passes parallel light from the subject to the camera lens, a successful procedure, called **focal-frame photography**. Here our camera is again focused at infinity or some adequate distance. The magnification of the camera image is given by the ratio of the focal lengths of the two lenses if the subject remains at the focus of the auxiliary lens, that is, if the supplementary lens is chosen to have a focal length equal to the lens–object distance:

$$\text{magnification} = \frac{f(\text{camera lens})}{f(\text{supplementary lens})}. \tag{1-15}$$

Supplementary lenses are usually simple (spectacle) lenses and have all of the aberrations of a simple lens. Their use is acceptable therefore

---

*We can frequently consider that $b = 1.0$. Highly unsymmetrical camera lenses are not usually suitable for close-up work.

only because the percentage of the optical work done by then is small; that is, their power is weak compared with that of the camera lens. A meniscus-type lens is best as a simple supplementary, with its concave face toward the camera lens, since this will least degrade the flatness of the image in the field. As the object approaches closer and closer and the power of the appropriate supplementary lens becomes greater with respect to the original camera lens, a lens of better and better quality must be chosen, such as a simple achromat, until at a magnification of 1X the two lenses should be equal in quality.

The procedure just outlined does bring the use of quite unsymmetrical lenses back to that assumed by the lens designer and does improve the results obtained with such lenses at "close-up" distances. However, camera lenses of almost symmetrical designs, such as the Tessar type, give excellent definition at moderate apertures from infinity to at least four focal lengths. With lenses of this kind the use of a supplementary lens would only degrade the definition.

The fundamentally best procedure is to select a lens designed for the use to which it is put, in this case for object distances of 10 focal lengths or less. An enlarger lens was made to work under just these conditions and is therefore excellent for the purpose. Because of the reversibility principle of light, we need only to see that the side of the lens that was designed for the shorter distance, either to object or image, is still used for the shorter distance. In this case the side that faced the easel in the enlarger should now face the object, since that is farther away than the camera image.

The illumination principles remain the same. We shall probably still want the illumination to come from the upper right or left; as the magnification increases, it becomes more and more important that the viewer of the picture know the direction of the incident light. Sometimes we may choose a front lighting and even that from all azimuths, for example, a ring illuminator, since it is becoming so important that no detail be lost in dark shadows.

Now we may have reached a magnification of 1X. Since the object has been approaching the lens, the image has been retreating, until at 1:1 they are four focal lengths apart and the system is entirely symmetrical. It is therefore not surprising that the best type of lens for this magnification is completely symmetrical. A process lens is such a lens, and a modern process lens, which is well corrected for color, is excellent for the purpose if one is available. It is important that the diaphragm, or stop, be in the middle of the lens. This is why placing two excellent camera lenses face-to-face will not give a lens equal to a good single lens of equal focal length; the limiting stop should be placed in the middle.

Note from the exposure formulas (1-13) that the time of exposure at $m = 1$ is four times that when m is very small and the object distance is great.

Entering the Domain of Photomicrography

If we now bring the object still closer, the image retreats and is magnified. We have entered the domain of photomicrography. The reversibility principle of light is pertinent here, since all the preceding discussion of optics still applies in reverse as we go up the magnification scale. A symmetrical lens, such as a process lens, should show some superiority from about $\frac{1}{2}$X magnification to about 2 or 3X. From about 3 through 20X an enlarger lens is again one of the best choices. Moreover, practical tests have confirmed the fact that modern enlarging lenses, such as the Enlarging Ektars, are preferable to the special photomicrographic lenses of the older formulas made for years for **photomacrography**, which is the term for this range of low-power photomicrography. Although its upper limit is often quoted as 10X, it can be advantageously taken as about 15 to 20X; its lower limit is likewise advantageously taken as $\frac{1}{3}$ or $\frac{1}{2}$X, since the optics and the technique throughout this domain are usually so similar. The lens producing the magnification is itself a microscope, since it is an optical instrument for magnifying small objects; it is a simple microscope, as opposed to a compound microscope. Of course, if we use an enlarging lens above 1:1, it must be turned around from the way it would be used as fractional magnification, if it is at all unsymmetrical, so that the end that faced the easel (longer conjugate) now faces the photographic film.

If ordinary camera lenses plus supplementary lenses are to be employed as the magnification becomes greater, the camera lens should be reversed so that its usual front faces the film. The supplementary lens should go on the image side. It is still on the original front of the camera lens, since the latter is now its rear. The ratio of focal lengths expressing magnification becomes the reciprocal of that of (1-15).

Finally, for 20 to 25X and higher magnifications with the simple microscope we should use the camera lens alone, since it will again be employed as it was designed to be if the original front of the lens faces the film. The image distance is now the longer conjugate and is reasonably large in focal lengths. Since the angle subtending the field and its image is quite small, a camera lens designed for 16 mm or even 8 mm cinematography is especially suitable. At 35X magnification and above superior photomicrographs will be obtained from these lenses over photomicrographs from lenses made for photomacrography, which are likely to have been designed with the assumption of a magnification of 10X.

As usual it is necessary to consider the assumptions of the lens designer when using a lens in a manner not strictly prescribed. Sometimes it is the mount of the lens and not the glass lens itself that is the limiting factor in its use in an originally unassumed manner. When a lens designed for infinity focus in a camera is employed in photomacrography, it may be found that the iris diaphragm cannot be seen with the eye in the image plane until it is closed down appreciably from the widest marked aperture. When the diaphragm is opened wider than the critical aperture for any magnification at which it is acting as the stop, the effect is far more serious than the fact that no additional image brightness is gained with further opening of the diaphragm. When the stop is transferred from the diaphragm to part of the outer lens mount, image quality will degrade and, moreover, the image brightness or exposure formula (1-13*b*), will behave as if a *b* factor appreciably different from 1.0 had been introduced.

Illumination (Reflected Light)

As we go up the magnification scale, we shall find the illumination arrangement dominated by two factors: first, the relative brightness of the image is decreasing with magnification as $(m + 1)^2$, unless we pour on increasingly more light, as we usually do; second, as the magnification increases, the working distance between the subject and the lens decreases until it indeed becomes the governing factor in the apparatus and technique employed. This mechanical factor causes so much apparent difference in methods that it is often overlooked that the fundamental principles may still be similar.

The simplest illumination method, in principle, although not always in practice, is that in which the light source shines directly on the subject, as it does outdoors and may in the studio or the living room with bare lamps. As the working distance gets smaller, the lamps must be smaller. They may range from large studio lamps to ordinary electric lamps (up to magnifications of about $\frac{1}{4}$X gooseneck lamps with photoflood bulbs are useful), then to automobile-headlight lamps, and finally to flashlight bulbs. This process of merely reducing bulb size is stopped primarily because we need increasingly more light at a high rate, and the smaller lamps give less of it. Instead of reflectors to fill in shadows, it becomes preferable to employ more lamps and a ring illuminator becomes exceedingly useful. (Illuminators are discussed in Chapter 4 and a photomicrograph made with such a device is shown in Figure 4-4.) It is obvious that this same discussion and general sequence apply to the **use of flashbulbs.** Since they are expendable, a great gain in the light from one bulb is obtained over those burned continuously. Therefore these lamps

are most suitable indeed for photomacrography, *including photomacro-*
graphs of still objects.

In another method of illumination, from studio photography through
all magnifications one or more beams of light with lenses or reflectors are
directed into the field from some distance outside. Studio spotlights are
in this category. The great advantage of this method is that, as working
distances decrease with magnification, the size of the lamp, with a
directed beam, need not decrease proportionately, hence can usually
furnish more light than can be obtained with the direct use of bulbs. The
problem is to control the illumination and its uniformity with easily
available bulbs. It is possible to gain uniformity and some control by
placing diffusing screens in the beam or by directing the beam against a
matte reflecting surface. This method is particularly useful for specimens
with round shining surfaces but is very wasteful of light. The best general
solution to the problem is the use of two condensers in which the second
condenser images the lamp condenser to form a uniformly lighted field
of any specified diameter. This is described in Chapter 4. Diffusion can
be added when advantageous.

When the magnification increases to such high values that only a very
small space is available between the object and the objective in its mount,
it becomes necessary to bring in the illumination beam through a condenser built around the objective by utilizing such special equipment as
the Leitz Ultropak or the Zeiss Epicondenser. Although the mechanical
difficulty of decreasing working space has forced a change in apparatus
and technique, the fundamental principles, including the mode of illumination, remain the same. This is the type of illumination with which
we view most objects and which seems most natural. It probably is not
chosen sufficiently often in place of transmitted light for viewing translucent objects. When the light is coming mainly from one direction, it is
important that the viewer of the photomicrograph be aware of that
direction. Notice that we are still depending on the specimen to scatter
the illumination, and thus to fill the whole aperture of the objective, and
that there is little or no trouble due to alignment of equipment. An
example of this type of illumination is shown in Figure 1-28 (see color
insert).

Specularly Reflected Light

There is one highly specialized but important field of photomicrography (i.e., metallography) in which the photomicrographer finds himself dealing with a specimen that in essence is a flat mirror. This involves
so-called **vertical illumination**, which could more generally be called

specular illumination. With these specimens we have entered a new phase of photomicrography.

Although we are still using reflected illumination, we now find that we must align the optics carefully so that the illumination and imaging optics form an integral system, and we must also deliberately arrange to fill the aperture of the objective lens to the extent that is desired. Note that these characteristics are usually associated with the type of illumination transmitted by the specimen. For the moment we merely remark that the exposure time will decrease and the reproduction of detail will increase as the aperture of the objective is increasingly filled by the illumination beam. The criterion to determine the amount of care required in filling the necessary objective aperture or aligning the optical system is the answer to the question: What percent of the light used to form the magnified image was scattered by the object? It is usually high for non-metallic specimens viewed by reflected light. A discussion of the illumination techniques for metallography appears in Chapter 4.

Transmitted Illumination

As magnification increases from very low values, there is always a tendency for the sake of convenience to resort to transmitted illumination as a standard method of microscopy. It overcomes the danger of excessively low image brightness with increasing magnification which occurs with oblique reflected illumination and which has already been discussed, since so much more of the entire original cone of light can now be gathered into the final image. Because of the limited depth of focus available with increasing magnification, microscopists have resorted to various devices for holding a specimen flat and making it thin. Making a thin section of the specimen is one of those devices that of course also serve other important purposes. Moreover, in transmitted illumination a short working distance between objective and specimen no longer limits the illumination available. In spite of these great conveniences, a microscopist should often ask himself whether his specimen cannot better be represented by the more familiar reflected type of illumination.

The simplest form of **transillumination** is that in which the specimen scatters so much light that it fills the objective aperture naturally. Some specimens do this, but in any case an alternative is available. It consists in putting a scattering medium *directly* behind the object, that is, mounting the specimen on opal glass. With this dodge, we return to the system by which the objective aperture is filled automatically and centration and alignment of the illumination are easy. Its potential disadvantages

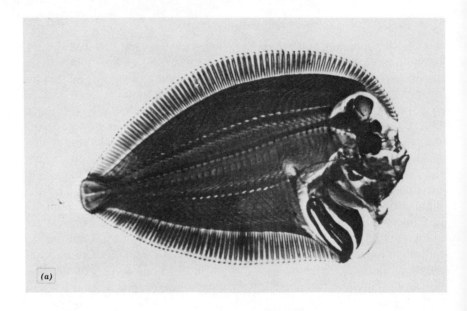

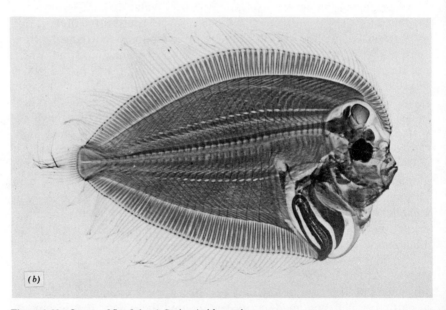

Figure 1-29 Larva of flat fish ×6. Stain: Acid carmine.
(a) Photomacrograph set for criticism, taken with inadequate aperture in illumination system; (b) slide placed on opal glass having a black paper diaphragm with 9/16-in. hole, ribbon filament, 72 mm Micro Tessar, $f/4.5$, 6-sec exposure, no filter, Kodak "M" Plate, D-41 developer.

48

are the lack of control of the aperture of the illumination and the ease with which flare light (i.e., light not proceeding from point to antipoint within the field *used*) can enter the field in the image plane to degrade the image. Therefore in this method it is *absolutely essential* that an aperture whose diameter is that of the intended field be mounted on the side of the opal glass facing the lamp. This mount can be a piece of black paper bearing a hole the size of the field or a clear aperture in a black-painted shield sprayed on from behind the opal glass. The aperture is easily made by using a temporary paper disk as a mask. This simple method is capable of yielding illumination superior to that of a poorly aligned system containing no diffusion or one with an inadequate aperture. Figure 1-29*a* is a photomicrograph obtained by a microscopist who used a condenser-objective system with no diffusion and inadequate aperture. Figure 1-29*b* is the result obtained with this same specimen by the method just described, that is, with the specimen on opal glass. A ribbon filament was focused directly back of the field. This method has the fundamental advantage of great versatility and at the same time of allowing complete control of field and illumination apertures. Yet we might have to examine the original photomicrographs to be able to detect the superiority of the results from this system. The method of mounting on opal glass, as the microscope slide, adequately diaphragmed, can be utilized from very low magnifications up to the point at which it is mechanically difficult to diaphragm it well at medium magnifications.

No further description of the illumination system is given in this chapter. As we go up the magnification scale in photomicrography by transmitted light, exactly the same considerations hold as in photomicrography by reflected light. We have already discussed the selection and use of the appropriate objectives; for instance, unsymmetrical camera objectives should be mounted so that the end that originally faced the film now faces the object when used to produce a magnified image. Figure 1-30 (see color insert) is a photomicrograph made with a 25 mm, $f/1.4$ cine lens, thus mounted. We may have flatter specimens for transillumination and therefore need sacrifice less for depth of focus. We have not discussed the principles governing the relationship between image definition and objective aperture. These principles apply at all magnifications but are more frequently the limiting factor at relatively high magnifications.

Chapter 2

General Optical Principles: The Compound Microscope

As the magnification becomes higher, objectives of shorter and shorter focal lengths with still higher apertures are required. This causes the working distances to become too small for much practical work, although in the very early days of microscopy Leeuwenhoek not only used simple lenses at high magnifications to make significant discoveries but preferred them to the compound microscopes of the time! As corrected optics were developed, however, the compound microscope was universally adopted for higher magnifications. It uses a primary lens (objective) of longer focal length to form a real magnified image that is treated as the object for a second lens (ocular). This system, with its valuable longer working distance, allows better optics of larger aperture to be designed by leaving the fulfillment of some optical design requirements to the ocular.

Objective and its Primary Image

The rule discussed in Chapter 1 that **a lens must be used as its designers assumed it would be**, for its optimum performance, **applies more and more strictly as magnification and aperture increase** until, with high-power objectives, variations of the object distance of very small fractions of a millimeter (or the same percentage of the image distance) cause noticeable changes of image quality. Optimum focus is assumed in each case. Not only is (1-4) applicable to describe the position and magnification of the objective and its primary image but also l and l' should be invariant for the reason just mentioned. Therefore a magnification, m_0, can be definitely assigned to each objective to describe the magnification of the

50

primary image. This value is usually engraved on the objective itself, which is now designated by this primary magnification in the catalogs instead of by its focal length, as done formerly. Objectives with both types of designation are in common use.

In a compound microscope the image formed by the objective is viewed and remagnified by an ocular, although in the common Huygenian type of ocular a lower "field lens" intercepts the beam just before it would otherwise form the objective image, as illustrated in Figure 2-1. In general, however, when used visually, the ocular acts as a simple magnifier. Its behavior is discussed in Chapter 1. For photomicrography a real image is formed in the camera and (1-4) also applies to the ocular. However, the "power" or magnification due to the ocular is designated by its visual use and so is fixed for each ocular. Let the designation of the ocular magnification be m_e. **The compound microscope is, then, a relay optical system** whose magnification M is

$$M = m_0 \times m_e$$

It will be found that when the image is projected it has this magnification, obtained as a product of those of the two components, *when the image is 10 in. from the eyepoint of the microscope.* If the microscope is focused to project a real image, as in photomicrography, the magnification will normally be this catalog (product) value, multiplied by the number of times 10 in. can be divided into the *bellows length*, which is the distance from eyepoint to final image.

The compound-microscope system usually begins with a 2X objective which has a focal length of 48 mm, or about 2 in. Therefore **there is an overlapping field of application in which either the simple or the compound microscope may be chosen.** The magnification range of the simple microscope is extended upward (from about 25X) by the use of reversed cine lenses of short focal length, as discussed in Chapter 1 and illustrated in Figure 1-30. The simple microscope is used when the size of the field is important for the reasons discussed on p. 32. On the other hand, it is not generally realized that **an objective made for a compound microscope can be used advantageously as a simple microscope when high central definition and contrast are more important than size of field.** This is partly because there are stern warnings in some texts *never* to use such an objective without an ocular. Actually, **within the central field the ocular can only degrade definition and contrast as it extends the area of the acceptable field!** With the 2 and 4X (48 and 32 mm) objectives, this area of markedly superior quality is not negligibly small. These facts are important to motion photomicrography, since the fields of the motion picture frame are quite small. The size of the superior central (paraxial) field diminishes rapidly with the decrease in focal length of the objective, but that of the 10X (16 mm) achromatic objective is still probably adequate to cover a 16 mm motion picture frame in a superior manner. On the whole, this does not apply to apochromatic objectives, except with monochromatic light, since part of its chromatic correction is made with the ocular.

Compound Microscope, a Relay System

This is probably the place to discuss the implications of the compound microscope as a relay system, although the basis for some of the argument is not discussed until later in the chapter. As a relay system, the objective forms a real image which is viewed by another lens that remagnifies it for the eye to view.

There can, however, be a number of variations of this relay system, all of which end by having somebody look at the final image directly through an eyepiece, by projection on a screen, or by having the eye view a photomicrograph, which is a record of the projected image. Moreover, **it** really **makes no fundamental difference how we vary this relay system mechanically.** The principal competing variation of this method is to record the primary image photographically and then to enlarge the photograph by projection, either directly on a screen to be viewed visually, or to make another photograph (enlargement) to be viewed later. In any case, **the primary image of the relay system contains all the elements of quality and the reimaging and recording systems can add no further definition.** Although they can and will degrade it, since neither perfect optical nor photographic systems exist, with proper selection and use of these systems this degradation of the image should be negligible with respect to the limitations of the human eye. Often such a photographic relay system is so convenient that it is the only practical method; usually this is because of the greater brightness of the image of less magnification, with its shorter required time of photographic exposure, or the lessened bulk of the apparatus. All cinemicrography must be included in this category.

Certain fundamental principles, however, offer guiding rules which have sometimes been neglected. The original objective and its correct use are the principal limitations to image quality, whether the objective is in a microscope or a small camera. We can consider these limitations together with the characteristics of the image that is finally viewed, neglecting the exact method of forming it or even the number of intermediate steps involved, except as they introduce image degradation.

The only factors to be considered are the objective aperture, the magnification of the *final* image to be viewed, the quality of the lens, and finally its appropriateness or its correct use. The last factor must be considered from a practical viewpoint and has already been discussed. The focal length is not normally one of the direct factors, since the same conditions can, as a rule be reached with any focal length. At the extremes of the scale it can enter as a practical consideration; most mathematical optical expressions that have eliminated it have done so by neglecting secondary terms that are usually negligible. Because of these principles, the requirements for objective aperture should be considered directly with respect to the *final* magnification of the image to be viewed, especially in regard to limitations of definition. If the final image is on a projection screen, the final magnification is determined by the ratio of the angle that some length in the picture subtends on the screen from the eye, to the angle it subtends when the intermediate picture projected is viewed in the hand. This is discussed specifically in Chapter 21 on cinemicrography.

When the statement is made that greater depth of field is obtained by intermediate use of a 35 mm camera than by use of a camera that produces the image quality at the final viewing magnification, it normally means that subjects are involved for which an appreciable depth of field is more important than the definition that otherwise could be obtained at the final image magnification. Such subjects abound in nature photography. This subject is discussed explicitly in Appendix 1.

Tube Length

As we noted at the beginning of this chapter, one of the advantages of abandoning the simple microscope in favor of the relay system, which is the compound microscope, is that *for any one objective* the object and image distances l and l' can be made constant to assist the lens designer. The primary image plane is Plane II of Figure 2-1 and l' is the **optical tube length** of the microscope and will necessarily vary with the objective. It is usually unknown except to the designer. To assist the user, manufacturers of microscopes have generally been able to extend the mechanical mounts that hold the various objective lenses in such a way that the location of the primary image is at a constant position below the top of the microscope tube. Thus the distance, called the **mechanical tube length**, from the bottom of the shoulder into which the objectives screw to the top of the brass tube on which the eyepieces (oculars) rest should be a constant one. Of course, the mechanical mounts of the lenses of the oculars must be made to focus this plane. Most makers of microscopes of the biological type have adopted 160 mm as the standard mechanical

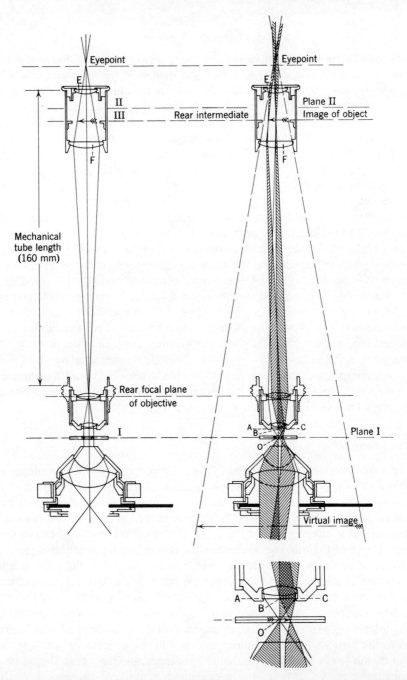

Figure 2-1 Compound microscope: (*a*) path of rays focused in the aperture planes; (*b*) path of rays focused in the object and image planes.

tube length; the E. Leitz Co. and Busch design for 170 mm. In many special types of microscope, including metallographic stands, it is impossible to keep the tube length that short. The objectives are then designed for the particular tube length; in the metallographic microscopes of Bausch & Lomb the objectives are designed for a mechanical tube length of 215 mm. Sometimes an optical dodge is resorted to, such as using a telescope lens between the objective and the ocular to take a parallel beam from the objective (which is "corrected for infinity") and then focusing the primary image by this telescope lens into the mechanically correct position for the ocular.

The practical question that inevitably arises is whether oculars from various manufacturers can be used indiscriminately with the objectives of any one single make or vice versa. The answer is not simple. It has been said, even by designers of microscopical lenses, that in practice the standard types of ocular from the various manufacturers could be interchanged freely without appreciable image degradation. Of course, everyone agrees that the *type* of ocular (Huygenian, compensating, etc.) appropriate to the objective must be used for excellent results. This is discussed in Chapter 3 and later in this chapter.

As shown in Figure 2-1, with no ocular in the tube the image formed by the objective would be formed within the tube at Plane II. The field lens of the ocular lowers the image somewhat to Plane III. **The position of Plane II,** at a distance below the top of the mechanical tube, **should be identical for all makes of microscope** if it makes no difference whose oculars were used. This is not accurately so, as discussed in some detail by Kurt Michel (4. 1967) (in German). Table 2-1 gives this value Δ of the distance of Plane II below the top of the microscope tube. Those of Reichert, Zeiss, and Zeiss-Winkel have recently been changed.

The existing level of the primary objective image is, of course, correctly utilized by oculars of the same make as that of the objective. For others the effect is that of using the ocular at the wrong tube length. The consequences are discussed later in this chapter. **Objectives corrected for infinity** (i.e., for a collimated image beam) **must be used with a microscope that contains a telescope lens** in its upper tube to focus the image in Plane II for the ocular. **Conversely,** if a microscope is made to be used with its objectives corrected for infinity, it will contain such a telescope lens so that **objectives of different tube-length design cannot be used with it.** Considering the largest discrepancies shown by the table, this error can be appreciable for dry objectives whose NA is greater than about 0.5 and for most immersion objectives.

A most important consideration is relatively new. The practical statement of the equivalence in use of all makes of oculars of the same type

Table 2-1 Tube Length and Lack of Parfocality in Objective-Image Distance

Make	Tube Lengths (millimeters)	Δ Top of Tube Plane II (millimeters)	Remarks
Bausch & Lomb Inc.	160	11	Biological
	215	11	Metallographic
	∞	11	Standard teaching
American Optical Co.	160	11.3	Biological
	180	11·3	Metallographic
	∞	11·3	Series 10 and 11 microscopes
E. Leitz	170	18	
	185	18	Ultropak
	215	18	Ore microscope
	∞	18	Metallographic
C. Reichert	160	13	
	190	13	Metallographic
Busch	170	13.7	
	190	13.7	
Zeiss[a] (Oberkochen)	160	10	Old value of $\Delta = 13$
Zeiss-Winkel[a]	160	5	Old value before 1948, $\Delta = 15$

[a]The objectives made for the MicroStar microscopes by the American Optical Co. and the metallographic objectives made by Zeiss are designed for an "infinite tube length" to make them insensitive to variation of the tube length *below* the telescope objective carried in the body tube. They are just as sensitive to tube length over-all and above the telescope lens that focuses the collimated beam from the objective to form the primary microscope image. This subject is discussed in Chapter 3.

was due to the fact that the great field curvature of all objectives of even fairly high NA was so great that this factor limited the size of the field before other factors, especially astigmatism, could show an effect. Now, with new glasses and the use of computers for design, objectives of superb central definition yet appreciably larger field within the tolerances for curvature have appeared. (Technically speaking, the Petzval sum has been made much smaller.) Some new-type wide-field eyepieces are available. Now the specific compensation of ocular and objective becomes important, especially for obtaining minimum astigmatism of the final image for the combination of objective and ocular. Therefore, **when**

objectives of newer design are purchased independently of a microscope, one should look for the manufacturer's recommendations for the oculars to be used with them. These are not always easy to get because of the common assumption of the companies that only their equipment should be used anyway.

Such special objective-ocular combinations include the following:

Objective	Ocular
Leitz Plano	Periplan and Periplan GF
Zeiss Neofluar	Compensating Plano, KPl
Zeiss Planachromat	Komplan
Zeiss Planapochromat	Komplan

Leitz particularly states the point that "use of the negative oculars for photography (Homals and Ampliplans) with plano objectives would be wrong."

Distance of Microscope Image. Now the microscope designer *could* design the ocular and the positions of its lenses so that the final image of the microscope would lie at any desired distance from the eyelens. In practice, except for some special types, **the optics of microscopes have been designed as if they were only to be used visually**, that is, **with parallel beams** emerging from the exit pupil (eyepoint) of the microscope **to form a divergent cone of light. Each beam** (having the diameter of the eyepoint at that location) **comes from one point in the object**, and the plane on which these individual beams of the bundle focus determines the location of the image. Therefore this statement concerning parallel beams is equivalent to saying that the **image is at infinity.** Note that this is entirely independent of the statement that the magnification of the microscope has its rated value at a projection distance (bellows length) of 10 in. or 250 mm; this is merely an accepted convention (see Chapter 1).

In order to focus to a real image at some shorter distance from the microscope, the eyelens of the ocular could be withdrawn a bit until the new projection distance is obtained. This would leave the objective, and also its primary image, undisturbed and as the lens designer intended. **If the microscope is refocused as a whole** (*fixed mechanical tube length*) to refocus the final image, **the objective will no longer be working at its prescribed condition.** The tolerance for this common practice is discussed in this chapter. **Any change in the distance of the ocular that represents a different tube length will not only degrade the image but also change the magnification.** The use of an incorrect tube length can be considered as introducing only spherical aberration in a well-centered system. **The first noticeable effect is a lowering of the contrast of the image**; its definition *might* even improve slightly. However, more spherical aberration will degrade definition and can make it very poor. The most sensitive test, and one that will identify pure spherical aberration, is the "star test," made with a pinhole in an opaque slide (qv). It will also indicate the way to change the tube length to reduce the aberration. **A tube length that is too short introduces undercorrection, a characteristic of a simple lens, whereas a tube length that is too long introduces overcorrection.**

There is an optimum tube length equivalent to the mechanical tube length of 160 or 170 mm, at which the optical characteristics of the object space are assumed to be correct. In practice there is some tolerance which can be expressed as a length on each side of the optimum position $(\pm \Delta L)$ within which the observer's eye cannot detect the degradation from the optimum. This **tolerance to change of tube length is affected only by the numerical aperture*** of the objective, in good objectives. Values for this tolerance, as determined by the star test and for dry objectives with the correct cover slip, are given in Figure 2-2, line A. Line B shows the tolerances allowed by immersion objectives of various apertures, **with or without a cover slip.** When the acceptability of an image of a specimen rather than the star test is the criterion, the tolerance may become about eight times greater. In a photomicrograph to be made with a 4 mm, 0.95 NA objective this allowable variation from the strict optimum would be about ± 8 mm. Because the tolerance graph is linear on logarithmic paper, the optical demand for exactly the correct tube length diminishes rather rapidly with decrease in objective aperture, that is, for objectives used at lower magnifications.

When classifying particles according to size by means of a reticle in the eyepiece, it is common practice to alter the tube length until the units of the reticle are equal to exact integral numbers. This is usually wise, but the allowable tolerance, as given in Figure 2-2, must be kept in mind.

The optimum tube length of a commercial objective may not be exactly the standard 160 or 170 mm with a standard cover slip (or without one, if so corrected). This is largely because the separation of some lens components affects the value of the optimum tube length so powerfully that the elements would have to be adjusted within microns of the specified separation. Therefore, **when optimum tube length is very important**, as it may well be at high magnifications and high objective apertures, the optimum tube length **should be determined for the particular**

---

*See Chapter 1 for a definition of NA. It is discussed in more detail later in this chapter. In this and the next few pages the NA value referred to is that engraved on all objectives.

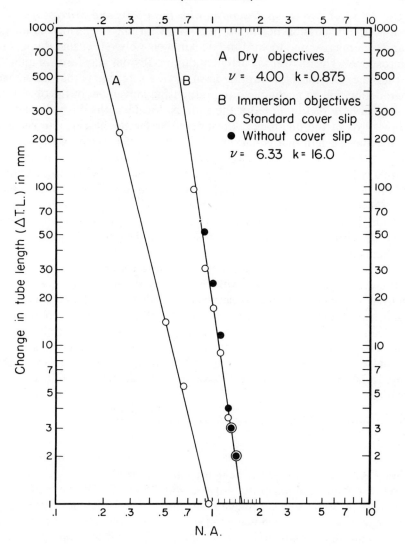

Figure 2-2 Tolerance to tube length change versus objective NA.

objective and, if possible, for the particular specimen and its cover glass. Although this is most sensitively done with the star test, any fine, well-defined structure will do; very fine particles, including dirt in the field, will do. The adjustment, in this case, is for maximum sharpness and contrast with no loss in resolution.

Use of Short Projection Distances with the Compound Microscope. To resume the discussion begun on p. 57, how far may we bring the image plane of a microscope toward the ocular by merely refocusing the whole microscope? The answer is partly conditioned by our standards of quality at the time. Obviously, from the discussion, a safe rule is that the image degradation will be negligible if the refocusing movement of the objective is negligible *in focal units.* For the shortest focal lengths this will be only a few turns of the fine-focus adjustment. **By far the best way to refocus for a bellows length that is too short is to pull out the eyelens of the ocular.** Oculars that allow such eyelens focusing are available. The next best artifice is to refocus by pulling out the entire ocular. The usual technique for an image plane that is too close is as follows: first focus visually directly through the microscope with a relaxed eye, and then refocus on the camera back with the eyelens of the ocular, or by pulling out the whole ocular. Finish to optimum focus with a slight adjustment of the fine-focus mechanism.

Obviously **an alternative technique is to insert a camera lens, in place of the eye, behind the ocular, with the film at its focus**, that is, a camera focused for infinity or any relatively long distance. The technique is discussed in detail under the subject, "Simplified Photomicrography." The camera lens will, of course, record the whole field, normally seen visually on the film, usually with an outer zone out of focus, as is the normal case with the eye. The magnification will be equal to the rated magnification M_r of the objective-ocular combination multiplied by a fraction:

$$\text{magnification of camera image} = M = M_r \times \frac{f \,(\text{millimeters})}{250}$$

$$= M_r \times \frac{f \,(\text{inches})}{10} \quad (2\text{-}1)$$

when f is the focal length of the camera lens.

When commercial apparatus for use with 35 mm film first appeared on the market, some devices placed the film a short distance above the ocular with a split-beam viewing telescope but no correcting lens between ocular and film. The degradation of the image was often apparent when adequate comparisons were made. Such conditions have now been corrected in the commercial apparatus known to the author. **Often a compromise has been made; the camera (or correcting) lens has a focal length two to three times longer than the distance between it and the film.** Thus some refocusing is necessary, but the out-of-focus zone is reduced or even eliminated, since only the central field is recorded. The magnification in the image plane of the camera will remain the same with the weaker lens if the height of the camera back from the eye point is unchanged, since this depends only on the height of the image plane with respect to the magnification standard distance of 250 mm. Equation 2-1 can be rewritten

$$\text{magnification of camera image} = M = M_r \times \frac{p}{250}, \qquad (2\text{-}1a)$$

where p, the distance from eyepoint to camera image is expressed in millimeters. This formula neglects two factors that are usually negligible: (a) change with distance due to refocusing from the best visual focus, and (b) some change in the optical distance if the glass of the camera lens is quite thick. The latter factor may reduce the magnification slightly.

Optics of the Object Space

Turning to physical and optical consideration of the object space of a compound-microscope objective, we find that many **limitations are much severer than those for factors in the imagespace,** such as tube length. This is because all lateral dimensions are divided by the magnification and all axial distances by the square of the magnification from the corresponding lengths in image space. At high magnifications this may indeed represent critically severe prerequisites for an excellent image.

Probably the severest restriction is that of depth of field, which is most galling when an uneven surface is being examined by reflected light. This problem is discussed in Chapter 4 and Appendix 1.

With most objectives for use with transmitted illumination it is assumed that the specimen will be under a cover slip (normally glass) of a standard thickness and refractive index.

Cover-Slip Thickness. American manufacturers of microscope objectives have designed them to be used with cover slips 0.180 mm thick. Manufacturers on the European continent have assumed the cover slips to be 0.170 mm thick. For many years a silly situation existed in which manufacturers designed their objectives for 0.180 or 0.170 mm cover slips, but in practice it was almost impossible to obtain them is quantity because cover slips are sold according to thickness class. Classes 1, 2, and 3 were on the market, but the specified thickness lay in between the thickness range of the No. 1 and No. 2 cover glasses. Recently cover slips of a new thickness class, No. $1\frac{1}{2}$, have appeared on the market. In this case the standard thickness around which most cover slips are grouped is 0.180 mm* and the variation is much narrower than

---

*The thickness-frequency distribution of some commercial slips of this type and from two commercial sources is given in Spinell and Loveland (4. 1960).

for the other types. How important is it to have a cover slip of the exact thickness assumed by the designer? What is the tolerance?

A cover slip is a parallel sheet of glass placed in the object space. If inserted when it is not assumed to be there by the lens designer, or if it is of a greater thickness than that assumed, spherical aberration will be introduced into the image. This, of course, is exactly the effect of variation of the tube length and also of variation of the distance between certain lens elements in the objective. Therefore deliberate variation of either of these conditions can be used to compensate for an incorrect thickness of a cover slip. Correction collars varying the distance between the lens elements are incorporated in most 4 mm apochromatic objectives. For other objectives, however, variation of tube length is a more generally available device. **A decrease in tube length is equivalent to a decrease of cover-slip thickness; consequently, to obtain minimum spherical aberration with a cover slip that is too thin increase the tube length. Conversely, with a cover glass that is too thick, decrease the tube length.** Criteria for these variations are discussed in Chapter 6 under "Star Test."

Spinell and Loveland determined the **tolerance to a variation of cover-slip thickness from the optimum**, depending on the visual inability to detect the degradation in the image within a certain range. The result is shown in curve A of Figure 2-3. It is a straight line on log-log paper when plotted against the numerical aperture of the objective. This means that **no other factor than the objective NA is involved.** Therefore it is not necessarily a function of the directly illuminated NA, as opposed to that of the whole objective. This is because the scattered light from the star test fills the objective. The test was repeated with photomicrographs of a critical specimen (a distribution of fine crystals), first to discover the tolerance for the first-detectable degradation of definition and second to determine a limit of acceptability, albeit a critical one. In each case the result was a straight line on the log-log graph as shown in Figure 2-3 and each line was parallel to the preceding one. Such a curve can be expressed by an equation:

$$\text{tolerance to variation of cover-slip thickness} = T(\Delta l) = k\,(\text{NA})^{-\nu} \qquad (2\text{-}2)$$

The value of ν is a constant for this function, but the value of k determines the position of the line on the graph; the curve is displaced to the right with increasing k. The parameter k, then, is a measure of the critical judgment required to determine acceptability or rejection of the image for definition. Values of these constants determined by Spinell and Loveland (4. 1960) for dry objectives were as follows:

Test	Criterion	ν	k
Star	Least-perceptible degradation	4	1.0
Photomicrograph	Least-perceptible degradation	4	2.5
Photomicrograph	Maximum-acceptable degradation	4	8.0

The last value is a practical one for critical work. By using it in Figure 2-3 we can read the tolerance for cover-slip thickness variation from line C. At 0.95 NA the cover-slip thickness should be correct to ±0.01 mm.

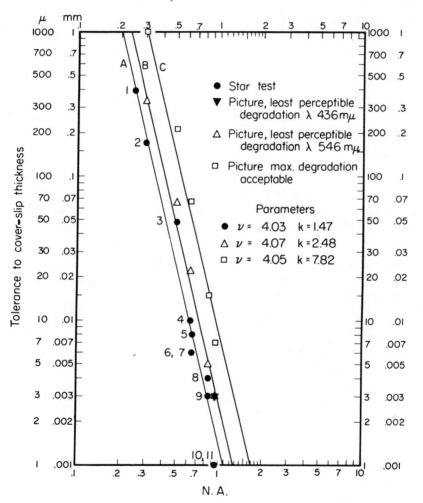

Figure 2-3 Tolerance to cover-slip thickness versus objective NA for dry objectives.

Such an objective is usually provided with a correction collar, *which should be used.*

However, 4 mm objectives of 0.85 and 0.65 NA are not usually equipped with correction collars. The tolerance values $T(\Delta l)$ are

	0.85 NA	0.65 NA
$k = 2.5$	$\pm 5\,\mu$	$\pm 15\,\mu = \pm 0.015$ mm
$k = 8.0$	$\pm 15\,\mu$	$\pm 45\,\mu = \pm 0.045$ mm

Since 85% of the slips of No. 1½ are within ± 10 μ of the 0.180 mm thickness, they can be considered satisfactory for these objectives, but for the 0.85 NA objective they are just so. The No. 1 and No. 2 cover slips should not be used for critical work except as they are selected with the aid of a micrometer.

The numerical apertures, for which the tolerance limit is equal to the thickness of the specified cover slip, are interesting, that is, where $T(\Delta l) = 180\,\mu$:

$$k = 2.5, \qquad NA = 0.35,$$
$$k = 8.0, \qquad NA = 0.50.$$

This means that with 10X (16 mm) objectives, which have a smaller NA than these values, it should be impossible to see the degradation in their images whether or not a cover slip is used and irrespective of whether the objective is corrected for use with one.

Refractive Index of Cover Slips. The refractive indices for cover slips assumed by the lens designers of two microscope companies are the following:

Manufacturer	n_D	n_e	V
Bausch and Lomb	1.5230	1.5252	58.6
American Optical	1.5244	1.5266	58.3

n_D = refractive index for $\lambda 589$ mμ (sodium).
n_e = refractive index for $\lambda 546$ mμ (mercury).

The refractive index of cover slips can be measured on an Abbe refractometer, for it utilizes the critical angle of reflection. It is wise to grind the edge of the cover slip with very fine abrasive paper at right angles to avoid a prism edge. Batches of cover slips obtained before

World War II had the same refractive index, in a group from one box, to within one or two units of the fourth decimal place. Now variations with third decimal place are common and slips from independent sources can vary by a unit in the second decimal place. An investigation[*] has shown, however, that this extreme variation is tolerable for the most critical use with "high dry" objectives (i.e., the 4 mm, 0.95) but just barely so.

Although there are many suppliers of cover slips and slides, it is useful to know that at the present time there are only three or four original sources of the glass used for slides and slips.

Manufacturer	n_D	n_e
Chance	1.5238 ± 0.0002	1.5266 ± 0.0002
Corning	1.5311 ± 0.0002	1.5339 ± 0.0002
"Commercial" (original source unknown)	1.5131	1.5159

Immersion Objectives. Development. Objectives of high aperture are designed to have the object space filled with a liquid, except when a cover slip is specified. There is an overlapping range of magnifications for which both dry and immersion objectives are available. Amici was the first to make the important advance to immersion objectives. However, "homogeneous" immersion, by the use of a liquid of a refractive index similar to that of the front lens and the cover slip, was first employed by H. B. Tolles in 1870 and then independently by E. Abbe in 1878. This technique puts the object point within the whole sphere of the lens so that an aplanatic image is obtained (Figure 1-16). The gain in the use of a spherical front lens is this way is great; this front element normally possesses all the magnifying power of the objective, except as it is incidentally modified by other succeeding lens elements which bring the diverging aplanatic beam to focus and correct its image for other than a single wavelength and for other qualities than aplanatism. The use of a liquid also reduces the flare that reduces image contrast with increasing apertures. Later in the chapter the more important contribution of the immersion liquid, that of allowing increased aperture and resolution, is discussed.

Cedarwood oil was chosen as the liquid to act as the liquid glass. It probably was the best choice of available oils. The refractive index n_D of the fresh oil of commerce is quoted below 1.5070. However, as it dries and oxidizes, its refractive index rises and it becomes yellower,

[*]Spinell and Loveland, "Optics of the Object Space" (4. 1960).

thicker, and stickier. The oil as bottled and sold for microscopy has been allowed to "dry" until its refractive index n_D^{25} is 1.515. The refractive indices of the hemispheres of microscope objectives are deliberately altered to some extent but probably are rarely less than that of crown glass, about n_D of 1.524, and have a lower spectral dispersion; that is, the disparity between immersion oil and the glass is even greater for blue light. The same discrepancy exists between the oil and cover slips also, which are usually made of the equivalent of crown glass. Many microscopists allow their cedarwood oil to "dry" further to raise its refractive index; it can be about 1.522, but is then very sticky.

Some years ago, Charles Shillaber did microscopists a great favor by compounding a synthetic oil, that is, **Shillaber's Immersion Oils*, A and B,** to replace cedarwood for immersion objectives. These two oils, differing principally in viscosity, are colorless and stable. Although even a little cedarwood oil on the bearings and the mechanical slides of the microscope finally became adhesive, Shillaber's oils actually are lubricants, although not necessarily the best for deliberate use as such. Shillaber assumed standardized cedarwood oil to be the optimum liquid optically. Therefore the refractive index of both of his oils at 25°C is $n_D = 1.5150$. However, Cargille's A has a slightly greater spectral dispersion, so that the two oils are not optically identical at appreciably different wave-lengths, say, in the near-ultraviolet region, in which the A oil has the higher index. Many objectives are now designed for use with Cargille's Oil.

Tolerance to Variation of Cover-Slip Thickness. Is this lack of accurate optical homogeneity between cover slip and immersion oil important? It is frequently stated that, with an immersion objective and its standard oil, it is immaterial whether a cover slip is used; certainly many microscopists act on this assumption. On the other hand, many of the best microscopists, for example, Belling, Shillaber, and Allen, have stated that for the best work it does make a difference. At the other extreme, many cases have been reported in which mineral oil has been substituted for the standard for the sake of economy or convenience.

Part of the answer can be obtained from Figure 2-4, which can be compared with Figure 2-3 for dry objectives. This is also a tolerance graph and the tolerance again is a straight line on log-log paper. Thus the tolerance can be expressed by the same equation (2-2) and the two constants have the same significance. For the star-test object

$$\nu = 7.2, \qquad k = 76.7.$$

Now marketed as **Cargille's Immersion Oil. Crown Immersion Oil is another satisfactory Oil. (see Table 3-1).*

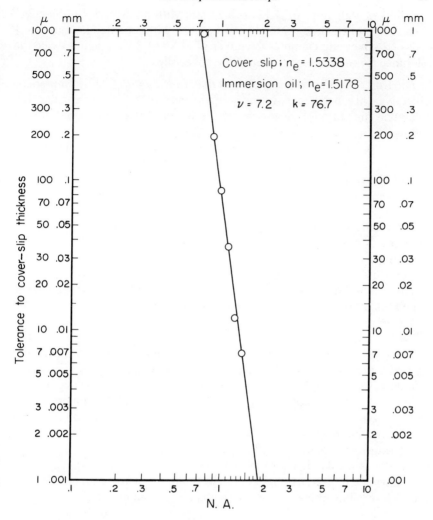

Figure 2-4 Tolerance to cover-slip thickness versus objective NA for immersion objectives.

As might be expected, the sensitivity of image deterioration to variation of cover-slip thickness is about $\frac{1}{50}$ that with dry objectives. Yet so powerful is the exponential form of this relation with NA that **at NA = 1.4,** which is that of available objectives, **the cover-slip thickness is no longer negligible, even with the immersion oil.** The line represents a determination obtained by the star test, which detected a thickness variation of $7\,\mu$. If it is assumed that the tolerance for best pictures or images is eight times greater, an image deterioration can be detected with a thickness variation of 0.06 mm, or 0.12 to 0.24 mm with an American objective. Cover slips of thickness No. $1\frac{1}{2}$ can be used without selection but No. 1 or No. 2 cover slip cannot. Objectives of NA = 1.3 have about twice the tolerance for thickness variation in the cover slips.

Tolerance to Variation of Refractive Index of the Oils. Immersion objectives are used in two ways: with and without a cover slip. The tolerance for variation of the refractive index of the immersion oil is as follows:

	ν	k_s(star test)	k_p(photomicrograph)
With cover slip	7.2	88.9	700
Without cover slip	7.0	38.7	310

The relation is shown in Figure 2-5 and is expressed by (2-2). Similarly, only the NA of the objectives governs the tolerance of image quality to variation of the refractive index of the immersion oil. With $\nu = 7$ it is less affected by NA than when $\nu = 4$.

There is naturally more than twice as much tolerance for the case in which about 60% of the object space is filled by a cover slip of constant index. The quality, however, is equally sensitive to variation of the refractive index of the cover slip.

At NA = 1.4 and without a cover slip, which represents a practical extreme, the tolerance $T(\Delta n)$, as represented by the star test, is 0.0004; that is, any larger variation of the index would give detectable degradation of the star image. Now, temperature affects Cargille's Immersion Oil to the extent of 0.0004 per 1°C. For the value of k_p for the photomicrographs, which is eight times larger than k_s, there is a tolerance of 8°C temperature shift before the effect becomes disturbing.

On the other hand, with the NA reduced to 1.0, $T(\Delta n)$ (even for the star test) is ±0.0075. It might be eight times larger for a picture. This is a very large allowable tolerance for a commercial immersion oil.

Resolving Power and Magnification

When magnification is employed to observe an object or its photograph, it is done to find more detail in it than could be seen directly. To see even more detail the magnification is increased. Now from the focal scale (q.v.) alone it is obvious that almost any magnification can be obtained by extending the image distance sufficiently and even this length can be kept convenient if the focal length of the objective lens can be decreased sufficiently. Beyond this point a relay optical system can be used, as already discussed. As everyone knows who has tried it, this method

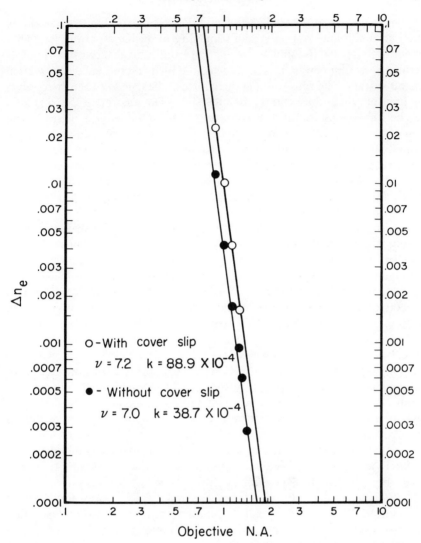

Figure 2-5   Tolerance to change in refractive index of immersion oil versus objective NA.

eventually breaks down; it is also obvious from the lack of photomicrographs which have much desired molecular detail. The images can grow larger but no more detail is added. Worse than that, under certain circumstances details called **artefacts** which are not directly representative of the object may seem to appear.

A poor lens or lens system can certainly limit the useful magnification at which detail can be added and may indeed be the limiting factor. Is the real limitation then only man's inability to design a perfect lens with which infinite detail could be seen? This ability to discern fine detail is termed **resolving power.** It is best measured by its reciprocal the **resolution** of the system. The resolution is the shortest distance between two points or lines at which they can be distinguished. The answer to this question of the fundamental limit to resolution has been the subject of study and discussion since the time of Fraunhofer, to which such men as E. Abbe, Helmholtz, Rayleigh, and Berek contributed outstandingly. A much more comprehensive discussion of the development of the theory of optical resolution can be obtained from any good text on optics such as that by L. C. Martin (2. 1930) or from articles by M. Herzberger (2. 1936) and H. Moore (2. 1940).

It turns out that the ultimate limit of the fineness of the structure of an object that can be discerned is set by the fact that light itself has structure and by its property of being bent or scattered by all optical edges or (equivalently) small particles. This is the property of diffraction discussed in Chapter 1. It is important to note that when we are dealing with objects of appreciable size we can deal with the geometrical optics of the system and consider only the shape and color of the objects; but when we are discussing the fine details of a specimen and the limitations of their discernment then, as Abbe and Helmholtz first pointed out, we must deal with the wavelength and diffraction phenomena. In both cases we assume that the ideal optical system would recombine all light coming from a point back into a point again, that is, its **antipoint.**

It is evident that if a structure is to be seen by its effect on a train of waves, when it becomes very small, with respect to the size or length of the waves, it will no longer have much effect on them and therefore will neither be seen nor imaged by any system. It is not surprising therefore that as wavelength gets shorter from radio waves, infrared waves, visible-light waves, ultraviolet waves, and finally x-rays the limit of resolution, the distance we call d, should become shorter. The resolution d, then, should be proportaional to the wavelength, which is almost invariably denoted by the symbol λ.

Again, in diffraction theory, as in geometrical optics, we can consider the object as a collection of points and then deal with only one point to simplify the discussion. Although the theory first considered only self-luminous points, first practice and, finally, theory showed that all imaged points show similar behavior. An ideal system would collect *all* of the light from a point to form its image point. Obviously that is impossible; only part of the emitted or diffracted light can get into the lens aperture. This should limit the quality of the antipoint and it does.

When the magnification of the image of a point is pushed high enough in any practical optical system, it will be found that **the antipoint is not a point but a disk*** and that it has structure. Moreover, the size and general structure is dependent on the size of the aperture, whether or not the aperture is filled with an excellent lens. A poor lens can, of course, degrade the quality of the antipoint. The upper row of Figure 2-6*a* shows the antipoint or Airy disk, as such images of a point are called, of a luminous point (a pinhole in an aluminum film) photographed at 1200X with an objective aperture of 0.63 NA. Figure 2-6*b* shows the antipoint of the same point and at the same magnification but taken with an objective NA of 0.15. In the lower row are representations† of a microdensitometer scanning through the center of the antipoints. It is obvious that a small aperture has increased the size of the antipoint and thrown more of the energy of the light beam into rings that surround it. With adequate aperture the rings have too small a percentage of the energy to make an impression in the printed reproduction. If the antipoints are relatively

*The antipoint is actually three-dimensional but as a rule only its intersection with a plane is observed. For its shape see Figure A1.2.

†The graininess fluctuations from the original negatives of the antipoints were eliminated in the reproduction of the traces. The asymmetry is probably due to the fact that the depth of the pinholes was not negligible nor were the pinholes exactly round with negligible size. The magnification of the pictures of the pair was 2800X. The image shown here is obviously the negative of the bright star seen in the microscope.

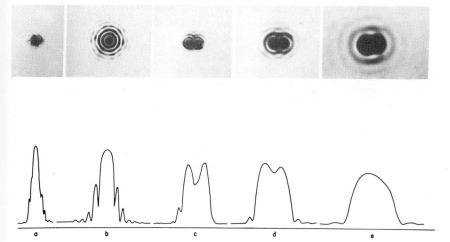

Figure 2-6 Effect of objective aperture on antipoint: (*above*) photomicrographs of pinholes in evaporated metal film; (*below*) microdensitometer traces of the negatives.

large, due to inadequate aperture, it will certainly be less easy or even impossible to distinguish or **to resolve** whether antipoints from two closely situated points are involved. This is illustrated by Figures 2-6c, d, and e with both the photographs and their corresponding microdensitometer traces. The original pinholes were 1.35 μ apart. At 0.63 NA (Figure 2-6c) they are definitely resolved. At 0.30 NA (Figure 2-6d) they are not reliably resolved, whereas at 0.15 NA (Figure 2-6e) they are not resolved at all. This phenomenon is discussed later. Returning to the single antipoint, we observe their general structure. They can be much more complex than this, for they can show the defects of an imperfect optical system. This, then, is the famous **star test** used to test lenses, including objectives, for quality.

Reverting to the limitations of excellent but practical lenses, we note from the preceding discussion that the **limitation to quality of the image**, as evidenced in the antipoint, was the **size of the aperture**. The lens occupying the aperture can do no better than bring the light given it from a point to the best possible antipoint, but the latter will always show this structure. The diffraction of the aperture throws more of the energy into the surrounding rings, the smaller the aperture is in proportion to its distance from the point; that is, the aperture should also be considered in focal units. **The aperture is best measured by the *angle* of light coming from the object point** to the aperture, that is, angle AOC in Figures 2-1 and 2-7 since it is a measure of the light that the lens aperture can take in and includes both this factor of distance and diameter. It was discovered by Lister and fundamentally utilized in Abbe's theory that **microscopic resolution is directly limited by the sine of the angle** rather than by the angle itself. Actually, the half-angle AOB, denoted as sine u (see p. 16), is used.

One fundamental characteristic of the object space has not yet been discussed in this connection, but we know it can vary; that is, the refractive index. The refractive index n_λ for any wavelength is 1.00 in vacuum, by definition, and nearly so in air, but **with increasing refractive index the wavelength becomes shorter**. With smaller structure of the light we should expect a smaller structure to be resolved, and this is so. We should expect also the factor (λ/n) to enter our final resolution formula and it does. For practical optical reasons, however, Abbe proposed that this factor n_λ be combined with that of the aperture to make a single term, and this has been adopted:

$$n \text{ sine } u = \text{numerical aperture} = \text{NA}. \qquad (1\text{-}8)$$

Let us consider three cases: (a) in which an object point O lies in air on a surface below an objective, (b) in which O is mounted in balsam and covered with a glass slip but with an air space between this and the objective, as illustrated in Figure 2-7 and (c) in which all of the intervening space between the object and the objective has a refractive index of about 1.53, although several materials, such as a resin, glass, and the immersion oil for the objective, may be present in layers. Such materials that are optically equivalent are most important in microscopy.

Considering case (a), Figure 2-7a, the largest cone of illumination coming from O that could possibly be used by the objective would have an angle of 180° ($u = 90°$). Since a practical working distance between the object plane and the front surface of the objective must be provided and some rim of the lens is shadowed by the mount, the widest cone that eventually actually passes through the microscope with dry objectives has an angle of about 143° ($u = 71.5°$) in the air space below the objective. The corresponding NA ($= n$ sine u) of this angle is 0.95, since the refractive index of air n is equal to 1.0; this therefore is the highest NA available in "dry" objectives.

Considering case (b), Figure 2-7b, we can see that the light rays cross a flat boundary from greater to lesser refractive index in the object space. According to Snell's law of refraction (q.v.), the angle of the illumination cone becomes greater as it enters the air from the glass or its optical equivalent. As previously discussed, if the inclination of the extreme ray of the cone becomes too great, it will be reflected back at the air-glass surface (i.e., when u is greater than about 82°). In practice, however, the objective can accept a cone of only 143° rather than 180°. This corresponds to about 77° in the glass, which is therefore the largest cone angle of illumination within the glass that is available if there is a layer of air in the object space. It is obvious that if the original illumination cone were coming down from the objective lens into air, as it does in a "dry objective" in metallography, this same limitation of about 77° in the glass

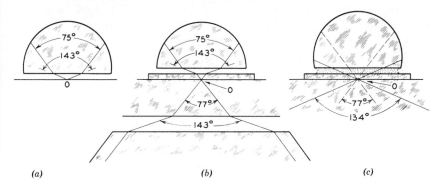

(a) (b) (c)

Figure 2-7   Path of rays in object space.

occurs, to correspond with the 143° limitation of the air space; both correspond to NA 0.95.

We can see that rays leaving the object point in the balsam with an angle of 77° ($u = 38.5°$) will apparently come from the point with an angle of 143° when they reach the objective. By the law of refraction the sine of the angle of the cone leaving the object point (sine 38.5°) is $1/n$th (or about $\frac{2}{3}$) of that in the air, so that in spite of the fact that the wavelength is $1/n$th of that in case (a) the resolution will be the same in both cases. Now it is also obvious that if all the object space has a refractive index of 1.53, as in case (c), and the angle of illumination collected by the objective from O is again 77°, the resolution will again be the same. With no intervening layer of air, however, it is no longer necessary to limit the cone to this angle; with a medium that is homogeneous with the glass, the rays go right through into the objective without any deviation and a cone of about 134° from the object point can be obtained in practice with immersion objectives. This cone obviously contains more light and has greater resolving power. With usual objective immersion oils this is equivalent to an NA of 1.4.

It can be seen that for practical purposes the value of n_λ, the refractive index at some wavelength in the expression of NA, must be considered as representing the lowest refractive index in the space between object and objective. If the illumination beam is to be adequate for an immersion objective, the air between condenser and objective must also be replaced with the same immersion medium. When an objective is used at the same working distance, so that the distance BO in Figure 2-1 is constant, as it usually is for a constant tube length and visual use (or with a fairly long bellows), the numerical aperture is equal to the effective aperture of the back lens of the objective divided by twice the equivalent focus. Therefore the relative proportion of the stated NA of an objective that is in use can be measured by observing the proportion of the back lens that is filled with light. Methods of measuring the NA of the objective that are actually used are discussed later in this chapter.

We are now able to combine the factors that we have found to affect, and indeed to limit, the resolution obtained with a microscope. The expression is

$$\text{resolution} = d = \frac{C\lambda}{\text{NA}}.$$

Although the value of the proportionality constant C could be obtained either by theoretical considerations or by actual test, the former method is preferable since we are considering a limitation imposed on even a theoretically perfect lens. For this we must make an assumption. By turning back to Figure 2-6c it becomes obvious that, because resolution is defined as the minimum acceptable distance to see two such points, the value of C will be set by expressing this distance. Since it was proposed by Lord Rayleigh, this minimum distance has been accepted as that in which the bright center of one Airy disk lies over the innermost dark band of the other, in which case it can be shown that $C = 0.61$. Our fundamental formula for the greatest resolution that can be obtained is

$$d = \frac{1.22\lambda}{2\,\text{NA}}, \tag{2-3}$$

where λ is the wavelength in vacuum and NA is the numerical aperture. Concerning empirical experience, it is a tribute to the quality of modern lenses that theoretical resolutions have been achieved not infrequently, and actually slightly better have been claimed. At the limit, either theoretically or empirically, it may always be a moot point, involving assumptions, just when it is no longer possible to tell of the existence of two points as distinguishable from one. Since it is a matter of practical importance for a microscopist to know when the limit of resolution may be involved in the quality of the image, it is also important that he calculate it, or utilize the pertinent tables, with no hesitancy. Therefore in this book the frequently quoted simpler formula* is used:

$$d = \frac{\lambda}{2\,\text{NA}}, \tag{2-4}$$

which assumes about 20% better resolving power than (2-3). This factor can easily be reintroduced into the final values whenever it is desirable.

It will be noted that neither the magnification of the image nor the characteristics of the eye enter the expression for the resolving limit; that is, that **a certain degree of detail exists in the image whether or not it has been magnified sufficiently for the eye to see**, as limited only by the wavelength (color) of the light and the numerical aperture. Eventually, however, the image or a photographic reproduction of it will be viewed. Then the limit of an increase in magnification that can furnish further detail will be when the smallest detail of the image, measured, as we have seen by the distance d, is equal to the limit of the visual acuity of the eye under the viewing conditions. Further magnification has been called "empty magnification." Obviously it should be used with discretion and with full knowledge of its existence and degree. On the other

---

*This is the formula as Abbe originally proposed it.

hand, some foolishly strong statements have been made about *never* employing it, as if it were almost morally wrong. Careful observation is required to discern the detail that is visible at the limit of resolution, when the magnification is such that the resolution limit is at the limit of visual acuity. If long periods of viewing are involved, and for other reasons also, bringing such detail up for easy or quick observation may be justified. It is the knowledge of the limitations that is important, partly to guard against artefacts. Discussion of the application of this formula, with examples, is given later in this chapter.

A knowledge of the difference between the resolving power and the **detecting power** of the microscope is also important. They constitute the two valuable characteristics of the microscope. It shares them with the telescope, but the detecting power of the astronomical telescope is probably more familiar. Although the resolving power enables the instrument to distinguish minute detail, the detecting power merely shows the presence or absence of points or groups of unresolved points. The shape of minute objects, including the representation of corners and the location of edges, is due to the resolving power. The two characteristics, resolving and detecting power, are independent of each other and are limited by entirely different factors. The factors limiting the resolving power have just been discussed; that is, wavelength, refractive index, and aperture. On the other hand, the detecting power is limited only by the brightness level and the contrast of the point against its background. The *difference* in the refractive index is important as a contrast factor, which is greatly enhanced by employing darkfield illumination, as is well known, so that the lower limit of the detecting power is greatly extended. The detecting power, however, is still operating and is important in brightfield illumination. Under optimum conditions (darkfield) it has been calculated that particles only a few hundredths of a micron in diameter can be detected. To resolve these particles would require the electron microscope, which might not be applicable under the circumstances.

Illumination Aperture and Resolving Power

To begin a discussion of the aperture of the illuminating cone, as related to resolution, let us consider a self-luminous specimen. Obviously no illuminating beam is utilized and therefore its aperture is not a variable. The light emitted from the specimen is assumed to scatter in all directions, hence to fill any objective aperture.

The analogous case has already been considered in which reflected light is used with a diffusely reflecting surface. Another analogous case, but one in which substage illumination may be employed, is that in which the specimen is photographed by fluorescence. In this method fluorescent light is emitted by the specimen on irradiation with ultraviolet which is not used directly by the objective. Here also we may expect any aperture to be filled.

Darkfield illumination (q.v.), in which the illumination beam is substage and the light is scattered by the specimen, might be considered an analogous case, but it is complicated by the fact that to utilize a high-aperture objective the aperture of the illumination condenser must be so high that the direct beam will miss the objective. The highly polar nature of this type of scattering may make this a different case, anyway. Excellent resolution as well as detection can be obtained, however.

The case in which the specimens are illuminated by transmitted light seems to have been considered almost entirely for biological specimens embedded in a fairly optically homogeneous medium, except for the extensive literature of the optimum illumination of diatoms provided largely by hobbyists. The conclusion is often advanced that these specimens are not at all representative of the "practical" sort. It will be remembered that Abbe considered that the number of orders of diffracted light that could be gathered in by the objective would determine how well the image would resemble the object. Diatomists would incline an illuminating beam to tilt the whole diffraction pattern to include higher orders and restrict the total azimuth of the cone to obtain contrast.

With the increase in industrial microscopy and photomicrography, the variety of types of specimen has probably increased, although the typical biological specimen may well still represent a majority of those viewed under the microscope.

Since resolution in the image at a given magnification depends on filling a specified aperture of the objective by light emanating from the points of the object, it is obvious that **any specimen that scatters a high percentage of the light will cause the required aperture to be filled with less dependence on the illumination** (condenser) **aperture.** This type includes granular specimens of high refractive index which are common in industrial microscopy. Practical results confirm this reasoning. For instance, during an investigation of microfilm photography it was found that when the granular silver image of a photographic film was examined by relatively high-magnification photomicrography, it was necessary to fill only about two-thirds or even less of the objective aperture theoretically required for resolution by the *direct* illumination beam to obtain a resolution as excellent as that obtained by directly filling the full aperture. Moreover, the contrast with the full illumination beam became too low to be acceptable for the purpose. The greater sharpness (edge gradients) of the picture taken with a less directly filled aperture made the picture quality better. Of course, an objective of adequate aperture is required for the diffracted light to fill it to the required limit. This is illustrated by Figure 2-8, although it is more obvious in the original photographs. For Figure 2-8a the illumination aperture was only one-half of that of the objective aperture, whereas for Figure 2-8b the objective aperature was 0.9 filled by direct illumination.

The more frequent case is probably the classic one in which the objective aperture must be nearly filled to the required value to obtain a resolution adequate for the desired magnification. It has certainly been found that filling the objective aperture completely to its rim lowers the contrast excessively without adding observed resolution. This has usually been ascribed to a degradation of the objective at its rim, and of course flare increases rapidly at high apertures. The phenomenon just described may play a part in the case of excellent objectives, even with specimens of normal contrast. It often occurs that adequate resolution is required to detect detail within an area of relatively low contrast, when there is high-contrast detail within the specimen and necessarily within the same microscope field. Here the procedure is to reduce the contrast that is excessive for the purpose, usually by choosing a mounting medium to match the high-contrast elements. The illumination aperture adequate for the latter, even for the same size of detail, will not be adequate for the desired detail.

The determination of the value of the required NA is discussed later. However, with that determined, the microscopist, having chosen an adequate objective, must *know* that his illumination cone fills the objective to the required extent. This is ascertained by looking directly at the aperture, that is, by looking down the tube with the eyepiece removed, preferably with a pinhole eyepiece in the tube or some other observation device. If the objective NA corresponds to that required, as it usually does, it normally should be filled to about 0.9 of the full objective aperture.

The microscopist should be quick to observe if there are noticeable diffraction rings around the edges of the image which usually denote a considerably inadequate aperture. Figure 2-9a represents a case in which the objective was filled 0.9 complete to give adequate aperture. Figure 2-9b shows the typical diffraction rings of inadequate aperture. The NA of the condenser was about 0.3 of that of the objective.

Darkfield, Stop-Contrast, and Phase-Microscopy Principles

Although consideration of illumination methods of enhancing contrast naturally belongs either in the chapter on illumination or in that on

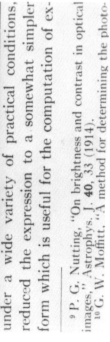

under a wide variety of practical conditions, reduced the expression to a somewhat simpler form which is useful for the computation of ex-

---

[9] P. G. Nutting, "On brightness and contrast in optical images," Astrophys. J. **40**, 33 (1914).

[10] G. W. Moffitt, "A method for determining the photo-

under a wide variety of practical conditions, reduced the expression to a somewhat simpler form which is useful for the computation of ex-

---

[9] P. G. Nutting, "On brightness and contrast in optical images," Astrophys. J. **40**, 33 (1914).

[10] G. W. Moffitt, "A method for determining the photo-

Figure 2-8 Documentary negative ×75. Objective: Zeiss Planapochromat, 16 mm 0.32 N.A. Ocular: 5× compensating. Illumination: λ 436 mμ (blue monochromatic). (*a*) Aperture 0.9 filled by condenser; (*b*) aperture 0.5 filled by condenser.

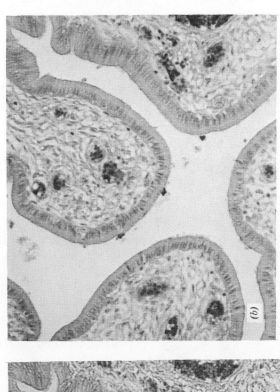

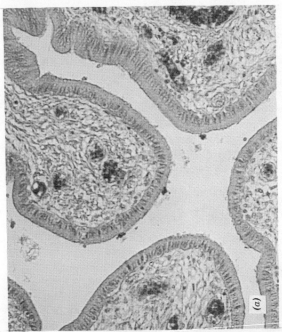

Figure 2-9 Human uterus, cross section. Mallory's stain 200×. Illumination: λ 546 mμ, from mercury arc. Optics: Planapochromatic objective, 16 mm 0.3 NA. Ocular: 10 × compensating. (*a*) Objective aperture 0.9 filled by condenser, photographic gamma = 1.0; (*b*) objective aperture 0.5 filled by condenser, photographic gamma = 0.6. Photographic contrast adjusted to compensate difference in optical contrast.

contrast, the principles involved are pertinent to the present discussion, and indeed both stop-contrast and phase microscopy were first employed as tests of the diffraction theory.

In a perfect darkfield illumination arrangement the objective would face into absolutely black optical space, with no direct light entering it but with one or more intense beams of light directed underneath and past it. Any material of different refractive index from that of the object space will, of course scatter light into the objective and the observing system to give a bright image against the black ground. **Its power to enhance contrast** (see Table 10-3) **is potentially enormous, but it is extremely wasteful of light.** The light from gross detail is deviated into the observable image by reflection or refraction and from fine detail by diffraction. **The detecting power for fine particles is the best of any system**, for the contrast of a well-prepared darkfield is supreme among the illumination methods. It was once thought that the resolution was less than the optimum for bright-field for a given objective aperture, but it is now acknowledged that it can be at least as good. The scattered light from the fine detail from illumination coming at a high angle to the optical axis completely fills the objective.

However, another type of darkfield — or bright contrast — forms images of lower quality. A high-power objective was sold at one time with an opaque mirror as a central disk lying on the back surface of the front lens, which was a hemisphere. Some of these objectives are still used. With such a design, when the specimen is illuminated by a very narrow axial beam, the beam is focused back on the specimen but the image is formed only by the annulus of the aperture surrounding the mirror. The occlusion is about 35% of the diameter. The result is an antipoint with the rings but with almost no bright central point. The result is also the formation of artefacts. This objective appeared at a time when micellular structure in fibers was being hypothesized and photomicrographs were made with this objective to bolster the theory. It was possible, however, to show micelles lying outside the fibers!

The fundamental problem in ordinary microscopy is that with illumination from below the stage and through the specimen, direct light, undeviated by reflection and diffraction, floods out the deviated light which is so essential to the formation of an excellent image, according to the theory. Wilska (4. 1953, 4. 1954) solved this problem by the method of **stop contrast,** which he advocates. He restricts the light source to a ring that is imaged in the aperture plane of the substage condenser and again in the aperture plane of the objective. The latter is normally in the rear lens of the objective or near one surface of it. An annular ring of soot is placed on the lens surface to cover the image of the light source, which, of course, is formed by the light rays undeviated by the specimen. This beam is reduced to about 10% of the original by the soot, thus allowing both the undeviated and the scattered light to participate in formation of the image. Moreover, with a ring light source, which is really formed by an annular diaphragm below the substage condenser, light scattered at even a slight angle is thrown clear of the annular density that reduces the intensity of the direct beam from the light source, making the method a very sensitive one. Details, such as the chromosomes in an unstained salivary gland of the fruit fly, which can now be seen clearly reveal excellent resolution. This method has been designated **anoptral contrast.***

For specimens of relatively high refractive index, which produce much diffracted light, Wilska uses a variation that was actually devised first. In this case the whole surface of the objective lens nearest the aperture plane is sooted over until the total transmission is reduced 50%. Then a clear annulus is wiped into this density to correspond to the image of the ring light source. Wilska claims that differences in contrast of detail within the specimen become clear. He also states that the young, well-nourished cells in a group of yeast cells stand out for the first time from the degenerating cells. This is one answer to the problem of showing delicate detail of low contrast within a specimen in which there is relatively high contrast in the whole field.

A more sensitive method of enhancing contrast was developed independently by F. Zernike and considerably earlier than the publication of the method of stop-contrast. It grew out of a diffraction test for telescope lenses to replace the famous Foucault test, but Zernike quickly realized its applicability to microscopy and by 1935 had applied for German patents. By the time of World War II Zernike's method, called phase microscopy, was known only to a few workers, and the system was developed mathematically along somewhat different lines by H. Osterberg in the United States. Zernike developed the method from the theoretical considerations of the diffraction theory and had applications of high resolution and high magnification in mind. It is now an important general method of enhancing contrast and has been used in the Kodak Research Laboratories very effectively at medium powers for quick identification of unstained animal tissues.

The illumination arrangement for phase microscopy is similar to that of stop-contrast in that the light source is a relatively narrow line in the rear focal plane of the substage condenser. In practice it is usually formed by a slotted diaphragm located there with a light filament focused on it. A slotted cross has been used, but a ring slot is more common, to il-

---

*Objectives of this type are now commercially available.

luminate all azimuths. With the condenser and the objective properly focused, light waves that originate at a point in this annular source will form a parallel bundle of rays when passing through the field plane (as shown by the solid lines in Figure 2-10*) and, *if unaffected by an object*, will be focused again in the rear focal plane of the objective (i.e., at *C'*) to give an image of the annular source. They will then be dispersed to cover the *whole* image field. However, with a specimen in the field, some of the light will be diffracted by its optical edges, including interior detail. This fraction of the light will deviate from the original path, as indicated by the dotted lines of Figure 2-10,* and will spread out from the object points as a source, although most of the energy from one point may lie within a small angle. It is diffuse in the focal plane of the objective and is focused in its image plane. With a specimen composed of mány refracting points, the whole aperture will be filled. Thus there is an extremely sensitive way of sorting the undeviated light waves from those that were deviated by the specimen, that is, by placing a mask in the focal plane of the objective that just covers the thin image of the annular source. As we have seen, for Wilska this mask was a density.

To explain Zernike's idea another phenomenon of diffracted waves should be considered. If the refractive index of the specimen is a little higher than that of its surround and the optical path of the diffracted light is a bit longer, the diffracted wave will be retarded one-quarter of a wavelength. This is illustrated in Figure 2-11,* in which the original undeviated wave is represented by curve *S* and the deviated wave by curve *D*. On the other hand, if the refractive index of the element of detail is less than that of the surrounding medium, the diffracted wave will be advanced by $\lambda/4$. Obviously only in the area of the image of the specimen, such as a particle, will both the *S* and *D* waves be present.

*From Bennet et al. (4. 1951) with permission.

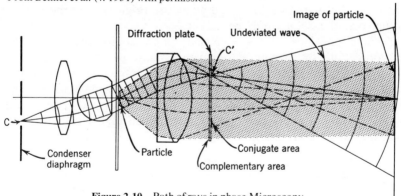

Figure 2-10 Path of rays in phase Microscopy.

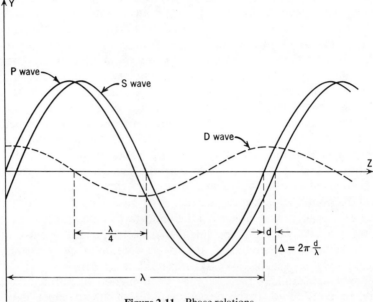

Figure 2-11   Phase relations.

The resultant wave, represented by P in the figure, assumes no absorption, that is, its amplitude is as great as that of the original S wave and nothing would be seen in the field, since the eye does not see phase differences. This wave would be advanced or retarded, according to the phase of the D wave; that is, $\pm \lambda/4$ from S. Referring again to Figure 2-11, if the general S wave or the diffracted wave D were retarded *or* advanced by $\lambda/4$ (equivalent to shoving a curve along the axis), the crests of the two waves would either coincide or be just $\lambda/2$ out of phase. In the first case the image would be brighter than the surround; in the second case, it would be darker.

Zernike, and later others who made phase equipment, accomplished this procedure by placing a transparent **phase plate** in the focal plane (at C' in Figure 2-10). By adding a ring of refractive material on top of the plate which is coincident with the image of the light source and of sufficient thickness to retard this undeviated light by $\lambda/4$, the S and D waves become coincident if the specimen had higher refractive index than that of its surround and **bright contrast*** is obtained. On the other

---

*A nomenclature that is independent of the relative refractive index of the specimen and background is to call one type, for example, retardation of the undeviated ray, "negative," and the alternative, "positive" phase contrast. This is discussed under the subtitle "Nomenclature" in Chapter 12.

hand, if the area of the annulus were ground down on the phase plate or refractive material were added to all of the area of the phase plate *except* over the image of the light source until the deviated rays were effectively retarded by a distance of $\lambda/4$, the two waves would then be exactly out of phase (crest to trough) and **dark contrast*** would be obtained with the same specimen. The amplitudes of the waves forming the image are determined by subtracting the amplitudes of the deviated rays from those of the undeviated rays ($S–D$, Figure 2-11). The observed intensity of the image light is, of course, the square of the amplitude of a wave.

As has been mentioned, however, with most specimens the amplitude (intensity)$^{1/2}$ of the diffracted light is very small with respect to the undeviated light, although a few lie at the other extreme at which most of the light is diffracted. Maximum contrast will be obtained when the intensities of both beams are equal. This can be done, as Wilska also discovered, by cutting down the more intense of the two beams with a neutral density. It might require a different neutral density for each specimen to bring its image to maximum contrast. Figure 2-12 illustrates how the density, usually an evaporated metal film, can be put over the annulus for undeviated light to bring up the relative intensity of the diffracted rays or over the rest of the objective to subdue the diffracted rays.

The type of contrast observed when the refractive index of the specimen is greater than that of the surround (the usual case) is also noted. Bright contrast designates the case in which the object is brighter than the field and conversely. Further consideration of this matter involves the nomenclature of phase microscopy which is discussed under "The Application of Phase Microscopy" in Chapter 12.

Objective Aperture: Practical Considerations

It can be seen that the aperture of the objective used in forming the image, whether with the simple or the compound microscope, is an

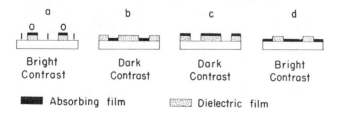

Figure 2-12 Types of phase plate.

important factor indeed in photomicrography. Let us recapitulate and list the effects that depend on it.

1. With a given lamp, the aperture determines the brightness of the image and therefore is a measure of the required photographic exposure. This varies with the square of the aperture, expressed either as f/α or as NA.

2. The detail in the image increases with the aperture, the lower limit of resolution being inversely proportional to the aperture.

3. The contrast of the image decreases as the lens aperture is increased and vice versa.

4. The depth of field decreases with the increase of aperture. Depth of field is inversely proportional to the size of the aperture.

5. The size of the field of good definition decreases with increase of aperture.

An image produced by a lens of very high resolving power, yet having adequately high contrast and depth of focus and requiring a short exposure for photography, would be highly desirable in nearly every case. We have seen however, that these individual qualities are achieved by opposite manipulation of the aperture. Obviously the actual picture must be the result of a compromise. The photomicrograph is taken in the first place to show something that cannot be seen directly with the eye with ease, if at all; the one quality that must not be sacrificed too much, then, is resolution of detail or the purpose of the picture itself will be defeated. One of the advantages of photomicrography over the mere visual examination of a magnified image is the fact that the contrast of the image can be increased during photography. Therefore **it may be advisable to open the aperture to achieve adequate resolution beyond the point at which the decreasing contrast of the image is pleasing to the eye.**

The size of the objective aperture is a value well worth keeping in mind at all times, whether one is using a simple or a compound microscope. In transparent illumination the extent to which direct light from the condenser fills the aperture should not only be known but deliberately chosen. The fraction of the aperture that is filled can be seen by removing the eyepiece after focusing the microscope and looking down the tube at the bright disk, which is the objective aperture. A pinhole eyepiece, obtainable from most of the microscope companies, is a convenient accessory and fortunately inexpensive.

Although the numerical aperture of the objective for a compound microscope is usually engraved on it, it is sometimes desirable to determine this value or that of the aperture which is actually filled with light. A device, called an *Apertometer*, is sold for the purpose of measuring the apertures of objectives and allows it to be done accurately. It is expensive and involves a slab of glass so thick that it is useless for apertures involved in a particular bench arrangement. The same is true of a method described by Shillaber (4. 1944) except that it is not expensive.

If a camera lucida is available, the objective numerical aperture can be measured quite accurately, although the setup must be calibrated, for the units are arbitrary. The illuminated aperture of the objective is viewed and measured directly, with the eyepiece removed from its tube and with a white ruler, which should lie on the microscope table, projected via the mirror onto the objective aperture. Alternatively, the aperture can be sketched on white paper and the sketch measured. The mirror and tabletop distance must remain constant throughout the measurements and calibration. This is done by measuring the aperture disk of an objective whose NA can undoubtedly be considered as the value engraved on it. If the apertures of a group of objectives are measured and the values plotted against the stated NA, those cases that lie off a straight line are easily spotted. In the author's experience this is a small minority, especially in modern objectives. A factor can then be calculated from the known case, with which the unknown NA's can be computed *after* dividing the measured value by the focal length of the objective or after multiplying by the primary magnifying power of the objective.

The numerical aperture of the objective can also be calculated with perhaps sufficient accuracy, after measurement of the diameter of the exit pupil of the microscope above the eyepiece (the eyepoint, Figure 2-1). This is best done quite carefully, with some support for the rule and hand and by using at least a 10X magnifier. The total magnification M at a distance of 250 mm (10 in.) must be known. The formula is (4. Beck, 1938)

$$NA = \frac{\text{diameter of eyepoint} \times M}{500}. \tag{2-5}$$

M is best measured at the time; the exact tube length is also involved. If the correct tube length is employed, the value of M should be that obtained by multiplying the magnifying power of the objective by that of the ocular; both values are engraved on the optics; that is, $M = m_1 \times m_2$.

Definition Factors

In any consideration of the importance of resolving power in making a photomicrograph the results of recent pertinent research should be included. Although the original aim was to improve aerial photography, these results are equally applicable to this field.

Let us assume that the best photomicrograph is the one with the best definition. To quote Higgins and Wolfe (1. 1955)

"**Definition** in photography is the quality aspect of a photograph that is associated with the clarity of detail. This is a *subjective* concept because it is an impression made on the mind of the observer when he views a photograph. Definition is the composite effect of several subjective factors, such as sharpness, resolving power, graininess, and tone reproduction. Definition will be considered to refer to the over-all appearance of detail. **Resolving power** will be restricted to its common meaning . . . and **sharpness** to the impression made on the mind of an observer when examining the boundaries of well-resolved elements of detail."

After an extended investigation with many observers, Higgins and Jones were able to develop an objective and a physically measurable quantity that correlated excellently with the subjective values of sharpness as judged by the observers. This characteristic has been named **acutance** and has become a very important one. On the whole, it has to do with the sharpness of edges in a photograph and it is evaluated as a density-gradient function across the edges of a picture after analyzing their microdensitometer traces. It is important to note that it is not the same characteristic as the contrast (including the photographic gamma) of the picture. Many observers were asked to grade several series of pictures (totaling a large number) in the order of their technical (not artistic) quality. The following important facts were derived from this experiment.

1. Within a certain gamut, resolving power and acutance can vary independently.

2. In many instances observers preferred pictures with high acutance and *relatively* poor resolving power.

3. When graininess and tone reproduction are constant and resolving power is adequate to reproduce all the detail that can be observed under the conditions of viewing, acutance correlates well with definition. On the other hand, when the resolving power is not adequate to reproduce all the detail that can be observed, it is also an important determinant of definition.

DETERMINATION OF THE REQUIRED APERTURE AND SELECTION OF THE OBJECTIVE

We have now considered the factors affecting image definition and the fundamental law limiting resolution in an image. Although most frequently presented from the standpoint of avoiding excessive "empty magnification," the actual situation is likely to be the converse and involves the selection of the objective. The desirable magnification for a given specimen is usually known; the question then becomes, what NA is needed? This the microscopist should know by preliminary mental calculation.

The limit to the detail that the unaided eye can distinguish is a variable that depends on a number of factors which include the brightness of the object, its distance, and its contrast. It can be assumed that a contact print will be viewed at optimum brightness and at the generally accepted "normal" viewing distance of 10 in. (25 cm). The contrast of the detail is most important; it can temporarily be assumed to be entirely that of black and white. In this case the average limit to vision, expressed as the angle subtended by the minimum discerned distance, is about one minute of arc, although it may average less for groups whose eyes have been corrected by glasses. In rare cases objects subtending less than 40 sec visual angle may be perceived [*Seeing*, Luckiesh and Moss (2. 1931)]. At 10 in. the distance subtended by an angle of 1 min. is about 1/300 in. or 0.085 mms (85 μ). It would be safer to consider 75 μ as the maximum distance which is usually *not* distinguished by the unaided eye, and since this value also gives a simpler factor for calculation it will be used. Some idea of these distances and the personal limit to resolution at the highest contrast may be gained from the patches in Figure 2-13, which have been printed from half-tone screens. The separation *between* elements is given in the captions. It is instructive to cover one-half of each patch with a transparent gray tint to lower the contrast.

It is evident that if a magnification of 100X is desired in a certain photomicrograph an antipoint of 75 μ in the image will be produced if the limit d of the microscope resolution of the specimen is 75/100 μ. Referring to the formula for resolution on p. 75, we find that the numerical aperture required is $\lambda/2d$. We assume the simple figure of 0·05 μ for the wavelength (which may be considered as the dominant one, visually, with the Kodak Wratten H Filter (No. 45)), the expression

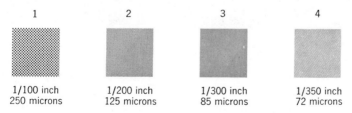

1	2	3	4
1/100 inch	1/200 inch	1/300 inch	1/350 inch
250 microns	125 microns	85 microns	72 microns

Figure 2-13   Antipoint size as related to visual acuity.

simplifies to: $NA = 0.5/(2 \times 0.75) = 0.33$. Accordingly, a numerical aperture of 0.33 is required at 100-diameters magnification if detail is to be as complete as the eye can see under optimum conditions. Conversely, a greater magnification than 100X with a NA of 0.33 merely increases the size of the antipoints further above the ultimate limit of vision.

Practically, this limit is often exceeded and rightly so. In the first place the smallest dimension that the eye can distinguish becomes considerably larger when the contrast diminishes; for example, when it is desired to examine small detail with other portions of the specimen as a background (Luckiesh and Moss, 2.1931). Moreover, there frequently is no reason why the desired detail should not be magnified until it is "easy to see."

In ordinary pictorial photography the photograph is commonly considered to be in good focus if the "circle of confusion" is no larger than 1/100 in. or $250\,\mu$. The photomicrographer, who must maintain stricter standards of definition, may then consider this to be the upper limit for the acceptable diameters of his image antipoints under normal conditions. In the case of 100X magnification the least distance resolved on the specimen must be $250/100\,\mu$ and the numerical aperture, determined as above, can be 0.10.

Let us formulate this for practical use. Let M represent the magnification of the print (including possible enlargement) and D stand for the acceptable diameter of the antipoints. Then

$$NA = \frac{\lambda \times M}{2D}. \tag{2-6}$$

Let λ be considered as $0.5\,\mu$:

Case 1. *Strictest Standard of Useful Definition.* If $D = 75\,\mu$, the visual limit of resolution under optimum conditions, then

$$NA = \text{magnification} \div 300.$$

Conversely, the highest acceptable magnification $= 300 \times NA$.

Case 2. *Lowest Acceptable Definition.* If $D = 250\,\mu$, then

$$NA = \text{magnification} \div 1000,$$

or the highest acceptable magnification $= 1000 \times NA$.

For other wavelengths these values for the NA required may be multiplied by the factor $2 \times \lambda$ *(specified in microns)*. It is easy to see that considerably less numerical aperture is required by the short wavelengths of ultraviolet for a given magnification.

These simple mental calculations, which set the upper and lower limits, should be made before undertaking the photomicrography of any specimen. The definition needed, of course, will depend on the detail of the given specimen. It is useful to set up some standards in which the definition, or the diameters of the antipoints, is known in each case.

Chapter 3

Apparatus

An individual is likely to enter the field of photomicrography by one of two principal avenues.

1. He may be a photographer, sometimes an expert amateur, who suddenly has a subject that requires magnification or wishes to extend his interest into that field.

2. He may be a microscopist—an amateur or novice—or he may have used a microscope for years as an expert in some scientific or technical field and now wants to record some fields or publish his results.

The photographic literature caters to those in the first group. The technical literature of the microscope companies tends to be directed to new visual microscopists and rather neglects the photographic aspects. In the author's experience students in schools, even medical schools, often get deeply into microscopy and later into photomicrography, without adequate briefing and supervision in these two subjects at any stage, and then go out into professional and technical life. Such a person may possess an elaborate and expensive photomicroscope about which no photographic knowledge is supposed to be needed. Trouble can be encountered here, too, although routine brightfield microscopy and photomicrography usually go well together and passable results are obtainable with other forms of illumination after some trials.

The treatment in this book for the two groups mentioned above, even the appropriate apparatus, should be different. Consequently some parts of this chapter and book will not be of direct interest to one of the two groups.

Types of microscope and microscopical apparatus have greatly increased in recent years. On the other hand, some highly desirable apparatus has been dropped from the market for commercial reasons, including increasing costs incurred when mass production, with minimum specific unit work by skilled labor, cannot be used. An example is the horizontal bench.

This chapter does not compete with the catalogs of the microscope manufacturers, which should be obtained before any purchase of apparatus is made. Before undertaking microscopy or even resuming microscopy with unfamiliar apparatus, one should **read the directions for the apparatus** provided by the manufacturer. Most manufacturers also publish a book with a title equivalent to *Use and Care of the Microscope*. Such a book is required reading for the novice. Both types of publication have priority over this one. In these first chapters the author will reiterate only what needs emphasis and discuss the points neglected in these pamphlets.

CARE AND CLEANING

Inadequate care will allow deterioration of the equipment which will eventually cause degradation of the quality of the results obtained. Microscopes are not lightly tossed aside but continue to be used often with images of needlessly lowered contrast. Actually, protection from dirt and mechanical damage is far preferable to depending on its being cleaned off, for most damage to optical surfaces is probably done by the cumulative effects of cleaning off abrasive dirt or from the corrosive action of solvents on surface interfaces. The subject of cleaning the microscope optics is discussed at the end of Chapter 6, and this portion of the book should not be overlooked or skipped.

If possible, the workroom should be kept dry and clean. A room that is especially dusty is bad, although few can obtain the ideal—a dustless room. The presence of certain corrosive chemicals, such as strong hydrochloric acid or ammonium fluoride on the shelves of a closed room will eventually and cumulatively deteriorate optical equipment, which is completely dependent on its mechanical parts for excellent performance. Preferably no vibrations should normally be present. Actually, the vibrations coming from sources inside or outside the room or made by manipulating the microscope and exposing the film may be the limiting factors to definition without the operator's being aware of the situation. This subject is discussed later.

QUESTIONS AND ANSWERS

Chapter 3 attempts to answer some of the questions of a person who might be considering the nature of the apparatus necessary for good photomicrography. To that end a summary is included here, near the beginning.

1. What kind of microscope should be used and what are the limiting factors?

In the first place, the author wishes to warn against the use of toy microscopes if there is to be any seriousness of purpose, even though some of them have been greatly improved in recent years. It may take some background to avoid some very poor buys. Even a very young person will eventually tire of not seeing more than is possible with a good hand maginfier, and the lack of standard tube size and the RMS thread will also eventually thwart the desire to progress. It would be better to look around for a second-hand standard microscope of relatively simple style. The agents for American microscopes may have some that were taken in during re-equipment of school laboratories. At the time of writing Bausch & Lomb also had some simple models for sale.

After one has observed a thrilling, or at least interesting, field, might one be disappointed in a well-exposed photomicrograph of it? The answer is yes. The lack of dynamic observation, with twitching fingers on the focusing knob, makes the selection of the field and the sharpness of detail in the picture most important. The field will certainly seem smaller, since one can no longer refocus for the periphery. The quality of the photomicrograph should be worth the trouble of taking it as a permanent record.

The quality of the microscope is obviously a limiting factor. In an optical instrument high quality is required in both the optical and mechanical components; yet the question of the quality of a microscope may not be quickly and completely answerable on inspection, especially by a novice, although some criteria and tests are provided here. The reputation of the manufacturer becomes important to the novice. The need for adequate aperture for the magnifications to be used has already been covered, and the nature of some of the components is taken up below.

Among first-class microscopes, with adequate aperture available in the objectives and substage condenser and almost equal in resolving power, the biggest difference is often the contrast or *crispness* of the image. This may be due to the presence or absence of a well-adjusted and well-diaphragmed illumination system or it may be inherent in the objectives and oculars.

2. For those considering professional equipment, under a limited budget, the author makes a plea not to confuse convenience with improved quality.

Today's microscope salesmen think that the public considers the first factor the more important. Convenience is not worth the sacrifice of quality in most cases. As an example that may cause argument, the zoom focus in place of changing oculars is primarily a convenience factor, though a nice one. A substage condenser of good quality that can be focused, is easily interchangeable with others, and divisible is definitely productive of an improved quality of image. Yet such substage equipment is often made available only on more expensive models after other models have been offered at added cost to furnish convenience factors such as zoom focus. With adequate demand, better substage equipment on simpler models would be provided.

Finally, the illuminating source obtained with the microscope may prove inadequate for photomicrography when that is undertaken. This is likely to be especially so for microscopes with incorporated lamps because some of the most effective types of illumination are so wasteful of light. Sometimes a picture is taken with a photographic material of extreme speed when that material would not have been chosen if there had been adequate illumination. Therefore a microscope should always have provision for a good mirror to be efficiently and easily added to the substage to allow the use of an outside light source. That makes almost all types of illumination available.

3. What type of camera is most suitable?

The answer may be a matter of convenience but it may also be dictated by circumstances.

There is little doubt that for the individual who is primarily a visual microscopist, who expects to sit over a microscope for hours, yet suddenly wants a record or an illustrative lantern slide of some field then in view, the **small camera, equipped to take 35 mm roll film**, is the most practical, since it can be placed almost instantly over the microscope and a picture taken. This is particularly so if someone else does the processing. This can be quickly and professionally provided by merely mailing it directly from the laboratory in a prepurchased mailer; it is also returned by mail. The author prefers an arrangement in which the camera swings over the microscope from a separate support and swings out of the way for visual use, but a camera adapter with a split-beam observation eyepiece that sits directly on the microscope is probably more frequently employed and is vigorously exploited commercially. This is the *Aufsitz Kamera* (attachment camera) of German literature. Both devices and a simplified method that is inexpensive to set up are discussed in Chapter 11.

Sometimes it is known that no more than one or two paper prints of the photomicrographs will ever be wanted. In this case it is preferable to make them directly in the photomicrographic camera by utilizing either **Polaroid Land materials or Kodak Super Speed Direct Positive Paper.** Both are made in relatively few minutes, the actual processing of the Polaroid film consuming only a fraction of a minute. The Polaroid material requires no darkroom but a special adapter, whereas the Super Speed Direct Positive Paper does require a darkroom, although its processing also consumes only a few minutes but the prints have much higher resolving power, being well below that of the eye, and are appreciably cheaper. Both materials do require a large camera as opposed to the "miniature," but in a laboratory as part of the optical equipment this should not be a disadvantage. When prints are wanted, the large camera utilizes the much more efficient system of using the bellows length of the photomicroscope directly as the enlarging equipment. **If this system is selected, it is best to get a camera that operates with sheet films and plates,** so that when prints for publication are desired they can be used. Both the Polaroid and Kodak Super Speed Direct Positive Film are suitable in this kind of camera (see Chapter 16).

Finally there is the **professional equipment** for those who want professional results, whether or not they are commercially engaged in this field. Such a person will need a photomicrographic camera that utilizes holders for sheet films and plates which will make permanent records available for an unlimited quantity of photographic reproductions. A negative material is capable of giving considerably better tone reproduction than is available with reversal positives principally because of the higher maximum density and the longer latitude for brightness reproduction (see Chapter 15). A much greater range of products is available from which to choose the contrast and spectral sensitivity desired for black-and-white materials and also the color balance for available light sources for color materials.

In order to provide for making 2×2 in. color lantern slides, a camera that utilizes 35 mm film will be wanted. This may be a completely separate photomicrographic 35 mm camera assembly or an **adapter back** that will fit the back of the view camera of a standard photomicrographic bench or stand. Such adapter backs for professional cameras are listed by photoequipment houses.* The most convenient and efficient type is that in which the assembly consists of a camera back to take 35 mm film which slides to one side and is replaced by a ground-glass back lying in the same plane as the film; this unit fits onto a camera back such as that of a Graflex Camera, and it, in turn, fits onto the back of the photomicrographic view camera. The ground glass is on top of a metal box to bring it to the plane of the film. Thus it is easily possible to measure the magnification and make photometric measurements in the image plane to determine the photographic exposure.

The extreme speed of some Polaroid materials is a definite advantage when one is restricted by an inadequate illumination level that may often occur from a built-in light source.

MECHANICAL PRINCIPLES

The optical principles of image formation in the microscope, both simple and compound, have already been discussed in Chapters 1 and 2, respectively, with Fig. 2-1 serving as the general illustration. The illumination optics are discussed in Chapter 4 for opaque specimens and in Chapter 7 for transillumination. Some comments on the essential parts of the compound microscope may be helpful.

Stage

The stage is merely the shelf for receiving the specimen, but it may vary greatly in complexity according to the sophistication and cost of the microscope. It should, of course, **be normal to the optical axis**, but used microscopes have been encountered in which this is not so. Accurate assessment of this characteristic may not be too simple; a tilt of the stage is probably best determined by the very effects it causes; that is, the in-focus field is narrower from one side to the other when the image is projected at high magnification to a camera back and the center of the in-focus field varies with the thickness of the cover and the thickness of the microscope slide.

The simpler microscopes have a square stage and use stage clips for the microscope slides, but a mechanical stage can usually be purchased as an accessory. The **mechanical stage**, which is the screw-and-worm or rack-and-pinion device for controlling the position of the microscope slide, is a great help to photomicrography and until recently has been considered almost essential to high-magnification photomicrography. The exception is due to the relatively new **glide stage** which involves a double platform with a top surface that glides over a parallel support with a smooth viscous motion. The stage must be horizontal, but the author has found this device to be quite satisfactory for most work, even at high powers. **Vernier position rulings**, however, in the two directions are ex-

---

*Among the largest of these houses, which also operate a mail-order business from a catalog, are Burke and James, Inc., 333 W. Lake St., Chicago, Ill., 60606; Willoughby, 110 W. 32 St (near Herald Square), New York, New York 10001.

ceedingly valuable in photomicrography, especially when more than one person is involved in the judgments of the selected field. As far as industrial microscopy is concerned, certainly, the assumption of the microscope manufacturers that the specimen will consist of a 1×3 or 2×3 slide for the biological-type microscope may become a severe limitation. Special **mechanical stages to hold other size slides are available**, including the petrographic slide of 45 mm length. **For industrial work a stage large enough to hold larger plates is often useful.** Lately some of the more expensive microscopes have square stages rather than the rotating circular stage. A **rotating stage** is exceedingly useful for quickly and conveniently aligning a field to the format of a rectangular photographic film or plate.

When there is much work with reflected light, as in metallography, it is almost a necessity to have the coarse focus adjust the height of the stage according to the thickness of the specimen. In many modern microscopes for transillumination the coarse adjustment for focusing operates the stage also.

Almost invariably in a microscope with a vertical optical axis the stage is a shelf supported at one side. Although this design does involve some convenience, the principle is really very poor from the standpoint of stability and freedom from vibration. Microscopes have been built with their stages in the form of a table, usually with tripod support, such as the former Bausch & Lomb research stand and a number of the older English microscopes. The author considers it to be unfortunate that this design has not been retained but must admit that some of the stage supports of the large photomicroscopes are now exceedingly sturdy.

Support

The microscope is usually supported from a flat surface by three "feet," even though the latter may consist of an area protruding only very slightly from the bottom surface of the microscope. For many years the base consisted of a horseshoe or large V-shape for nearly all models except a few English microscopes that utilized the tripod. With the advent of the built-in illuminator, this form of base is becoming the exception rather than the rule, which is often unfortunate for photomicrography when the built-in source is inadequate and a mirror is inserted below the stage to reflect in the illumination beam from an outside lamp. This substage mirror may often be too high to allow a substage condenser to be interchanged without throwing it completely out of alignment. As discussed later, for photomicrography the author prefers an arrangement by which the foot of the mirror is below the microscope and is independently supported to ensure good alignment.

The desirable photomicrographic camera on a horizontal bench, discussed later, has disappeared as a catalog commercial item, but it may still be set up. It generally used a microscope set at right angles to a horizontal optical axis around an inclination joint. Such microscopes themselves have almost disappeared from the market. Many are, however, still in use. The extension of the body tube beyond its support was a great source of instability. Utilizing the Leitz Model B Microscope, the author had the microscope reversed on its inclination joint so that when pulled to a horizontal position the horseshoe support lay under the horizontal body tube to give a much more stable system. When a microscope has no inclination joint, a stable support system for the horizontal microscope which keeps the center of gravity low and below the microscope proper can be devised without too much difficulty.

Having **the base itself large and heavy enough to keep the center of gravity low** is an important consideration for a microscope, whether the tube is vertical or horizontal. Fortunately this is usually done. Another factor is important. For microscopes that may be moved about there must be a rugged part, usually the arm, by which the microscope can be both conveniently and safely carried. The writer knows of an expensive measuring microscope with an inherent serious error because the microscope had been lifted by auxiliary personnel by the only part convenient to grab, yet which allowed the heavy base to hang by a bearing.

Vibration Considerations

The author is sometimes amazed at the indifference to the possible presence of vibration that is exhibited by some people engaged in photomicrography. Relatively recently the author took a slide that was being used for photomicrography on a bench in a factory, back to his own laboratory. Here the large photomicrographic bench, seen in Figure 3-15, was used; it is located on "isolation piers" in the basement of a building. The improvement in definition and microcontrast was startling, yet probably the optics were equally excellent in both laboratories. Apparently quality which has not been seen from a given setup is not missed unless the results obtained are obviously poor. **Vibration can be the limiting factor to the definition**, including resolving power, **obtainable in a photomicrograph**. If so, the photomicrographer should at least know it.

Detection of Vibration. It is not difficult to set up an arrangement sensitive to vibration and with which its presence is easily detected. Shillaber (4.1944) describes the familiar test of dust floating on water below a microscope objective. For this test place a ring cell, cemented to a microscope slide, on the stage of a microscope. The cell should be filled with water until it bulges up almost to overflowing. A few drops of some dust, such as carbon, are sprinkled on the surface of the liquid and observed through the microscope. This test is exceedingly delicate; Shillaber believes it to be too delicate.

The author likes a simple device that can be set on any surface that is to be tested for the presence of vibration. This device is made around a round-bottomed porcelain evaporating dish which contains a liquid that does not wet it. For this purpose the inside of the dish is rubbed with grease or a silicone and water is used for the liquid. A beam of light, reflected from its top surface, is created by shining a spotlight on it, as from an A-O Universal lamp, and the beam is allowed to hit the ceiling to form a long light lever. An image pattern should be focused on the ceiling; it may be merely the lamp filaments, and this image should be on a mark on the ceiling, or it need be only a crack or stria. The dish itself must be supported on legs or other means so that *it* will not wobble on the surface being tested.

The methods mentioned above are adequate for *detecting* vibration. The author learned much that was pertinent and useful from the use of a General Radio Sound and Vibration Analyzer on several occasions. This electronic device does much more than merely detect the presence of vibration, although it is very sensitive to it. The polarization of the vibration can be most important; there can be bad vibration in one plane and almost none in another. **Often the source of vibration can be determined with its aid.** In one case the source proved to be a motor and fan fastened to a wall in another department. Once, when the bench was supported by solid rock, there was no detectable vibration at all except when a truck passed a point on a nearby street. High definition photography was done by suspending operations when such an event occurred, as told by watching the meter. The effects of these vibrations are damped out more quickly in the test device than with the two liquid systems.

Vibration may be set up by the use of the microscope or camera. In one case in which very high definition photography was being done there was no camera with a focal plane shutter made by any manufacturer that did not cause vibration that made its use for this work impossible.

Elimination of Vibration. Some of the principles involved in the elimination of vibration are fairly sophisticated. The handbook by General Radio Co., issued with their meter, is excellent. Charles E. Crede (4. 1951) discusses the matter.

On the other hand, some of the principles are simple and well known. **It is important to have the center of gravity low.** It is also important to have **sufficient mass to absorb vibrations** of low energy. Therefore the *base* of the microscope and photomicrographic stand should be somewhat massive. **The supports themselves should be well separated.** A bulge near the middle of a flat support need be only 0.001 in. high to provide a fulcrum for vibration. Years ago Lord Kelvin showed the stability of wide-spaced support at three points of a triangle.

Good sponge rubber under high compression will absorb much vibration. The author has used a pile of three rubber bath sponges under steel plates, each of which lay beneath a leg of a heavy support, to isolate a bench effectively from vibration coming from outside. If natural sponge rubber is not used, the individual should know something of the elastic properties of the synthetic material chosen.

The microscopic board, discussed in Chapter 6, lends itself to a support to isolate it from vibrations.

A rather dramatic case in the author's experience concerns a couple of horizontal photomicrographic benches which had to be set up on an upper floor of a building tied to what were then the world's largest compressors. All methods of vibration isolation tried had failed—among them standing on rubber and suspension from springs. A rather heavy wooden platform was made for each bench and lifted from the floor by softly inflated inner tubes of truck tires. The tubes were loosely wound in canvas for protection. Photomicrographs of high resolving power and sharpness were made at 2500X magnification in routine while the liquid in the bottles on nearby shelves vibrated continuously and obviously. If loss of air caused one corner of the platform to touch the floor, the image in the camera was instantly sent into wild vibration.

The special case in which the vibration is caused by the shutter used in making a photomicrograph is discussed under *Shutters* in this chapter.

Mirror

Until quite recently all microscopes had a mirror that directed light from an outside light source into its optical axis. In metallographic microscopes it might be incorporated as a prism or a cover-glass type in the vertical illuminator, but with all other stands it was in a round frame at the bottom of the substage and carried by a universal joint involving a yoke that allowed the center of the pivot to lie in the mirror for tilting it in any direction. With the advent of the built-in light source for biological stands, this accessible mirror may not be part of the equipment. However, if there is considerable photomicrography to be done, a substage mirror for outside light source should be available and usually is. Some methods of illumination are exceedingly wasteful of light, and an outside light source may be wanted for adequate intensity or to give the preferred quality.

It seems strange, but the substage mirror, which is definitely part of the optical system of a microscope, has often been the poorest part of an expensive microscope. Added to this, individuals without training or who have done no adequate reading of directions have often used the mirror incorrectly. The standard "mirror" in reality consists of two independent mirrors, that is, independent in function. One side is concave; that is, it is a catoptric lens with a definite focal length and aperture. It should be designed for illuminating the specimen viewed with the lowest power objective with which the microscope is equipped, usually the 2X (48 mm). It would thus receive parallel light, often from the sky, focus it on the specimen, and fill the aperture of the objective. Therefore it is used without a condenser lens in the substage. For all other purposes the plane mirror should be used. To be really efficient and convenient the mirror should have the following characteristics:

1. The optical axis from the lamp through the microscope formed by the illumination beam should hit the center of the mirror, whose central point of swivel or tilt should lie at this point on the reflecting surface (see Figure 3-1).

2. The mirror should consist of a neutrally reflecting single surface, that is, a first-surface mirror.

3. The mirror should be low enough below the microscope to allow changing substage condensers without hitting the mirror and upsetting its adjustment.

Sometimes even with expensive microscopes all three of these desirable features may not apply. Under some circumstances flare is sometimes introduced by the relatively thick second-surface mirror because of the double image of the light source, as illustrated in Figure 3-1. The first-surface mirror is perfectly practical these days. The simplest procedure seems to be to obtain two substage

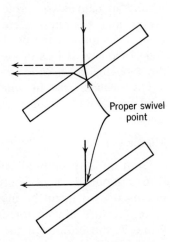

Figure 3-1 Substage mirror alignment.

mirrors for the microscope, which are inexpensive, and then to have an aluminum mirror overcoated with silicon monoxide evaporated on each.* One is required as a standby, for eventually the exposed one does deteriorate, principally during cleaning.

Unfortunately, substage mirrors are frequently *not* in good optical alignment, as mounted in the microscope; that is, the optical axis of the microscope does not hit its reflecting surface in the middle of the round mirror and the axis of swivel or tilt of the mirror does not pass through this intersection point, so that a further tilt of the mirror displaces as well as tilts the reflected beam. That this is an inconvenience rather than a defect of any good centration can be seen from the following consideration. Assume that the mirror is infinite in extent. Any movement in its own plane is inconsequential. Since the beam from the lamp is tilted anyway and can always be redirected to center on the **intersection of the mirror and the axis of the microscope,** any vertical or horizontal displacements of the mirror are equivalent in that each merely changes the point of intersection of the axis by the mirror. However, in each case, since it changes the tilt of the lamp, it will require a new tilt of the mirror. Several back-and-forth determinations of the lamp tilt versus the mirror tilt will be required. If the mirror is nearly correctly tilted in the first place, touching up its tilt is relatively uncomplicated. However, it is most convenient and time saving when the microscope axis intersects the round mirror in its center, and this can be provided by a support, independent of the miscroscope stand.

One of the advantages of a separate board support for the microscope is the ease with which the substage mirror can be supported low and independent of the microscope when there is an open horseshoe support or its equivalent. The mirror usually has a central post support that can be stuck vertically into a hole made below the microscope in a metal plate that is centrable. It is, of course, necessary to have a hole or slot cut in the board, as shown in Figures 6-2 and 6-3, with a temporary mark in the center of the mirror; the mirror is easily centered by looking down the microscope tube through a pinhole ocular, as described in detail in Chapter 6.

The Substage and Substage Condensers

Except for inverted microscopes, including the metallographic type, the substage of the microscope includes a condenser for illuminating the specimen, an iris diaphragm for controlling the aperture of the illumination, and sometimes an auxiliary spectacle lens for providing a larger illuminated field for use with a low-power objective with the assembled high-aperture condenser. Then there should be a shelf or ring to hold

*This is done, for instance, by Evaporated Metal Films, Inc., Ithaca New York.

auxiliary color filters; preferably it should take disks 33 mm in diameter.

The substage is usually the biggest sacrifice made for inexpensive microscopes. What is worse is that the microscope models may become more expensive and sophisticated in other ways before the substage is adequately improved so that other optics and mechanical features achieve a level beyond that of the substage and its condenser. **Except in the most inexpensive microscopes, one should be able to remove a substage condenser and replace it with another type.** For photomicrography, at least, one should be able to focus the substage condenser, preferably with a rack and pinion. Some of the most sophisticated models have a substage iris diaphragm that focuses independently of the condenser; this is highly desirable, even though some condensers, such as the achromatic type, come with their own incorporated iris diaphragm. In the better models it should be possible to displace the substage diaphragm to one side and to rotate it to obtain oblique illumination from any azimuth. Ideally, it would be nice to have the focal length and the aperture of the condenser match those of the objective in use, plus having its optical corrections match those of the objective. Devices are sold by which a second microscope objective can be used as a condenser; these devices hold a spectacle lens to adjust the corrections of the objective for its "tube length" when used as a substage condenser, as discussed in Chapter 6. Although sometimes most desirable, this is, of course, rarely done. It is, however, poor microscopy to allow the substage condenser and objective to be too badly mismatched, particularly in focal length and aperture. An auxiliary negative spectacle lens below the substage condenser, to provide the equivalent of increasing its focal length, is an excellent convenience for using the low power to search the field for high-power work but a definite sacrifice when used in routine for low-power examination in photomicrography. One may have a battery of condensers in which the better corrected condensers of high aperture provide means of unscrewing one or more components to obtain longer focal length and lower aperture. Some of these are listed in Table 7-1 and are, of course, discussed with their focal lengths and apertures in the catalogs of the microscope companies.

The quality of the substage condenser makes more difference in the quality of a final photomicrograph, according to the author's experience, than is usually assumed. It is interesting than in theoretical discussions by authorities in the general field of optics quality of illumination of the specimen is usually considered unimportant, whereas in discussions by practical microscopists it is emphasized as important, sometimes to an extreme extent. This is probably so because the theoretical authority assumes that the image-forming light is scattered by the object points, effectively originating from there, so that the resolving power of the image is unaffected by the mode of specimen illumination. In practice it is easy to show that with a poorly corrected substage condenser and a source of appreciable size it is impossible simultaneously to fill the aperture of the objective uniformly and obtain even reasonably uniform illumination in the image plane. As discussed in Chapter 15, it is now known that although resolving power of the image is fundamentally important other factors such as "microcontrast," which can be independently variable, contribute greatly to the judgment of good quality and good definition in the photograph. Light flare in the image field can be an important factor in image degradation to which some images are much more susceptible than others. Therefore, in the writer's opinion, among models of microscopes, as the quality of the whole instrument with its objectives and eyepieces becomes better the quality of the substage condenser should also improve. This will require customer demand. This discussion, of course, applies to brightfield microscopy in photomicrography and is complicated by the advent of phase microscopy and interference microscopy, as discussed in later chapters.

The two-element Abbe substage condenser is of ancient design, but it is still numerically the most commonly used condenser for the compound microscope. A cross-sectional view is given in Figure 3-2a through the courtesy of Bausch & Lomb. Maximum numerical aperture quoted for this condenser is usually 1·25. Because it is made of one kind of glass, usually ordinary crown, there is no chromatic correction. The top element is a truncated sphere with its lower spherical surface calculated to be of such radius that the object point will lie at the interior aplanatic surface (see Chapter 1) if the bottom of the slide is homogeneously immersed with the plano upper surface of the condenser. The lower element is necessarily of high power with an extremely strong lower surface and with some attempt to make the interior surface of the lower element aplanatic. The strong surfaces introduce much optical degradation to the image light source, and since the condenser is rarely immersed the air-glass surfaces at its top introduce more large aberrations spherically and comatically. Since only spherical aberration, which is a purely axial aberration demanding a point source, is corrected, in practice the coma and other zonal and field aberrations are so great with practical light sources of appreciable diameter that little improvement can be noticed in illumination quality with immersion of the slide condenser interfaces. This is not considering the increase of aperture obtained by immersion of the condenser, as discussed in Chapter 2. The flare light introduced by the use of this condenser for practical light sources must be very great, but its degrading effect on the quality of some images is discussed in subsequent chapters. The statement has been made that two-element

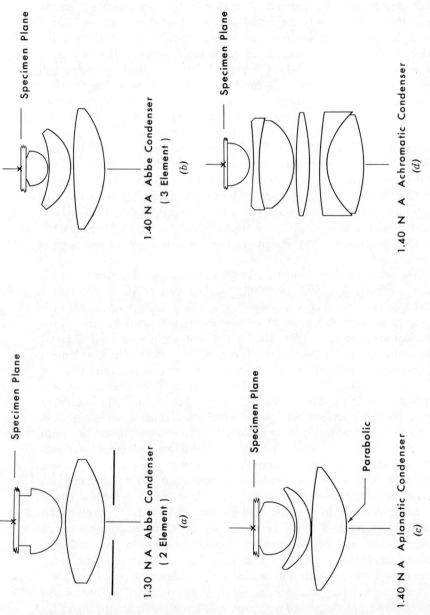

Figure 3-2 Types of substage condenser for brightfield illumination: (*a*) two-element Abbe condenser; (*b*) three-element Abbe condenser; (*c*) aplanatic condenser; (*d*) achromatic condenser.

Abbe condenser is unfit for photomicrography. This statement seems foolish to the author. Specimens differ enormously in their demands for good image quality for all the characteristics that go with that expression, and the operator may be content with the optical image as he sees it formed on projection.

Very substantial improvement can be made with the **three-element Abbe condenser.** A cross section of this is shown, also by courtesy of Bausch & Lomb, in Figure 3-2*b*. The truncated spherical element at the top may be the same for both. However, the very strong lens with high curvature of the two-element condenser can now be weakened by splitting it, which is a great help in itself, but the top surface of the middle element can be made concave; the exit rays will be almost normal to the surface and thus no additional spherical aberration will be introduced. Such a design is often opened up to NA 1·40.

Although a great improvement, the three-element Abbe still contains large spherical aberration and no correction for coma. Therefore an **aplanatic condenser**, usually having an aspheric lower surface, is still a greater improvement. A cross section of this design is shown as Figure 3-2*c*. This condenser, usually of NA 1.40, is a worthwhile improvement although it still has no chromatic correction. For black-and-white photography with color filters and often monochromatic illumination the author employed an aplanatic 1.40 NA substage condenser most satisfactorily for many years. An excellently sharp image of the lamp condenser and field diaphragm for Köhler illumination (see Chapter 7) is provided in the object and image plane of the microscope by this condenser when a color filter is being used. The **achromatic substage condenser** is a great improvement over any of the previously discussed models. On the whole this design corresponds to that obtained when design of an achromatic objective is blown up in focal length with possibly a few other economies made. Its advantages are discussed in more detail in later chapters.

However, the better the optical corrections for a substage condenser, the more rigid the specifications for using them. This is a truism for all lenses. In designing an aplanatic or achromatic condenser, the designer must assume the distance for a light source, and this assumption should be fulfilled by the user. This matter is discussed in Chapter 7, and the specifications for a number of condensers are given in Table 7-2. **If it is necessary to use the light source at a different distance than assumed by the designer, it can be brought,** *optically*, **to the correct distance by adding a corrective spectacle lens**, as discussed in Chapter 7. This simple lens may bring some degradation in the use of the achromatic condenser but relatively little and less than the neglect of this factor. The principle, of course, is to make the light rays enter the lens at the angle assumed by the designer. Darkfield condensers are discussed in detail in Chapter 12.

The Focusing Mechanisms

The focusing mechanisms, coarse and fine adjustments, of a microscope are purely mechanical components but such that **their perfection greatly influences the convenience, stability, and reliability of the whole instrument** and its performance. An experienced microscopist can usually judge the satisfactory nature of both mechanisms in a relatively short while if he is allowed to examine some critical object. **The quality of the coarse mechanism is quickly determined by its feel.** This is usually a rack-and-pinion or a worm train and rack. In the best design the rack-and-pinion teeth are cut **in helical form** so that several are in engagement at the same time. **There should also be some mechanism for taking up wear** as the instrument gets older. Reference to "Care and Use of the Microscope" may indicate that a method is provided and give directions for doing it. Some screw may be tightened or the coarse focusing knobs rotated firmly in opposite directions. In some of the newer instruments no provision is made for the user to do this; it may be asserted that it will be unnecessary. However, the experts in the shop repair of particular instruments have usually made this type of renovation when the focus became loose or was slipping; for instance, in the newer Leitz adjustment the shop may tighten the end screws of the knobs or insert a new flannel-type washer in the train which acts like a brake.

The design of the fine-focus mechanism can vary a great deal, especially when all models, past and present, and the various manufacturers are considered. Most, except the most recent, are discussed and illustrated in textbooks on microscopy and by B. O. Payne (4. 1957). The fine-focus mechanism, which usually moves the objectives with respect to the stage over a range of about 2 mm, should have a positive stop at each end so that it cannot jam. **An excellent fine adjustment should be stable when focused at a given level, show no variance (play) when reversing its motion, and be exceedingly sensitive and free of vibrations.** It should also be unaffected by appreciable changes in temperature. It may be important that it be a micrometer type, that is **linear over its range** with respect to the rotation of the actuating knob. The various designs of fine adjustments will differ in these qualities and the need for them will vary with the use to which the microscope is put. There have been considerable design changes throughout the industry in recent years so that many types are now in use. Certain uniform trends are apparent.

Previously almost all fine focus mechanisms were almost **entirely** separate from the coarse adjustment and carried the entire adjustable structure, either the stage or the whole body tube, including the incorporated coarse adjustment. This form may still be used; it can be very satisfactory. It is significant that this applies to the Leitz Ortholux and Panphot, the two "top models" of the Leitz line. With both instruments the coarse adjustment and fine adjustment rotating knobs are separate, as are the entire mechanisms. However, in the new Zeiss (Jena) microscopes the two knobs are concentric, although the mechanisms are separate. There is a definite trend toward concentric shafts for the fine and coarse adjustment knobs, a definite convenience.

The trend toward **ball-bearing ways for the focus adjustment**, especially the fine movement, seems complete. The ball-bearing slide helps wear and can prevent a serious tendency of the objective (or stage) to move laterally, besides furnishing the familiar smooth, almost effortless translational movement. E. Leitz has utilized a double ball-bearing slide in which one set of balls rolls between the stationary track and an unattached plate. Another set rolls between the plate and the moving member. This free-floating plate is used on their laboratory microscopes, including the Labolux and the SM. The writer has tested and used this device, as represented by the Leitz Model B microscope (now unfortunately obsolete) in which the tube was moved. In this case the movement was very smooth and it was impossible to detect any play in the fine focus, even at high magnification. On the other hand, a projected image was exceedingly susceptible to lateral vibration (in the plane of the free plate); when the microscope was used horizontally, a single tap at the side caused the image to vibrate for a relatively long time.

For the fine adjustment the problem is to reduce the ratio of motion between the rotation of the fine adjustment knob and the motion along the axis of the objective or specimen on the stage **without allowing appreciable lost motion** when the direction of focus is reversed because of a succession of linkages. An advance of 100μ for a rotation of the knob is desirable in the better microscopes, **allowing graduation to a micron;** this graduation can still be done with a rotation that represents a 200-μ advance by using a larger drum, as does the American Optical Co. For many years the preferred method was the use of a lever with a high lever ratio. The weight of the tube (or the stage) was carried on a pin that was lifted or lowered by the short movement of the lever. It could be carried on either side of the fulcrum. The long length of the lever could be moved by a worm or gears. Backlash, especially in horizontal microscopes, was removed by springs that pressed in one direction. The problem of the lever was that the movement was not linear throughout; that is, it could not be used as a micrometer as is sometimes desirable. The lever was also especially affected by temperature variation, an important consideration if a well-focused image at high magnification is to be maintained for long periods, as in cinemicrography.

Just before World War II Carl Zeiss abandoned the lever in favor of a **train of gears**, similar in appearance to those of a clock. They were, in fact, stamped out, and the original motivation was the economy of factory space and labor. This gear movement has been kept by Zeiss and was adopted by Carl Zeiss in Oberkochen after the war; it still is the mechanism of their fine adjustment. It shows no appreciable wear with time and furnishes accurate micrometer movement. Zeiss claims that because of the push of the spring it should show no variance at any magnification. All slides, coarse and fine, are on ball bearings and so offer little holding friction.

The Zeiss (Jena) microscopes, such as their Ng and Nf models, also have a rectangular box that rests on the base containing the entire focusing mechanism, with concentric focusing shafts and aligned knobs. Each side of the box has slide ways. The rear side carries the arm, hence the objectives; the height of the arm is adjusted by the coarse focus. The fine-focus mechanism acts on the height of the stage. Graduations are to 2μ. The author would expect that care would have to be exercised in lifting the microscope by its arm.

In general however, the lever has been abandoned in favor of the cam for reduction and regulation of the focusing movement. **The cam is reliable, can give a strictly linear relation between the rotary knob and the translated member and can be made to show little wear.** It can furnish almost any ratio of movement, although in practice one of reasonable size is limited so that **some of the reduction of motion** is likely to be made **by a worm screw and its gear.** The heart-shaped cam is famous for its linear control. Leitz, Inc., utilized a cam many years ago and their Ortholux and Panphot models still do. In each of these two models the coarse adjustment has a helical rack-and-pinion and ball-bearing ways. The fine-focus knob turns a worm screw; its nut bears an end collar which is a cam on which the follower is a roller on an arm from the block that carries the the coarse adjustment, the stage, and the substage assembly. The ways also have ball bearings, of course, but not the double ball bearings with a floating plate. One turn of the knob advances the stage 100μ toward or away from the objective. One division corresponds to 1μ. A spring, working upward against gravity, relieves some of the weight of the stage assembly and specimen.

The most recent tendency of focusing mechanisms is to **have the fine adjustment act *through* the coarse adjustment so that there is only one carrying mechanism.** Coaxial shafts and aligned focusing knobs would be intrinsic features for which the coarse adjustment has had to be modified so that no variance (play) would be encountered in it and so that it would be adaptable to the system. This **has usually restricted the motion of the coarse adjustment** to 40 or 45 mm, a potential disadvantage for some industrial microscopy and photomicrography.

The Leitz adjustments on their SM and Labolux models have a unique feature. There is only **a single focusing knob and shaft**, which is **for** *both* **coarse and fine focusing**. For coarse focusing we employ it in the usual way; it regulates the height of the stage support which also carries the sub-stage condenser rack. For fine focusing we back off a little. The knob now **acts as a fine focus for about one-third of a revolution**. This feature is achieved as follows:

> The shaft from the focusing knob bears a worm screw, but this is mounted loosely on the shaft. The flat end of the worm screw bears two pins, each projecting parallel to the shaft. When the knob is rotated, a pin, projecting perpendicularly from the shaft, hits one of those from the worm screw and causes the screw to revolve also. When the knob is suddenly rotated the other way, the pins become disengaged. The worm screw causes its nut and a gear on the same shaft with it to revolve. The gear mesh into a rack (on the left side with the microscope facing the operator) and thus change the height of the stage. To actuate the fine focus there is a ball bearing at the other end of the worm that rides against a slightly inclined plane. Rotation of the knob moves the worm screw along the axis a little, and thus causes a slight rotation of its nut. This system does not seem quite so desirable to the author for photomicrography as that of the Ortholux and Panphot, but he has not used it extensively. The SM and Labolux models have the double set of ball bearings with the free-floating plate.

In the Bausch & Lomb Dynazoom and Dynoptic and the A-O Micro-Star Series 10 microscopes the fine-focus mechanism also acts as a micrometered force to actuate the coarse-focus adjustment. All have some features in common. In the B&L instruments the stage support is moved by the focus mechanisms; in the MicroStar the objective support is moved. In all cases **the larger coarse adjustment knob turns a hollow shaft that contains the freely turning inner shaft of the fine-focus knob.**

In the Dynazoom and Dynoptic microscopes a worm gear is mounted on the outer shaft. Its gear, which is fairly large, turns a small pinion as its own shaft and teeth at the end of this pinion push a rack bearing the stage support up and down. The teeth and rack are helical. The worm, which is free on its shaft, is really caused to rotate by a nut protruding through a longitudinal slot in the outer shaft. The nut is on a fine micrometer screw thread on the inner shaft. Rotation of the shaft by the fine-focus knob translates the nut, hence the worm gear, between the limits allowed by the slot. This causes the rest of the mechanism to change the focus by just the allowed bit. Pushing the worm from one side is used to eliminate variance. Actually the helical gear on the worm rotates freely on the shaft of the pinion. Therefore the pinion shaft follows only because

of a clutch, one plate of which is fastened to the shaft; the other is the face
of the gear. A Mylar washer lies between; a spring washer holds the
assembly together. Thus the microscope could be turned down against a
slide without damaging an objective. The tension on the clutch is easily
adjusted from the back of the microscope by withdrawal of a screw. There
is no other provision for wear, which it is claimed should never be needed.
The spring plus gravity takes up wear in the gears.

The mechanism used for the focusing of the MicroStar is somewhat
different. The worm which rides freely on the outer shaft is a larger
special design with a single large groove that also is an inclined plane as it
turns. See Figure 3-3. The vertical arm of a sturdy right-angled lever
rides on the worm and has a split ball at its lower end (in the groove) to
prevent variance. Rotation of the worm, as the outer shaft rotates, raises
the lever with its axis and rotates it. A shorter side arm of the lever lifts a
block that bears the support for the objective changer. This allows a
range of $1\frac{5}{8}$ in. The worm is caused to rotate with the large knob and
outer axle shaft by a pin from the follower nut on the screw thread of the
inner shaft attached to the fine-focus knob. Rotation of the fine-focus knob
translates the follower nut along the axis of this differential screw thread
and thus causes the large worm to translate along the axis also. This
moves the whole focus mechanism slightly. Gravity and the split ball are
said to remove all variance throughout the life of the microscope. The
American Optical Co. declares the action to have been designed so that
the fine focus is linear over its range. This form does lend itself to great
stability of focus which is especially important for cinemicrography. Testing the mechanism at both low magnification (small image depth of focus)
and at high magnification, the author found the focus to be unaltered by
pounding the table *after* the refocusing required for the first pound. (This
illustrates the importance of a jar to stabilize focus for a single picture.)
Actual vibrations touching the tube or stage were transmitted to and
amplified in the image, as is the case with all microscopes tested.

The Body

In the past the body almost invariably consisted of a brass tube of
sufficient diameter to clear all image rays **whose function was to hold the
objective and the oculars in good alignment and at the proper optical
distance.** Modern design has proliferated it into many forms, but it must
maintain these functions. In the European models especially the old tube
has almost disappeared, except for the long narrow one into which the
oculars slip, and various alternative bodies can be slipped into an adapter
ring held at the end of the arm. The lower end, which holds the objective

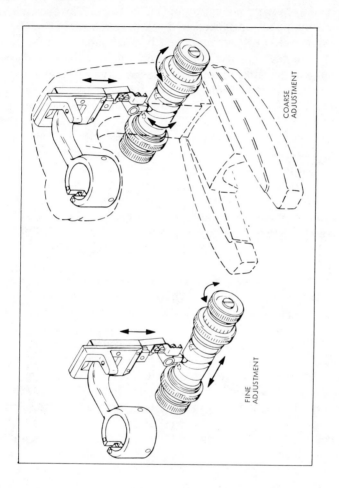

Figure 3-3 Focusing system of Microstar Microscope, Series 10.

COARSE
ADJUSTMENT

FINE
ADJUSTMENT

or set of objectives, **must provide a way of easily changing the objectives. The rotating objective holder is the most convenient in use when the objectives are parfocal.** Frequently a set of objectives will be purchased with the microscope, centered into the optic axis on the rotating changer at the factory, and used with the microscope at all times. However, a laboratory that is well equipped and expects to do extensive work will not be satisfied with a single set of objectives, even by the same manufacturer, since especially valuable objectives are scattered among the manufacturers. **A mechanism for bringing these objectives into central alignment with the optical axis is definitely required.** All of the objective carriers of professional microscopes utilize the RMS thread. This thread, originally specified by the Royal Microscopical Society (4. L. W. Nickolls, 1955), has become the international standard. The American standard (ASA Bl.11-1958) is interchangeable with it but modified principally in having greater truncation to the peaks of the threads. [Figure 3-4, which was taken by courtesy of the Royal Microscopical Society from p. 378 Vol. LVI (1936) of its journal, shows the specifications of the old British Standard, since the illustration of the ASA publication appears to be more complicated at first glance. A British RMS thread gage would be satisfactory to test the threads of either the ASA or the British specification but not conversely.]

When there are a large number of objectives, the centrable, sliding-clutch type of objective holder is best, such as that Bausch & Lomb use in their petrographic microscope. This holder was put on the author's special photomicrographic microscope. It is easy to use, but very firm and stable. For a smaller number of objectives the revolving centerable objective carrier for four objectives, made by Leitz, was found to be most convenient and fitted several microscopes, including those of other makes. Other revolving objective carriers are centerable, but of those noted, the Leitz was most easily adjustable by the user with special wrenches and yet was stable.

In Chapter 2 **the importance of the corrected tube length** between objective and ocular, as related to the cover-slip thickness and other constants of object space, was discussed, as were the advantages of being able to control the tube length. There is real difficulty, even in the most sophisticated microscopes, in obtaining this control in present day microscopes. When the tube length should be lengthened, it can be done by merely pulling the ocular out and providing some mechanism for holding it there. For objectives corrected for infinity the same technique is applicable. They have a telescope objective whose distance from them is not critical, but the distance of the ocular above the internal telescope objective is just as critical as with other type objectives and can be manipulated for tube length.

IV.—R.M.S. SPECIFICATION A.

SCREW THREAD FOR OBJECTIVE.

Form.—Whitworth, i.e. a V-shaped thread, sides of thread inclined at an angle of 55° to each other, one-sixth of the V-depth being rounded off at the top and bottom of the thread.

Pitch.—36 threads per inch = 0·02778 in. approx. (= 0·7056 mm.).

Length of thread.—0·125 in. (3·175 mm.).

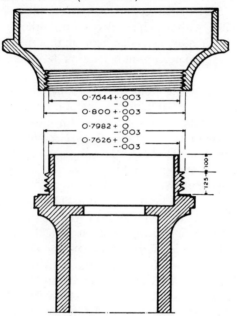

	1	2	3	4
	Maximum.		Minimum.	
	in.	mm.	in.	mm.
Full diameter	0·7982	20·274	0·7952	20·198
Effective diameter	0·7804	19·822	—	—
Core diameter	0·7626	19·370	—	—

NOTE.—The objective is to screw home properly to shoulder.

Plain Fitting (Pilot) above the thread of the objective :
Diameter is not to exceed 0·7626 in. (19·370 mm.).
Length of pilot : 0·1 in. (2·54 mm.).

Figure 3-4 Specifications of RMS screw thread for objectives. [Old British Standard, L. W. Nickols, *Trans. Royal Micros. Soc.* **75**, 58 (1955).]

TYPES AND CHOICE OF APPARATUS

The types of apparatus commercially available have increased. On the other hand, some desirable types are no longer commercially available, principally for economic reasons. An example is the completely horizontal photomicrographic bench discussed elsewhere. The basis for the decisions that may be made within a certain magnification range between the use of the simple and the compound microscope has been discussed; for instance, if at 50X magnification we wish to fill an 8×10 frame, we can do so acceptably with a simple microscope, although we would have to be content with a much smaller field at this magnification for the compound microscope. The central resolution, however, would be somewhat higher.

It has been customary to hold the photomicrograph objective of **the simple microscope** on the front of the camera. Smooth focusing over a relatively extensive range is necessary for convenient and reliable work, especially at the higher magnifications. Bausch & Lomb's rack-and-pinion focussing mount to hold the Micro Tessars is satisfactory for very low magnifications but has enough wobble in the back-and-forth movement to make it aggravating and even inefficient for the objectives of smaller diameter that need more accurate centration. The objective mount on the Bausch & Lomb L Camera for low-power work is more satisfactory. The author set up a photomicrographic bench with the simple microscope shown in Figure 3-16 on which **the support for the objective was independent of the camera.** For this purpose the body tube with its rack-and-pinion was obtained separately (Bausch & Lomb Unit No. 31-29-01-01) and the original tube replaced with a wide tube that carried an adapter plate threaded to take Micro Tessars and other photomicrographic lenses. These lenses were usually enlarger lenses reversed in their mounts. The entire unit has proved exceedingly satisfactory, being both stable and smooth working.

Assembly with Compound Microscope

Since World War II there has been a proliferation of **simple laboratory compound microscopes** and **student microscopes** on the market, frequently by companies not in this field before the war. No specific comments will be made, since the general principles have been discussed under evaluation of microscopes. When an acceptable image is obtained, a photomicrograph can be made. This is fully discussed, particularly in Chapter 11. As already stated, the author prefers a horseshoe base because it **allows the use of a mirror below the substage separately supported** by a board, as discussed at the beginning of Chapter 6. It is most worthwhile to obtain a microscope **substage condenser that is demountable and focusable.** Again the author suggests investigation of the availability of second-hand, possible reconditioned, microscopes. There is a fairly lively market in these instruments, although their availability is usually quite tight.

In a discussion of the **professional microscope** stands we can no longer entirely defer the subject of **"flat field"** to the section of the accessories, objectives, and eyepieces. This is an attractive selling feature that is much emphasized. Until recently the statement was true that objectives and oculars of different makes could be interchanged with impunity, provided that the tube length considerations discussed in Chapter 2 were observed. This situation allowed the picking up of especially desirable objectives for specialty types from the different manufacturers and was a very great advantage.

The advent of the relatively new wide-field (or plano-) optics invalidates this assumption. An extreme example is the Bausch & Lomb Dynazoom (see Figure 3-5) and Dynoptic microscopes. With these two microscopes a lens that may be said to function as a part of the objective is incorporated in the lowest part of the body tube so that only the special plano-objectives that come with the microscope may be used with it. Both the Dynazoom and Dynoptic microscopes may be obtained without the flat field component lens so that standard objectives, including those of other makes, may be used with them. The Dynoptic model contains a two-element relay lens system in its body tube which gives color compensation about equivalent to that of the Hyperplane eyepieces; therefore ordinary Huygens eyepieces may be used up to the point of which full compensation is required. The plano-objectives of both Leitz and Zeiss may be screwed into any microscope of appropriate tube length, but in each case the oculars specified by the manufacturer should be used. In the case of Leitz this is the Peri-planatic, and with Zeiss the Komplan (Kpl) eyepieces must be used for excellent results. However, the old assumption is still true for the older optics that are still in greater use; collections of objectives may be obtained from the several manufacturers without buying new oculars if *one possesses good oculars of the appropriate type for the objective;* for example, compensating eyepieces for high-power achromats and all apochromats.

The new "flat-field" objectives and accessory optics are disccused under *Objectives* later in this chapter.

Body Design

Although there is an unfortunate tendency to get away from international or even intercompany standardization to allow interchange-ability of components, most of the microscope companies have increased

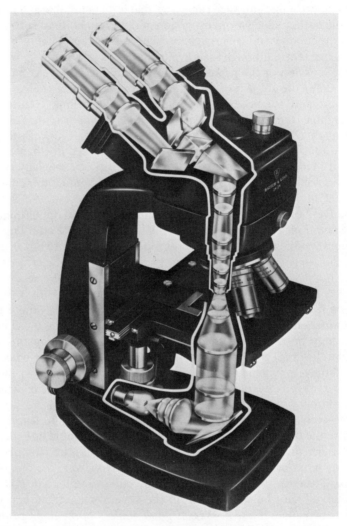

Figure 3-5   Cross-sectional view of Bausch & Lomb Dynazoom Microscope.

the possibility of interchangeability within their own equipment by unitized components. Many of the stands of the European companies have a sturdy arm ending in a horizontal portion above the microscope stage which ends in a ring that allows what Carl Zeiss calls the "circular dovetail system" to carry any type of objective carrier below it and any type of body tube above it. A sturdy ring clamp may be used instead of the dovetail. This system is illustrated in Figure 3-6 by the "exploded"

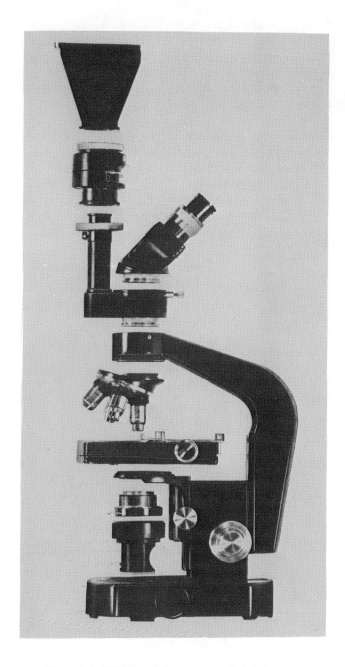

Figure 3-6 Wild M-20 Microscope, exploded view.

view of the M-20 Research microscope of Wild (Heerbrugg, Switzerland). The same system can also be seen in Figure 3-7 of the A-O microscope and in Figure 7-20 which shows a Leitz Model FM.

The MicroStar series of the American Optical Company illustrates a complete system of unitized parts that assists in the intelligent selection of a microscope to fit the purpose. These microscopes also use the short sturdy U-arm and circular dovetail linkage described above for the European stands. The author could wish for some more alternatives in the brightfield substage condensers. He does rejoice in the retention of the U-base (equivalent to the horseshoe), as illustrated in Figure 3-7 of a MicroStar Series 10 microscope, which allows the use of a substage mirror that could be located below the microscope, as described in Chapter 6, and that also easily allows the use of any light source. This

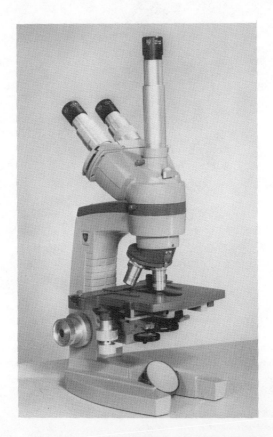

Figure 3-7   A-O MicroStar Model XM10TG-QW.

does not prevent the inclusion of a "built-in light source," several models of which are provided, and which slips into the base (see Chapter 8). The figure shows the trinocular head, now provided by most microscope companies, which normally would be chosen when both visual and photographic work is to be done with the same microscope. However, a special monocular head allows the tube length to be varied.

Unfortunately, with the tendency in microscope design to leave as little as possible to the operator in the adjustments, some of the most expensive models provide little opportunity for the microscopist to adjust the tube length, the importance of which was discussed in Chapter 2. The unitized construction makes it simpler to add a specially constructed component without great expense, as evidenced above. The special tube of variable length offered by A-O should be most valuable for critical photomicrography. It should be especially useful with monochromatic near-ultraviolet, which has proved so valuable for particle-size work and other fields that utilize black and white films, as discussed in Chapter 14.

Some objectives, for example, those of the American Optical Company and the metallographic objectives of Zeiss, are "corrected for infinity." From the discussion in Chapter 1 this must mean that the objective is calculated for the specimen to lie in its front focal plane which gives parallel bundles of light from each object point and a diverging cone of rays from the exit plane of the objective. This image cone needs an achromatic telescope lens in the body tube, usually in a position about opposite the supporting arm. The image beam is therefore brought to the focal plane of this lens, of course, except as intercepted by the field lens of the microscope ocular. A variable distance between the objective and the telescope lens will not deteriorate the image—hence vertical illuminators and other components can be interposed—but neither can variation of this distance be used to improve the image. On the other hand, above the telescope objective any variation of this distance to the ocular will affect the conjugate distance in front of the objective, when it is in focus, and so can be used to adjust the tube length as before. In fact, in the MicroStar Series 10 microscope the telescope lens is mechanically linked so that its height is adjusted to compensate for the different interocular distances that may be chosen. With the special monocular tube, whose length may be varied, the telescope lens is at the bottom of the attachment and the normal mechanical tube length is 282 mm.

Unitized construction allows change of body tube between binocular and monocular for photomicrography. The trinocular body tube can now be obtained from all of the microscope companies, thus allowing photomicrography to be done at any time. In fact, in order to make photomicrographs with either the Bausch & Lomb Dynazoom or Dynoptic microscopes, a trinocular body must be specified. When a microscope is obtained primarily for photomicrography, the author feels that microscopists should obtain one that allows the use of the beam through a monocular or trinocular body tube, unhindered by the presence of a prism or other split beam arrangement. The simplest arrangement with a trinocular is to have a mirror or prism that flips in or out of the selected optical beam.

The much used Leitz Labolux may be specifically mentioned because of the diversity of components accessible for it. It has a completely built-in light source and illuminating system discussed in Chapter 7 and normally uses the 6-V, 15-W LINOP bulb, also discussed in Chapter 7. However, a substage mirror can be substituted under the substage condenser. Photomicrographic accessories, such as the KAVAR observation eyepiece for 35 mm cameras, discussed in Chapter 11, are provided for this microscope, or it may be used with the Aristophot illustrated in Figure 3-8. It can also be fitted with the Orthomat automatic exposure device discussed in Chapter 17.

The relatively recent Leitz research microscope, "Orthoplan," seems especially impressive for photomicrography because of its emphasis on stability and freedom from vibration, the importance of which has just been discussed, although the author has no experience with its efficiency. This model is shown in Figure 3-8. It also carries a built-in illumination system that operates with a 12-V, 60-W lamp. This same bulb is used in the Leitz Panphot. The Orthoplan features 30 mm eyepieces (28-mm field of view), together with bigger prisms and the larger body tube that is required to handle the wider bundles of light rays in order to utilize the largest possible field offered by the newer plano-optics. The eyepieces are the "wide-field" Periplanats. This microscope probably could not utilize a vertical illuminator.

The Camera and Its Support

A 35 mm camera can, of course, be used with almost any microscope. The common commercial system is to support the camera by the microscope itself with an intermediate split-beam observation tube, the so-called "attachment camera." Because of the mechanical shock of the exposing shutter, primarily, the author still prefers to support even this small camera separately from the microscope but by some method which it is quickly and easily superposed or thrown out of the way. However, many successful photomicrographs have been made with attachment cameras and observation eyepieces. This matter is discussed under *Eyepieces* in this chapter and in more detail in Chapter 11.

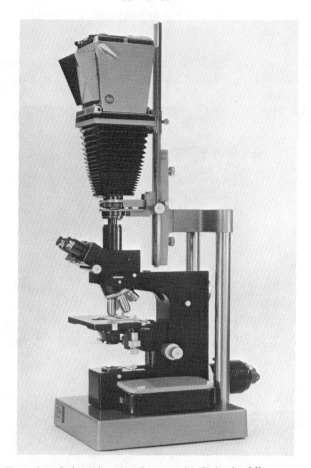

Figure 3-8   Leitz Aristophot Camera with Orthoplan Microscope.

Since we already have a projected beam when a microscope must be set up for photomicrography, the enlargement, which is a subsequent and independent operation with 35 mm film, is incorporated in the first picture and its negative with the larger format, thus allowing contact prints or immediate direct positive prints, for example, Polaroid or Kodak Super Speed Direct Positive, to be made. The supporting member for the camera over the microscope must be very sturdy, much sturdier than apparently seems necessary to some people, judging by the supports the author has seen. It is an advantage to have a stand that allows a microscope to be set under it for photomicrography and then to be withdrawn when there is extended visual use elsewhere. Sturdy stands of this type are commercially available. The Bausch & Lomb L stand shown in Figure 3-9 has been made quite adaptable for photomicrography with a simple microscope.

The sturdy post for the Wild stand is shown in Figure 3-10. A supporting bracket for the Wild Universal Lamp can also be seen. Leitz

Figure 3-9   Bausch & Lomb L Camera Stand.

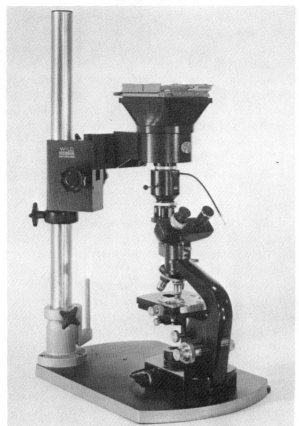

Figure 3-10 Wild Multipurpose Camera Stand with M-20 Microscope.

utilizes a double support for its Aristophot photomicrographic camera shown with the Orthoplan microscope in Figure 3-8. It is quite possible to set up a vertical photomicrographic camera of one's own; the author did just that very satisfactorily. The vertical post used for the view camera had a diameter of about 3 in. Strong stop rings (or "collars") below the clamp rings for the camera gave much greater reliability from accident and allowed it to be swung around with impunity. Some careful planning is required to give adequate support that is not susceptible to vibration, yet has sufficient and easy vertical movement.

The Large "Universal" Photomicroscope

This large, yet compact and necessarily expensive, type of microscope is designed to furnish any kind of illumination, whether by transmitted or reflected light, including metallography and petrography or even their combination and usually phase contrast and polarized light. The author has not seen interference microscopy included. Moreover, it is supposed to require the least skill and knowledge by the operator by becoming almost a "pushbutton" type. The illumination is necessarily built in. Such a device lacks flexibility, for many decisions are made for the user. The writer agrees with E. P. Shillaber (4.1944) that rather than being suitable for research it is more suitable for service photomicrography operated by technicians who are not necessarily experts. In the author's experience the first "universal photomicroscope" was introduced by Reichert (Vienna) and was a very unorthodox-appearing instrument with the camera at the bottom and an inverted objective pointing up at the microscope stage on which the specimen was laid from the top. This is their Me F Model with which the emphasis is now mostly on its use for metallography. For general photomicrography Reichert now offers the **Zetopan**, which it claims to be a "truly universal microscope" and to furnish all forms of illumination from its built-in or attachable light sources. However, this now appears similar to the more conventional microscope with a trinocular head, which has an adapter for a 35 mm camera and one for a camera taking 4×5 sheet or Polaroid films. A 200-W mercury vapor arc can be attached to the rear of the microscope with a subsidiary 30-W tungsten filament lamp for fluorescence microscopy.

The corresponding universal microscope by Leitz has been their **Panphot**, which seems to be especially compact and stably constructed. It does not have the inverted stage for metallographic specimens but it is still stressed as a metallographic microscope as well as a petrographic stand (Panphot-Pol). The universal aspect that involves transillumination is stressed. It normally carries either the XBO 150 xenon arc or a carbon arc with a tungsten illuminator, LILOM, which uses a 12-V, 60-W lamp with a 3×4 mm rectangular filament grid offset to one side of the lead wires and binding posts. The lens of the illuminator appears to be a simple aspheric lens, with the lamp side made somewhat matte. An 8-V, 0.6-A lamp (LISEY) is available for visual use which is considerably cheaper. All forms of illumination may be switched in by appropriately flipped mirrors and buttons. The three types of camera are each available, the largest format for sheet film being 4×5 in. This might be a size limitation for some photomacrography. The cameras for 35 mm film include the Leitz Orthomat which is a device that incorporates cassettes for 35 mm film with an electronically controlled shutter through a photocell that automatically exposes the film from $\frac{1}{100}$ of a second to half an hour or more, according to the speed setting of the film and the accumulated light. This type of device is discussed in Chapter 17.

E. Leitz now feature their **Ortholux** microscope which, like the Reichert's Zetopan, can be outfitted also to handle all forms of illumination and cameras; hence its total system might be considered to be a universal microscope with all sources either built in or attachable. It is a little simpler to add an external mirror for an outside light source than the Panphot. It has built-in illumination, of course, but features the Leitz Universal Light Source 250, already discussed, or merely the simple tungsten illuminant; reflected illumination is obtained when needed by attaching the light source to a tube adapter that extends through the arm of the microscope directly to a vertical illuminator or to the Leitz Ultropak for reflected dark-field illumination used for textiles or other non-metallic surfaces. A variety of cameras may be used, as for the Panphot, 35 mm directly or the sturdy Aristophot camera stand. With the Aristophot there is the same limitation of a 4 × 5-in. format as the largest contact negative than can be made. Figure 3-11 shows the Ortholux with the Orthomat 35 mm automatic camera (see Chapter 11).

The Panphot will take any accessories that fit the Ortholux, but, as a dealer has pointed out, if one obtained most of the accessories for the

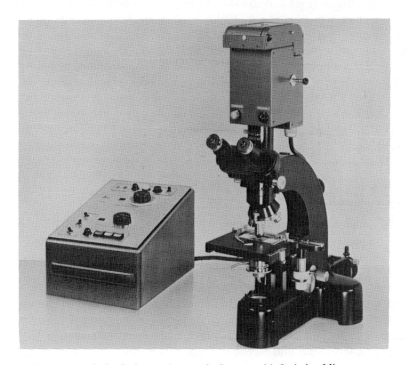

Figure 3-11   Leitz Orthomat Automatic Camera with Ortholux Microscope.

Ortholux it would cost as much as the Panphot, and "one might as well have the latter compact sturdy stand." It does seem to the author, however, that the complete Ortholux-Aristophot system would have a bit more versatility.

Carl Zeiss (Oberkochen) now features two photomicroscopes, the Ultraphot II and the Photomicroscope. The Ultraphot II, which is shown in diagrammatic cross section in Figures 3-12 and 13, is described as "a streamlined and comprehensive design." The "photohead" of the camera is interchangeable. The standard head employs plates and sheet film of 9 × 12 cm format. Photoheads for other sizes may be obtained. The Ultraphot II has a three-stage auxiliary magnification of the image beam of the objective to render the change of eyepieces superfluous. Trans-illumination and reflected light darkfield (called epi-illumination by Zeiss) are easily obtained as desired, as are phase illumination and polarized light. To make a picture one merely pushes a button to open the camera shutter. "Exposure time is then adjusted automatically to the correct value, in dependence on the intensity of the illumination, after which the shutter closes itself." There is no indication that the spectral sensitivity of the photocell is related to that of the photographic material, a feature discussed in Chapter 17. A 12-V., 100-W tungsten lamp is normally incorporated but provision is made for the substitution of other illuminants, including mercury and carbon arcs.

The Zeiss Photomicroscope is somewhat similar in design but lighter and has cassettes to hold 35mm film in its large arm. Exposure of this film is also automatic. The Photomicroscope is equipped with a built-in tungsten lamp of 6 V, 15 W, discussed in Chapter 7. This can be briefly overvolted to give considerable illumination. The 35 mm roll of film is wound backward, emulsion out and against its natural curl, in the camera. This has occasioned some difficulties with a few types of photographic film.

Inverted Stands.

The efficient photomicrography of metal surfaces (i.e., metallography) takes a well-prepared metal surface and excellent illumination and resolving conditions. By the turn of the century LeChatelier had designed a microscope with a horizontal stage but with the microscope objective looking up through it so that only a single surface of the metal had to be prepared well. This has been called the inverted stand, and because of its convenience it is used for all large professional metallographic equipment. **Fundamentally, no better results can be obtained with it than with an upright microscope and vertical illuminator with equally adequate illumination.** Because of the importance of metallography, including the

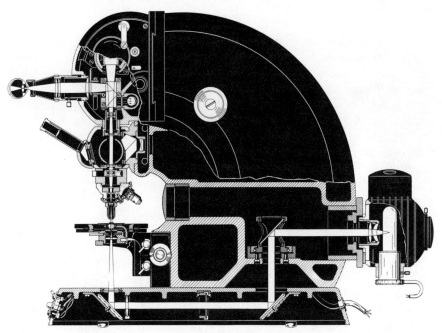

Figure 3-12 Zeiss Ultraphot II, cross section, transmitted light, 35 mm roll film camera.

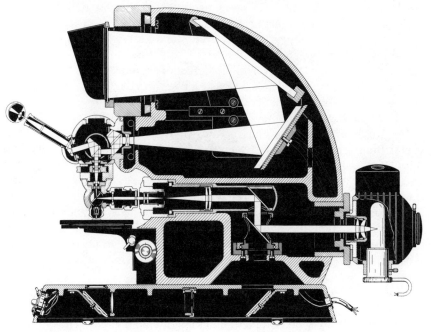

Figure 3-13 Zeiss Ultraphot II, cross section, reflected light, sheet film camera.

fact that the steel companies can afford to pay for excellent equipment, competition, plus the price that can be charged, has brought metallographic equipment to a state of high efficiency and convenience. As in the automobile industry, competition sometimes drives design to features that will seem attractive, whether or not they are of primary importance. The new metallographic equipment, although employing the same principles, will take the form of desk-type enclosed apparatus in which the operator flips switches or moves knobs with little direct observation, aside from reading labels, of exactly what operational change he is performing. However, these new metallographic microscopes will produce excellent results. Sometimes decisions are already made which the operator would do differently, for instance, the exposure photometer might be a cadmium sulfide cell which under certain circumstances could lead one astray in the exposure determination (see Chapter 17).

The principles, design, and operation of metallographic equipment are discussed near the end of Chapter 4.

Sometimes the inverted stand becomes especially convenient with transillumination, as when growing tissue is observed, looking up through a cell with the cover glass open. The B & L Dynazoom Metallograph has such an inverted illuminating device for brightfield, separately supported, available as an accessory, and some metallographers who used this model found this feature occasionally most useful.

Models of microscopes with inverted stands for biological work with transillumination which have appeared from time to time, include the Reichert and Leitz "Universal" Microscopes mentioned before, and Unitron (Newton Highlands, Mass.) have been selling a model that seems to have considerable distribution, especially for tissue culture work. A photograph of their latest model, BN-13, is shown in Figure 3-14. Its light source, which is the low-voltage lamp EL-lB with a rectangular grid described in Chapters 1 and 7, seems to be very efficient. The lamp condenser is well corrected. The maximum aperture offered by their objectives is Na 1.25. Unitron states that their substage condenser is not designed to be immersed. The reason given for this is that it is not needed with phase microscopy (the annulus is within NA 1.0) and that the average microscopist closes his iris diaphragm down below this point anyway with ordinary brightfield illumination. The Model BN-13 is equipped for use with both phase and polarized illumination.

Cameras of various sizes of format may be obtained for the BN-13; the standard is sheet film of $3\frac{1}{4} \times 4\frac{1}{4}$ in. An attachment for Polaroid or for 35 mm film may be obtained, including a cine camera. The exposure meter is built in.

Figure 3-14 Unitron BN-13 Inverted Photomicroscope.

The Horizontal Photomicrographic Bench

The author considers the "truly universal" micrographic stand, in terms of its versatility and convenience of use, to be the horizontal optical bench with separately mounted components. A completely horizontal bench, including the camera, is easy to align accurately and, on the whole, easier to use when it is applicable. In the author's laboratory both horizontal and vertical benches are available, yet the group tends to use a horizontal bench whenever possible. This is not possible for specimens that require a horizontal stage, and the latter may be more convenient for most types of reflected illumination. When space is available and funds permit, possession of both a horizontal and a vertical bench is a real time-saver when much photomicrography of miscellaneous character is to be done. Unfortunately, the completely hosizontal bench is no longer commercially available. Primarily it went out for economic reasons. It was expensive to make a good horizontal optical bed with adequate length, and unless it is optically good it has lost its advantage. Because the optical elements are at the top of rod supports often at least 9 in. high, a relatively wide bench to support the bed clamps is required to prevent slight tilting magnified by lever action. Unfortunately, the optical beds in use did not all meet this requirement.

Improvised Horizontal Optical Bench. Some of the advantages of a horizontal bench can be obtained with an improvised arrangement if a table or bench with a smooth level top is available with sufficient overhang to allow the use of C clamps. A long board, such as a $2 \times 4''$ of sufficient length, is laid along the edge of the table or laboratory bench and clamped (or screwed) to it. The inner edge must be made smooth and very straight by selection or planing. The optical bench supports for individual components are merely blocks of wood with carefully planed sides. It is wise to give them a three-point support on the bottom with small pieces of adhesive tape. (A little talcum powder allows them to glide easily.) Deep C clamps that reach over the edgeboard can be obtained to hold them in position. Support design may be variable; a hole in the wooden clamp that will just take a support rod is one way. The vertical rods should have collars (stop rings) that limit their vertical movement and should be kept from rotating by a set screw.

Horizontal optical benches, often sold for educational purposes, can be used for photomicrography, especially for the illuminating system with a vertical microscope (Chapter 6, Case A).

Use of Lathe Bed as Photomicrographic Bench. After using commercial horizontal photomicrographic benches for years, the author installed a special one that has been so successful and advantageous up through present use that it is described here. He learned that an old Hendey 10-ft lathe bed had been reposing in the salvage department for some time. It is also frequently possible to purchase such abandoned lathe beds second-hand. For our purposes its age was a great advantage. The castings certainly had come to equilibrium. Its surface consists of flat and raised V-tracks 8 in. apart. The bed was touched up on a large lathe bed to the minimum degree; this may be unnecessary. Its geometrical and optical straightness was tested by utilizing an autocollimating telescope for one end and a movable mirror that ran along the bench. Here a vehement warning must be given. A slight warp was put into this bench during its transportation by being held improperly from the swing of a crane. It seems impossible to convince many people that objects can be very heavy and apparently rugged and still require "delicate" handling. Such a bed can warp itself from its own weight if slung from the middle.

Special bed clamps were designed and made by casting. The foundry patterns consisted of three parts. The iron bed clamps kept the center of gravity low and allowed centration of the $\frac{5}{8}$ vertical rods they were equipped to accept. They are illustrated in several figures, including Figure 6-5. They could be clamped from underneath at any location of the bench. Frequently this was unnecessary for single components, since they tended to be stable. Heavy chromium plating was given to the bed clamps and its parts with no special preparation of the surface. This proved very satisfactory. The matte finish is attractive and there has been no corrosion. Note in Figure 6-5 that the support rod has been offset so that it is near one edge of the clamp and thus may get close to some other bench component. The design allows the top parts to be swung around after loosening so that the support rod could come at the other edge of the bed clamp, for the clamps themselves are not reversible on the bench. The optical elements held by these bed clamps are completely centrable, vertically by means of the rods, laterally by the clamp on top of the bed clamp, and along the bench by movement of the clamp itself. The vertical rods are always used with a collar with a set screw so that they can be rotated with no danger of vertical movement. Unfortunately, these collars were not in place on the vertical rods when the picture for Figure 6-5 was taken.

It is worthwhile to design the clamp to take a rod whose diameter is that of some commercial optical bed. In the case described a $\frac{5}{8}$-in. rod was chosen because it is the size used by Bausch & Lomb for their bench supports. The Gaertner Company optical bench takes a rod of $\frac{3}{4}$-in. diameter. A rod size equivalent to that used by Zeiss might be chosen.

The Bausch & Lomb metal photomicrographic camera was utilized but the back of it was held uniquely. Two narrow optical beds were raised vertically on each side to make a U from the clamps on which the camera was supported. However, a metal expansion screw support was provided. When the side clamps were loosened, the height of the camera could be easily adjusted vertically, with the expansion screw acting as a jack. Then, with the side clamps tightened, the central support was loosened so that it was not supported from the center by even a thousandth of an inch. The metal front board was also separately supported from the sides, and provision was made for the easy withdrawal of the bellows for the occasional circumstances in which this is useful. Aside from the microscope itself, all bench components were separately supported, as shown in Figure 3-15; one component was a simple table for supporting water cells, liquid color filters, or square glass and gelatin filters. When the lamp condensers are themselves mounted in a tube, as they are, for instance, in those from Bausch & Lomb, this tube can be simply supported in an outer tube, as shown in Figures 3-15 and 8-7, and focused by frictional adjustment. Condensers of very short length can be held at the back of a relatively long tube so that they can be thrust into the lamphouse near the lamp. Other bench components are discussed under *Accessories* in the next section. The microscope itself is discussed below.

Figure 3-15   Large horizontal bench for a compound microscope.

This bench was so successful that some years later a similar bench was set up for the simple microscope. See Figure 3-16. The optical arrangement is that of Case 7 of Chapter 5. Each component is separately supported, except as a shutter may be screwed to the iris diaphragm. An iris diaphragm is also screwed to the lamp side of the lamp condenser which is independently centerable. The stage is separately supported and carries the field condenser which is centerable by means of the bed clamp. The photomicrographic lens is carried on the compound microscope rack discussed at the beginning of this chapter. In Figure 3-16 the housing of the lamp is put aside. The lamp is discussed in Chapter 8.

Special Compound Microscope for Horizontal Bench. Because a great variety of photomicrographic work was done in the author's laboratory and because this laboratory was unusually well equipped with microscope accessories, including optics, from many companies, it was decided to obtain a special microscope to allow greater versatility of use and yet retain a horizontal bench. Great versatility in an instrument, especially a microscope, is usually obtained at the expense of stability. Two famous special microscopes had already been built, one for F. F. Lucas of the Bell Telephone Laboratories, the other for L. C. Graton and E. B. Dane at Harvard University. Both had cost more than $30,000. In both cases these individuals had given a set of specifications to a microscope company who had met them to make a good instrument. Some time after 1940 the author collaborated with H. Kurtz and G. Galash of Bausch & Lomb in the design of a special microscope that would be versatile yet stable *but* would chiefly use parts that were sold by Bausch & Lomb or had been made by them so that the blueprints and tools would be available. The result was a successful practical instrument which is still in active use but which cost less than $5000. It was called their 42-18-01 photomicroscope; a few were made on special order after that.

A basic cast-iron base was made, which sat on the flat base plate of the lathe-bed optical bench mentioned before. A sturdy pillar of the casting carried the coarse-and fine-focus adjustments. For the latter there is a dial, $2\frac{1}{2}$ in. in diameter, graduated with 100 divisions, each representing 0.001 mm horizontal movement. There is a vernier for this. The ball-bearing fine adjustment is stable, there is a very slight variance. Sometimes a curved spring is used from the focus controls to give remote control from the camera back, but this is bad for that type of bearing. Therefore a geared remote control was used.

Various body tubes could be set in firmly or removed with great simplicity. The stage focused on a horizontal movement bearing a vernier millimeter scale so that magnifications could be accurately reproduced.

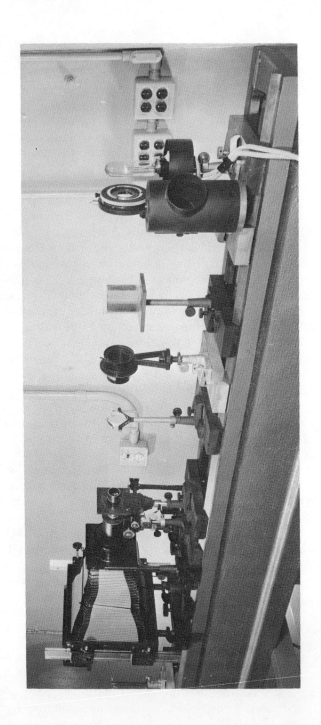

Figure 3-16 Large horizontal bench for a simple microscope.

A number of stages were available, again with stable firm positioning, yet with simple interchange possible.

The objective carrier is the large centering nosepiece used for petrographic microscopes. Once centered, an objective on its changer came back into good alignment reproducibly after it had been removed.

The substage rack is a short optical bed on which several substage elements can be carried. Thus a centerable yet stable iris diaphragm can be racked into the true aperture plane of the lower power condensers. A large ring on this track carries a metal disk firmly. This disk is centrable and in turn carries the substage condenser. Thus almost any substage condenser of any make can be used.

Although the longer substage rack is advantageous and versatile for many purposes, its one disadvantage appears on the occasions when one wishes to get a lamp, or its condenser of very short focal length, closer to the substage iris. This can be done by use of offset clamps but is less convenient at the time.

Nothing need be said of the components actually used with this microscope, since they can be chosen from many that are available. A certain standard set does tend to be employed in routine.

PHOTOMICROSCOPE ACCESSORIES

If we consider, by microscope accessories, all components of a stripped-down microscope stand that might be obtainable with some variation and often from various manufacturers, there are a very great number of then. Many, such as the various graticules used for quantitative microscopy, are almost entirely useful for visual studies with the microscope and are not considered here even though they may be important items in the total microscopical work. In order to be helpful, items made by specific manufacturers are discussed, but it must be understood that more changes may be made from the time of publication of this book in this field than in almost any other. Specific reference to up-to-date catalogs from the manufacturers therefore is a final prerequisite.

Objectives

For many years microscope objectives have been classified primarily according to their degree of chromatic corrections. As discussed below, some new classes must be added. Here, however, a general distinction is of fundamental importance.

Simple **achromatic objectives** overwhelmingly constitute the majority of objectives in use and are made by all microscope manufacturers. There is an especially complete line of them from the lowest power, usually the 2X (0.8-mm equivalent focus) 0.08 NA, up to a 60X (3-mm e.f.) 0.85 NA, which are used "dry," that is, with air in the object space plus (in the case of objectives for "biological use") a carefully specified cover slip, as discussed in Chapter 2. Added to these are the immersion objectives, going up usually to 97X (2-mm e.f.) 1.25 NA, which use an oil that is almost homogeneous optically with glass in the object space, also discussed in Chapter 2. **Achromatic objectives** can be said to be **chromatically corrected at two wavelengths but spherically corrected at only one intermediate wavelength in the yellow-green.** Photomicrographs made with achromatic objectives of 10X (0.25 NA) or of higher aperture should therefore be used with a restrictive color filter, as discussed in Chapter 9, if good definition is important; at least, an ultraviolet-eliminating filter, such as the Kodak Wratten Filter No. 2A, should be used.

Apochromatic objectives have been chromatically corrected at three wavelengths in the blue, green, and red, respectively, **and spherically at two wavelengths** (see Chapter 1). They furnish quite good images with white light and so are most useful for color photomicrography. Most of the large microscope manufacturers offer them. Their power usually ranges from 10X (0.30 NA) upward. with somewhat better NAs for their magnifying power than for achromats. Immersion apochromats often include the following and may be confined to them:

60X (3 mm),   1.40 NA,

90X (2 mm),   1.30 NA,

90X (2 mm),   1.40 NA.

Since the resolving power from an objective is dependent on its numerical aperture, the 60X 1.40 NA is an especially useful one because it has a somewhat longer working distance (e.g., 0.12 mm) than an objective of equal aperture but higher initial magnification. The very small working distance of the latter is sometimes troublesome. The author used the 60X 1.40 NA objective for cinemicrography for years; it gave superb images.

This high degree of chromatic correction is obtained at a sacrifice of two characteristics. First, its *primary image* has high lateral color which gives a rainbow effect at all edges formed within the field (see Chapter 2). This effect increases as the focal length decreases [see Figure 3-17 which

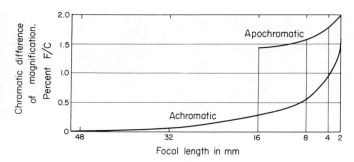

Figure 3-17 Chromatic difference of magnification as shown by achromatic and apochromatic objectives.

is due to L. V. Foster (4.1934)]. Obviously, apochromatic objectives show much more lateral color, although the achromats of highest power are almost as bad. Fortunately, this effect can be corrected by using so-called compensating eyepieces, which are made to exhibit the same phenomenon in reverse chromatic order.

The other sacrifice in apochromatic objectives cannot be so easily compensated. This is the high field curvature with apochromats which causes the field in excellent focus in the plane of a ground glass or photomicrograph to be considerably smaller than that of a corresponding achromatic objective.

Although fluorspar in the form of the mineral fluorite is normally used in the construction of apochromatic objectives, there is a series called **fluorite objectives** whose chromatic correction is intermediate between the two preceding series. They are appreciably less expensive, yet their quality is closer to that of apochromats than achromats. This is especially so of some redesigned fluorites which utilize synthetic fluorspar offered by Zeiss-Winkel. A complete line of these objectives is offered, from the 6.3X 0.2 NA through 100X 1.30 NA. The author found the quality to be superb. A 40X (4.3-mm) 1.0 NA oil immersion objective (Bausch & Lomb) has proved itself especially useful for darkfield. It gives superior images, partly because of the lack of flare with the oil immersion.

Flat-Field Objectives. A major advance in microscopy, especially as applied to photomicrography, has been made with the advent of objectives of fundamental redesign which have been able to overcome some of the field curvature limitations of microscope objectives. This has been done somewhat differently by the several companies who offer these objectives, but in all cases the redesign is fundamental and part of the improvement is obtained by utilizing thick component elements of the

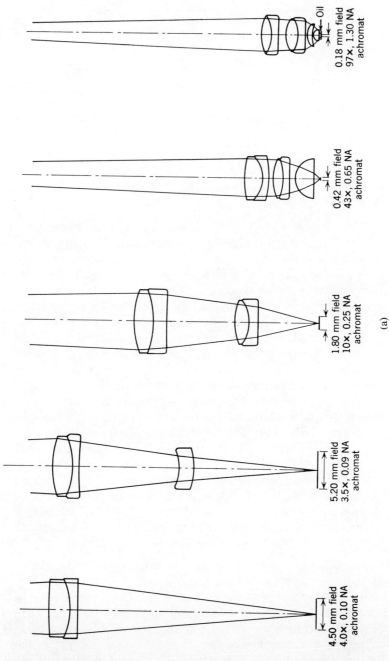

Oil
0.18 mm field
97x, 1.30 NA
achromat

0.42 mm field
43x, 0.65 NA
achromat

1.80 mm field
10x, 0.25 NA
achromat

5.20 mm field
3.5x, 0.09 NA
achromat

4.50 mm field
4.0x, 0.10 NA
achromat

(a)

Figure 3-18 (a) Diagrammatic construction of typical achromatic objectives; (b) diagrammatic construction of Bausch and Lomb Flat-field objectives.

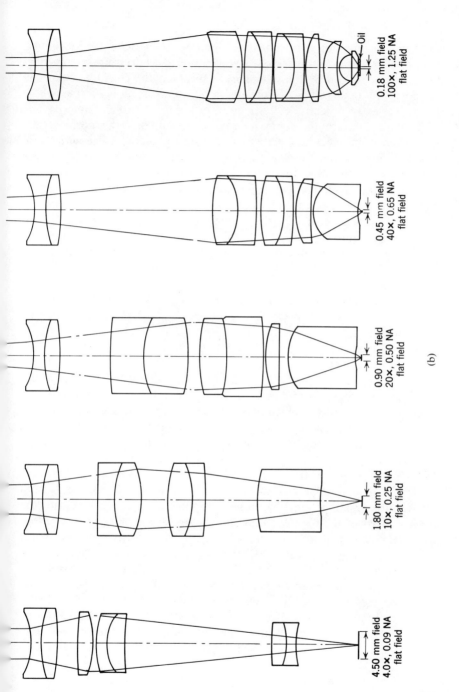

0.18 mm field
100×, 1.25 NA
flat field

Oil

0.45 mm field
40×, 0.65 NA
flat field

0.90 mm field
20×, 0.50 NA
flat field

(b)

1.80 mm field
10×, 0.25 NA
flat field

4.50 mm field
4.0×, 0.09 NA
flat field

141

total microscope objective lens. See Figures 3-18*b* and 3-19. Moreover, **lateral color has actually been introduced into the lower power objectives to equal that shown by those of shortest focal length.** Thus an ocular can be designed that will give correct color compensation for the whole power series of objectives. Leitz and Zeiss offer "flat-field" microscope objectives that are utilizable in any microscope of the correct tube length but require oculars made by the same company for use with them. The solution offered by Bausch & Lomb is different. In their case the upper lens of the objective always was an achromatic meniscus lens. Therefore this objective lens component was incorporated at the bottom of the body tube mechanically independent of the objective. This means that the microscope, that is, the Dynazoom or Dynoptic, equipped with this feature, must be used with the factory-installed objective which precludes substituting other objectives, even from other companies, for special work. The imagery produced by these objectives seems excellent to the author.

A power series of achromatic plano-objectives is offered by Leitz. Leitz also has a 40X fluorite plano-objective and a 100X plano-apochromatic objective.

Zeiss (Oberkochen) has a full line of planachromats and a lesser number of planapochromats in which a 10 and a 100X are included.

The American Optical Company introduced a somewhat different system with their MicroStar microscopes, especially Series 10. As already mentioned, the objectives for this microscope are corrected for an "infinite tube length," that is, for the specimen to lie in the front focal plane, with a telescope objective at a definite distance below the oculars which are now a "wide-field type" with the focal plane below and outside the eyepiece; the primary objective image is formed, of course, in the focal plane of the telescope lens. This lens lies near the bottom of the upper part of the body-tube component which lifts off with the oculars. However, this telescope objective is a complex achromat, that is, a lens that can be considered part of the objective system, since certain corrections for coma, astigmatism, etc., have been shared between them. In this way the whole system becomes completely corrected, including color-wise, and color compensation by the eyepieces is no longer necessary; the same eyepiece may be efficiently used through the magnification series. It is also the new wide-field type. At the time of this writing only achromatic objectives are available. The writer understands that apochromatic objectives will become available for the Series 10 microscope.

The introduction of these plano systems represents a major advance in photomicrography and has required fundamental alterations in the

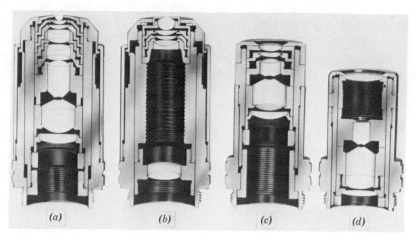

Figure 3-19 Leitz Plano Objectives seen in cross section: (a) Plano Apo 0:1 100× NA 1.32; (b) Plano Achromat 40× NA 0.65; (c) Plano Achromat 10× NA 0.25; (d) Plano Achromat 4× NA 0.10.

optical design of the objective and imaging system. This may be illustrated by reference to a study made some years ago in which the author compared the size of the field obtained when an excellent 4 mm apochromatic objective was used, first with a 15X compensating ocular and then with the Zeiss Homal III, also of 15X secondary magnification, but which is a negative ocular supposed to furnish a larger field plane of good focus. Magnification was 1000X in both cases. The test object was comprised of exceedingly fine crystals in excellent plane and so was unusually critical. The size of the field in excellent focus was judged to be 11 cm in diameter in each case (0.11 mm in the object field). However, with the negative ocular, the field became out of focus at its edges much more gradually, which means that if the detail had been coarse, as with larger crystals, the Homal would have appeared to furnish a considerably larger field of good focus.

When a magnification of 100X was set up with a Bausch & Lomb 10X apochromatic objective and 7.5X B&L compensating ocular, the field size of excellent focus was less than 3.5 in., whereas, with the Zeiss 10X planapochromatic objective and 8X Kpl ocular, the area of the 5 × 7 in. photomicrograph was more than filled with the field.

When the photomicrographic bench was set up to make the photomicrograph shown as Fig. 10-2, the fields were examined first by white light because of the color picture to be made. With the apochromat and its corresponding compensating ocular, the central definition obtained was superior indeed and absolutely no lateral chromatic aberration could be detected in the field. On the other hand, with the planapochromat and the Kpl ocular, the definition was good over the picture but not so excellent as the central definition in the preceding case. There was notable chromatic aberration (lateral color) increasing from the center of the field and giving the effect in the final color picture as if the components of the print from the two negatives were not *quite* registered. This can be seen on inspection of Figure 10-2. On the other hand, when any other ocular was substituted for the Kpl, the observed lateral color fringing became marked. In making a choice of the photomicrograph to be used, the planapochromat was chosen. With its appreciably larger field, it was thought that in practice this was the one that microscopists would choose for photomicrographs if both were available. The author must admit that this would be a matter of personal judgment.

Specialized Objectives. All companies offer what may be considered to be specialized objectives, although in some cases the use may be as great as for any other. The simplest are the **water-immersion objectives** for which the call is limited, so only a few are offered, usually 40X, although Leitz has a series through 90X, both achromats and phase objectives. These objectives can be very useful in industrial work; the author has found then useful in cinemicrography in which they can be dipped directly into the preparation carrying organisms.

Objectives **for use with polarized light** on petrographic microscopes are listed separately but usually constitute achromatic objectives selected to be strain-free. Because of the fluorite component of apochromats, they rarely can be used satisfactorily with polarized light.

Objectives made **for use with phase illumination** constitute such an important class, definitely specialized in design, that they are considered in Chapter 12 and not discussed here.

Separate objectives are made **for use at medium and high powers with obliquely incident light**, that is, reflected darkfield illumination. These require the condenser to be incorporated around them and the objective mount to be as thin as possible. Zeiss calls this epi-illumination and has designed special equipment for it. Likewise, Leitz has a complete set of equipment called the Ultropak illuminator which covers the whole range of magnifications for the compound microscope. The Tri-Vert Illuminator of Bausch & Lomb is somewhat more limited in range. This type of apparatus is discussed in Chapter 4.

A type that might definitely be considered specialized, but is so important and so common as to be a major item, is the **objectives for metallography**. These objectives are not only corrected for use without cover glass but are built into short mounts with the rear element as close to the top as possible to subtend an adequate aperture in the presence of the reflector of the vertical illuminator just above them. They must also be calculated for a longer mechanical tube length to allow for the space occupied by the vertical illuminator. Bausch & Lomb and Leitz utilize a tube length of the 215 mm, whereas the objectives of the American Optical Company and Zeiss are corrected for "an infinite tube length." Aside from these design specifications, metallographic objectives are available in the complete line of achromats, fluorites, and apochromats discussed above.

Situations do occur in which it is desirable to use an objective corrected **for use without cover slip** for transillumination. If there is sufficient working distance in the object space, this can be accomplished by oiling the correct cover slip to the front of a dry objective, being very careful in doing this, and later in cleaning it off with a solvent, that the objective is not injured by oil getting behind the front lens, since it was not designed for that purpose. However, it is usually sufficiently protected. Special objectives corrected for use without cover slip for transillumination can usually be obtained. Both Leitz and Zeiss market objectives corrected for use with no cover slip for fluorescence microscopy (see Chapter 20).

General Considerations

The importance of utilizing the cover slip of correct refractive index and thickness and effects of variation from these conditions are discussed in detail in Chapter 2. The American manufacturers have adopted as standard cover-slip thickness the value 0.18 mm (180 μ), whereas the European standard is 0.17 mm (170 μ).

The standardization of the tube length, which standardizes the image distance l' of the objective, really is for the purpose of standardizing the object distance which is critical because of the high aperture that may be used. Therefore any immersion fluid used should be that assumed by the designer of the objective, if image definition and contrast are important. The following immersion oils are assumed by the designers:

Table 3-1 Immersion Oils Assumed by Designers of Objectives

Manufacturer	Oil
American Optical Company	Crown Immersion
Bausch & Lomb	Shillaber's A
E. Leitz	Cedarwood (processed by Leitz)
Zeiss (Oberkochen)	Cedarwood (processed by Zeiss)

Many of the best objectives are now "spring-loaded," that is, the objective proper is held in its outer tubular case by a coil spring above it which pushes it down without great force. This is valuable, for sometimes even the careful microscopist may encounter a specimen whose level below the cover slip is difficult to locate.

Eyepieces

It is amazing that the first microscope eyepiece to replace the crude simple plano-convex lens, which was designed by Christiann Huygens about 1685, was so good that it is still the design most commonly used. The **Huygenian eyepiece** consists of two plano-convex lenses with the convex sides of both turned toward the objective and both made of ordinary crown glass. The focal length of these lenses will vary with the wavelength, but the effect is in opposite directions with respect to the image that lies between them. Therefore lateral chromatic aberration and coma can be eliminated by manipulation of their relative focal lengths and separation. Huygens showed that this is achieved when the sum of the focal lengths, divided by two, equals their separation. However, these constants (focal length and separation) are adjusted in commercial Huygenian eyepieces to show 0.25% lateral color and with the focal length for blue being greater than for red. This just balances the lateral color defect of the 10X (16 mm) objectives [but see the discussion of flat-field objectives]. With 2X (48 mm) objectives that show no appreciable lateral color, the final image with a Huygenian eyepiece will show 0.25% overcorrection. There is usually some spherical aberration and axial chromatic aberration, but these effects are usually small because of the small numerical aperture used by the eyepiece, which is that of the objective divided by the magnification. Some astigmatism is introduced but this actually helps a little to combat curvature of field. Huygenian eyepieces are usually called achromatic. The following ratios have been used in making them:

focal length (field lens) : focal length (eye lens) : separation :: 3 : 1 : 2.

The weak field lens lies a little below the plane in which the image from the objective would have been formed, bringing it down a little to the plane of the diaphragm within the eyepiece. Thus the field lens lies sufficiently far from the field plane so that fine dust on its surface cannot be seen. Usually a series of Huygenian oculars is available, which varies in power from 5 to 15X through a series of magnification steps of roughly $\sqrt[3]{2}$.

As stated in the preceding section, the primary images from apochromatic objectives suffer from bad lateral chromatic difference in magnification, and the blue images have noticeably higher magnification than images of longer wavelength, such as the red illustrated in Figure 3-17. The reverse effect is built into **compensating eyepieces** into which an overcorrection of about 1.5% is incorporated. This could be done, at least largely, by manipulation of the variables of the Huygenian eyepiece, but in practice it is chiefly done by utilizing a flint-and-crown doublet for the eye lens. Thus with more variables to control better control of all aberrations can be obtained. Compensating eyepieces are generally furnished in a series of magnifications that rise to 25X. As can be concluded from Figure 3-17, the images from high-power achromatic objectives also require the use of compensating eyepieces. In 1954 Bausch & Lomb introduced a new series of compensating eyepieces with marked improvement in both definition and contrast of the image.

Bausch & Lomb and Leitz have both had a **series of eyepieces designed to give about half the compensation that is built into "compensating eyepieces,"** that is, about 0.8% lateral color. Bausch & Lomb call their series "Hyperplane" and Leitz calls its eyepieces "Periplanatic." Both series are supposed to offer slightly larger fields, but this kind of improvement is negligible. They are useful with achromatic objectives not yet needing full compensation of the apochromats. The specific recommendation for the selection of the appropriate eyepiece is given in Table 6-1.

Negative lenses for image projection, such as the *Homals and Ampliplans*, are considered later under "Projection and Camera Systems."

Wide-Field Eyepieces. When using a stereo binocular microscope, it is advantageous to have an unusually large field of view through the eyepiece. The fact that the closely spaced pair of objectives of a stereo microscope is restricted to less than the usual aperture, hence to less than the usual resolving power, allowed the diaphragms of the eyepieces to be opened up exceptionally wide with less sense of loss. Somewhat later an eyepiece of more complex design, involving an achromatic doublet for an eye lens, was devised which really allowed a field of view about one-third greater than normal. Both negative and positive oculars were made, although sometimes the draw tube into which they fitted had to be of larger diameter. **By negative oculars is meant one, like the Huygenian, with a field lens below the objective image and an internal focal plane; a positive ocular lies entirely above the image with the latter in its focal plane.** Although developed for the stereo microscopes, such wide-field eyepieces have been offered for the standard compound microscope for some time.

With the advent of flat-field objectives, wide-field oculars were again redesigned to make matched sets between the objectives and oculars. The oculars were made to have negative lateral chromatic magnification error of the compensating oculars. **Then lateral chromatic magnification aberration was actually introduced into the low-power achromatic flat-field objectives so that the compensation** of the oculars was correct for the series and **did not vary with focal length of the objectives.** (See Figure 3-17). Thus they need to be used with the flat-field objectives for the best results. They can also be used as compensating oculars of flatter field than usual with standard objectives that normally should take compensating oculars. At the present writing Bausch & Lomb has only one ocular (of 10X power) of this wide-field type. Zeiss has a series, denoted as Komplans, with the designation Kp1 marked on the eyepieces. The Leitz series is denoted as GF on the eyepieces. Bausch & Lomb does list an ordinary Huygenian 5X and also a 10X ocular for use with their wide-field objectives but that is to satisfy the economic demand, since they are $10 versus at least twice that sum for the new wide-field type. The latter really should be used for photomicrography unless the combination discussed below is employed. Actually the corresponding ocular from any manufacturer would be appreciably better than a Huygenian for use with any flat-field objective.

Some of this discussion is not applicable to the system introduced by the American Optical Company (See Index, *flat-field objectives*).

Zoom Systems

Bausch & Lomb Inc. have introduced a set of three achromatic doublet lenses within the microscope body that by means of cam action (see Figure 3-20) will vary the magnification of the primary objective image continuously from 1 to 2X. Therefore their 10X wide-field eyepiece can be made to function continuously as a 10 to 20X ocular. This is the Dynazoom line of microscopes (Figure 3-4). An equivalent line, but without the zoom feature, is offered as Dynoptic microscopes. Some optical compensation, chiefly chromatic, is required for this zoom set of lenses in the ocular and therefore a somewhat different set of oculars is used for the Dynazoom than for the Dynoptic microscopes.

The Unitron Instrument Company offers a zoom eyepiece by which continuous eyepiece magnification from 10 to 20X is obtained by turning the "zoom collar." The images are said to remain in focus as the magnification is changed to give "a dramatically useful effect for studying a field." There is a locking screw to hold the eyepiece in place and a graduated scale to show positions of the collar for 10, 12, 15, and 20X. The eyepiece

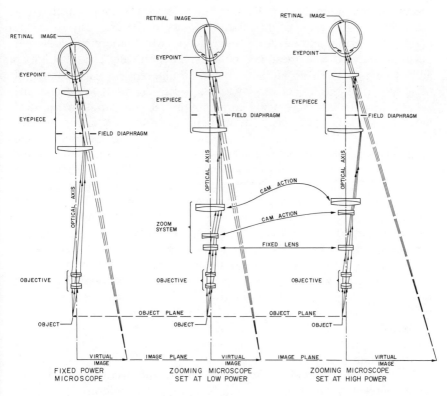

Figure 3-20 Diagram of principle of Zoom Microscope (courtesy of Bausch and Lomb Inc.).

is stated to fit any standard microscope and be readily adaptable for photography. The chromatic classification type is not stated but is assumed to have some chromatic compensations, because use with oil immersion objectives is specifically recommended. The author has had no direct experience with it.

Projection and Camera Systems

If, instead of focusing the compound microscope visually, as assumed by the designer of the objectives, a real image is projected at some near distance on a camera film, both greater limitations and some freedom of design become applicable. About 1922 Zeiss brought out its series of **Homals**. Each of these was a negative system of several components but **functioning as a single negative lens that carried the primary image** of the objective **to the final projection plane** for its first imaging, at the same time correcting the sine condition and making chromatic compensation for the objective characteristics. Obviously there was no eyepoint that represented the exit pupil of the microscope above the draw tube, the exit pupil being virtual and inside the microscope, which, therefore could not be used visually with these components. Theoretically the focal length and placement of these Homals should be different for each change of focal length of the objective. Therefore a set was made to bracket the several focal-length increments corresponding to the variations in focal length of the objectives. Some time thereafter Bausch & Lomb brought out their Ampliplans which had the same function. Both the Homals and Ampliplans require special adapters, since they are larger than the normal draw tube size. The Ampliplans were later replaced by the Bausch & Lomb Ultraplane which did fit the usual draw tube but involved some sacrifice. On the other hand, they did give superior images. The principal advantage of these negative "amplifiers" over eyepieces was supposed to be a larger flat field of focus. **The gain for fine detail that lay in one plane was negligible but because the curved field did not drop off so rapidly** as with positive oculars **coarse detail seemed to stay in focus over a larger diameter**, which was some gain. (See Index, *flat-field objectives* for a measure of the field size with Homal III.) The Homals seem to be still offered by Zeiss (Jena) but not by Zeiss (Oberkochen). Since the introduction of the flat-field system microscopes Bausch & Lomb have dropped the Ultraplane but have kept the principle in their Photographic Amplifier for the Dynazoom and Dynoptic microscopes.

As discussed in detail in Chapter 11, if the microscope image is projected onto a film or other imaging surface close to the microscope and the position of the objective has to be appreciably changed from the position at which it was used for visual focusing in order to focus the image, the quality of the image will be somewhat degraded. It can be restored by interposing a lens above the complete microscope that will project this image in focus on the plane when the objective is focused as for visual use. This principle was first applied to allow photomicrography with roll-film cameras (Loveland, 4.1944), especially the 35 mm; nearly all microscope companies brought out a device for 35 mm cameras. The principle is now also applied as "attachment cameras" for sheet films, including Polaroid, but the discussions of Chapter 11 are just as valid.

The case of the "Photographic Amplifier" sold by Bausch & Lomb for photomicrography with their Dynazoom and Dynoptic microscopes is somewhat different. This unit, which sits on the trinocular head of these microscopes, can carry any one of the three available cameras directly or can be used to project the image to a more distant and larger camera, as on the L stand. The unit does not use a separate ocular, since it is, in itself, a four-component negative lens system of the same type as the

Ampliplan and lowers a lens component below the primary image in the microscope to focus it in the plane of the film. A ring can be turned to focus this image from the near distance of the 35 mm camera to that of a larger, more distant camera with a scale available for calibration, as determined by focusing through the binocular viewing eyepieces. This is not so precise, however, as direct focusing of the image in an open-image plane with a magnifier, which can also be done. The magnification of the primary image is as follows for film held in the standard cameras:

Camera Format	Secondary Magnification
35 mm	2.5X
$3\frac{1}{4} \times 4\frac{1}{4}$ in.	7.5X
4×5 in.	10X

This negative lens system is not color compensating in itself. Therefore, when the flat-field apochromatic, or even fluorite objectives are used, for which compensating oculars would normally be needed, a special optical color compensating unit should be inserted into the Photographic Amplifier.

The American Optical Company has an attachment camera (their 1053A for a 35 mm camera) that utilizes the same optical principle. The assembly fits into the hole for the trinocular tube of their Series 10 MicroStar microscope above the telescope lens. The assembly consists of a negative lens, a shutter, an auxiliary cylindrical tube for image projection, and the 35 mm camera. The camera may sit almost on top of the shutter for a 2.8X projection magnification or at the top of the auxiliary tube for 5X. A lever flips to focus the lens to suit the magnification. The lens consists of three achromatic doublets and intercepts the image beam to relay the primary objective image onto the film plane. An alternative camera at 5X projection distance utilizes Land-Polaroid film. The shutter is elastically mounted "to absorb its vibration." The author made pictures at $\frac{1}{50}$ sec, others at $\frac{1}{5}$ sec with the shutter, and then at 2 sec by momentarily removing a block to the illumination beam below the microscope. These tests merely required neutral densities of 1.0 and 2.0, respectively. Those made by time exposure were very sharp; those made at $\frac{1}{5}$ sec were noticeably less sharp; those made at $\frac{1}{50}$ sec were fuzzy in comparison. A number of attachment cameras have their lenses similarly mounted. **The author strongly recommends that a similar test be made with fine grain black-and-white film whenever the shutter is mounted on the microscope**, though the shutter may be in the lens of an attached camera. An independent interruption of the light beam should be compared with a short exposure by the shutter. Fractional-second exposures may be required for good results with color film.

The American Optical Company does offer a stand by which the camera and shutter may be held independent of the microscope and should be very efficient with the trinocular type. This is their No. 1070 camera and stand, which may include a 35 mm camera, a $3\frac{1}{4} \times 4$ Polaroid Camera, or a 4×5 Graflok Camera (also allowing use of Land-Polaroid Sheet Film). This also permits the alternative of a calibrated conjugate focusing through the binocular eyetubes, as does the attachment type of camera, and individual focusing and photometry of the image plane when desirable — and it may become very desirable. However, the use of a dummy camera to accomplish this with a roll-film camera is discussed in Chapter 11.

Jackson Tube-Length Corrector

The importance of using the tube length of the microscope which gives optimum definition for a particular objective and conditions of the optics space was discussed in Chapter 2. It is unfortunate that the convenience built into most modern microscopes, with reduction of versatility, has eliminated the variable draw tube and thus the ability to control the optical and mechanical tube length between objective and eyepiece. W. Watson & Sons Ltd. sell the Jackson Tube-Length Corrector, which is a small tube that can be screwed in between the body tube and the objective on the objective carrier. Rotation of the knurled ring controls the separation of the component lenses, which in effect provides variable distances between the objective and eyepiece, ranging from 100 to 300 mm. The author has not used the Jackson Corrector with white light and color photography, but it works excellently in black-and-white photomicrography with markedly improved definition when such results should be expected following change of tube length. Some decrease in contrast was observed in comparison with the direct use of the correct tube length without the Jackson Corrector. However, the author was able to obtain a special unit from the manufacturer in which the optics had been given antireflection coatings for the specified wavelengths. This device proved especially helpful in some photomicrography with near ultraviolet. It is also most useful in diagnosing whether a given objective is being used at optimum tube length for a given case, being much faster in such determinations than direct change of the tube length. The numbers on the scale of the Tube-Length Corrector form an inverse scale series to those of the actual tube lengths. A lens bench examination showed its power to vary from -6.5 diopters to about $+3$; there was no power at a scale reading of 14.5. A straightforward calibration of the scale versus actual tube length was worthwhile. The equivalent tube length could then be set up, when desirable, without the Corrector. The equation for the Jackson Tube-Length Corrector calibrated is

$$TL = -0.94\,S + 306,$$

where

TL = tube length in millimeters,
S = scale reading on the Corrector.

The rounded off form for this equation is

$$TL = -S + 315.$$

The author does not know how well this would apply to other units of the Watson (Jackson) Tube-Length Corrector.

Shutters

One of the big and unsolved problems of photomicrography is the inefficiency and inaccuracies of the shutters that are so frequently used for controlling the photographic exposure. **One of the difficulties is that the problem is usually not recognized**; few people realize how poor the shutters may be. If the exposure control is very inaccurate, as discussed below, the poor exposure received by the photographic material may be blamed, for instance, on reciprocity failure. Rarely is a shutter in photomicrography placed in a true aperture of field plane. All else is "vignetting space." If the shutter opens up to an aperture from the center and then closes down from the outside, it could be considered as fixed halfway open when the full exposure time is used up by its opening and closing When the aperture is very large, such as 2 in., no commercial shutter has negligible opening and closing time for the small fractions of a second. When such a shutter is imaged in the field, even out of focus, there will be uneven illumination with a central bright spot which has often been blamed on the illumination system. If it is in a true aperture plane, the aperture will not be effectively filled during the exposure, in spite of the optical arrangement. Both underexposure and poor resolution may be the result. Shutters with rubber blades have fast opening and closing speed but are perfectly transparent to the infrared. Light steel blades have proved satisfactory for the latter. The cocking shutters are on the whole more accurate than the automatic self-cocking type. After a different exposure time has been dialed on the shutter, it should be clicked once or twice, with the film protected, before the actual exposure is taken; the exposure will be more accurate.

Shutters are now calibrated photoelectrically for the integral amount of light transmitted during the exposure time. This is good, but it assumes that the whole aperture is to be filled with light. Consider the case in which a fairly large shutter is placed at the eyepoint of the microscope, which is a true aperture plane. Most of the time of opening and closing will not affect the light beam; therefore the calibration will be affected if the opening and closing time is an appreciable part of the total exposure. When placed behind the eyepoint, where it cuts the individual pencils of light going to the antipoints of the field, the central portion of the field will receive the greatest exposure, as mentioned above. We have been able to get some very small camera shutters for use behind the microscope, but the manufacturers are loath to distribute shutters for this purpose because an excellent small shutter is like a fine watch, with its works normally well protected by the glass lenses of the camera. **An unprotected shutter should be popped into a box or covered when not in use to prevent the dust from getting into its works. It may be corked** if kept in place. A favorite of the author has been the Acme Shutter No. 3 or the Universal Size 3 made by the Ilex Optical Company who will sell the shutters independent of optics. The No. 3 has a clear aperture of $1\frac{3}{8}$ in.

The author has had a "time dwell switch," built in his own laboratory, by which any circuit can be held open or closed for a duration of time from 1 to 10,000 msec, as dialed on the face of the instrument. The problem is to obtain shutters that could use this capability. **It is useful to think of exposure duration in milliseconds.**

If the illumination beam is first opened and then closed in its exposure to the film from the same side of the beam, the efficiency of the shutter is no longer of much importance for a still subject. It is not even important whether the shutter is in an aperture or a field plane, and the deleterious effect is less susceptible to departures from these planes. A number of individuals have taken advantage of this point to utilize **sector-wheel shutters.** In a laboratory in which space is no great consideration, especially for a horizontal bench or when the illumination system is on a horizontal bench, this method has great advantage. In recent time it has been easy to obtain electronically controlled motors which can be set simply and accurately over a range of speed. For the extreme range possibly one or two simple gear changes might be desirable. With the speed of the motor as a variable, plus the angular opening of the sector wheel, both calibrated in time units, the method would seem quite attractive. The sector wheel should be used to determine the actual photographic exposure given, whereas an auxiliary shutter would protect the film and be open long enough to catch one sweep of the open sector. This could be arranged by utilizing a standard flash shutter and either a mechanical or photoelectric auxiliary switch to the sector disk shutter. A secondary light source, offset from the axis on the other side of the shutter, could trigger a photocell switch that opens the protective shutter. These photocell relay switches are now commercial items.

Illumination Bench Accessories

Assuming that a rather wide variety of illumination conditions is encountered and unless one has a rather elaborate "universal" microscope equipped with ring illuminators, etc., for low power work, a variety of **clamp-stand type of accessories** become invaluable. As stated previously, the writer favors a horizontal bench for photomicrography, for the illumination system at least. The components then would be supported by solid rods which should be a commercial standard, $\frac{5}{8}$ in. diameter for Bausch & Lomb accessories, $\frac{3}{4}$ in. for accessories of the Gaertner Scientific Corp. Some of the accessories for reflected light are discussed in Chapter 4. In any case, **offset supports and clamps** are most useful to bring optical elements closer to desired planes than the optical bed clamps would otherwise allow. This usually means a vertical rod held parallel to the original supporting rod but offset in front of the first one to hold the actual component. It is worthwhile to keep their weight to a minimum by making them of aluminum. It is also quite simple to make the clamps so that it is unnecessary to do more than bore holes or possibly make a few sawcuts, since standard machine screws and nuts are otherwise used. It is important to drill the holes fairly accurately for the size of the rod so that they can be squeezed smaller; in fact, the diameter of the hole can be very slightly less than that of the rod. It is much easier to work with this type of clamp than with the V-groove clamps that wobble badly whenever they are loosened (see Figure 3-21).

If much work at low magnification is done, a supply of **extra lenses**, kept neatly in a cabinet according to focal length, is helpful. Such components are not expensive and can be picked up from many sources including The Edmund Scientific Company (Barrington, N. J. 08007) and A. Jaegers (Lynbrook, N. Y.). **Mounting the individual lenses on the optical bench** or as an accessory to a stand is simple but must be done with some preparation to be satisfactory. The simplest method, but one that is satisfactory, is to use a Y clamp. If the lens has a thin edge, such clamps are commercially available (e.g., from Central Scientific Corporation or Gaertner Scientific Corp.). If the lenses have an appreciably wide

Figure 3-21　Offset extension rods and clamps.

flat edge, it is preferable to mount them on a Y made from a narrow section of brass right angle held as a V by a support rod. Black electrical tape over the top to hold it to the clamp has proved preferable to some commercial accessories; it is not too sticky, yet adequate. A V notch must be made in the top of the support rod and the brass right-angle component fastened in, either by a screw or very easily by some modern plastic cements. When the lens is already mounted in a cylinder, as are some condensers, it is usually possible to find a tube that will make an outside fit with very little extra work so that correct focus will be obtained by sliding friction; a set screw can be used.

There are many reasons why it may be convenient to utilize **neutral densities** to control the illumination level, and thus the exposure time, to one that is convenient or is prescribed by the photographic material being used. The characteristics of these important accessories to photomicrography are discussed in Chapter 9. The important conclusion is **that there is no one type of neutral density that is best for all purposes;** the truly neutral filter, important for color photomicrography, that is, the inconels, may be more expensive or inconvenient than would be used for all purposes. The Kodak Wratten No. 96 lacquered gelatin film filters, although not **photographically neutral**, are most convenient and are available in a quite complete series for illumination beam attenuation.

Therefore they are the most frequently used. They cannot be kept in the intense beams from the strongest lamps, however, such as the arcs.

Filter cells for water or other liquids are often convenient and sometimes required. Their use is discussed in Chapter 9, followed by a discussion of the cells. **The heat-absorbing glass filters** are discussed in the same chapter.

The use of many accessories for the alignment of the horizontal bench with its illumination or for the illumination portion of a bench is discussed in Chapter 6. **Pinhole eyepieces**, which contain no lenses, only the central small hole for centration of the microscope, are obtainable from most microscope companies. We have found it worthwhile to make **pinhole objectives and pinhole condensers** which are merely brass accessories that screw or fit into the places held by the components for which they are termed. In making them on a lathe it is important not to take the piece out between the time the outside dimensions are worked and the central pinhole is made.

The accessories made by the Polaroid Corp., including **analyzer, polarizer, and retardation plates**, can be most useful. They are sold by several microscope companies and are completely described in the catalog on accessories of Bausch & Lomb. Possibly insufficient warning is given that the polarizer, which may be in the illumination beam, must not be allowed to get warm.

Chapter 4

Photomicrography by Reflected Illumination

The term **incident illumination** is sometimes used to oppose the term *transmitted illumination,* since the light is incident on the face of the specimen that is viewed. Whichever term is selected, reflected or incident, it should include the scattering of light by small particles, since this type merges into reflected illumination with increase in particle size. However, this illumination is usually obtained with modern darkfield condensers, which in turn require a technique more akin to that for bright field transmitted light. Therefore the discussion of darkfield illumination is postponed until a later chapter.

The general principles of illumination by reflected light have already been discussed in Chapter 1 under the section, *Use of Lenses: Up the Magnificent Scale.* It is assumed at this point that this section has been read or its subject matter is already known.

Reflected light is the kind of illumination that gives the most familiar appearance to surfaces that can also be seen with no magnification or, in such cases as insect eggs, resemble objects that can be seen without magnification. As has been noted, when the illumination comes at an oblique angle to most surfaces, it gives the enormous advantage that the required aperture is filled automatically by scattered light. In Chapter 2 we have seen why this is important. On the other hand, if the illumination is directed so that the surface, or small elements of the surface, reflects the light specularly like a mirror into the objective, the second type of reflected illumination, appropriately called **brightfield reflected illumination**, is obtained. Its principal use is in metallography, in which the specimens are good mirrors, and it is usually obtained by an axial beam that is vertical to the specimen (i.e., **vertical illumination**). The nonspecular or oblique illumination renders planes that are perpendicular to the axis of the objective as dark areas, and thus the image appears as a negative of its appearance by specular illumination and conversely. Oblique illumination can be obtained by equipment also used for vertical illumination, that is, metalloscopes, by illuminating only through the outer zone of the objective to give **conical illumination**; moreover, part of the azimuth may be blocked to give an unsymmetrical relief appearance. The two photomicrographs of the Texas meteorite illustrate the vertical versus conical types (Figure 4-*1a* and *b*). A controlled mixture of oblique and specular illuminations is sometimes useful if one of the two is definitely subordinate. Specular (vertical) illumination is discussed on p. 187.

CAMERAS AND OBJECTIVES

Gross-specimen Photography

As discussed in Chapter 1, the principles, techniques, and even the apparatus are about the same for photography at fractional magnifications from about 1/5 X (usually included in **gross-specimen photography**) through 1/2 X, 1X, and finally up to about 15X. Above 1X it is usually called **photomacrography**, and a simple microscope, which may consist of the original lens of a standard camera, will normally be used. The domain of flower photography through insect photography is included here, but also included is the photography of a host of other specimens, those of industrial origin probably being the most miscellaneous. Anything is included that one might desire to examine closely, possibly with a hand magnifier. This last statement leads to an important principle of technique. Often it is wise to examine the specimen first, probably entirely off the base on which it is to be photographed, using the appropriate hand magnifier and frequently holding the sample in the hand, turning it, and subjecting it to various kinds of lighting. A wide range of light from flat to contrasty and at many azimuths can often be tried out in a relatively few minutes and frequently in light available from lamps in the room. **The optimum lighting is then decided on and the arrangement made to obtain it with the somewhat less flexible arrangement near the camera.** This procedure is, of course, not always feasible, but it is usually best.

The German literature is particularly prolific with books and articles in this domain, often well illustrated. The files of the *Journal of the Biological Photographic Association* contains quite a few very helpful detailed articles; those on industrial specimens are found mostly

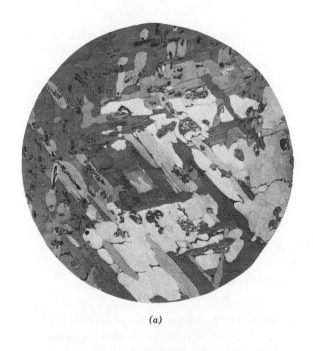

(a)

(b)

Figure 4-1   Odessa, Texas, Meteorite: (a) vertical illumination; (b) conical illumination.

160

in trade and photographic journals, but it should be frequently possible to find the desired help from articles written by authors with very different applications in mind. The German articles are chiefly concerned with the use of cameras for 35 mm film. The literature in English (from both America and England) is much more likely to discuss the adaptation of a larger camera to this purpose. The difference in magnification desired for placing the image of the subject in a 35 mm frame or in a considerably larger one is most significant here. It must be remembered, however, that in judging quality, including depth of field, we must consider the pictures (prints or screen images) that are finally viewed.

The 35 mm camera is often much more convenient for taking the picture, and if the specimen is living or evanescent it may constitute a major advantage over a large professional outfit. In this case, also, the use of flash lamps (see chapter 13), including a ring of flash around the lens, may be outstandingly useful. With such subjects, probably the focal-frame device, mentioned in Chapter 1, is the most effective method of quickly framing, focusing, and taking the picture. Moreover, this method can utilize the very common type of hand camera instead of the more expensive reflex cameras or the more cumbersome view cameras.

Focal-Frame Photography

It will be recalled that "focal-frame" photography was defined as use of an arrangement by which a frame, slightly larger than the object field imaged on the film in the camera, is held by a support extending from the camera almost against the field to be photographed. If the frame is in the focal plane of the supplementary lens and the latter is against the camera lens, focused at infinity, the plane of the frame will be in focus (see chapter 1). It is more practical to hold the frame about 10% of the distance short of the field plane. Actually we can generalize this definition and refer the term "focal frame" to the general practice of **providing a movable frame at a known focal distance from the camera in order to obtain a desired magnification**, which may still be small (including $1/10 < m < 10$). Usually a supplementary lens will be used, but even this may not be so. Focal framing and focusing for a nature subject is simpler and faster than the reflex view camera unless the illumination is very bright.

Supplementary lenses are rarely designed as well as camera lenses, and their limitations have been discussed in Chapter 1. **They should be used only at relatively small apertures**, but fortunately at these magnifications the need for more depth of field also demands small apertures.

It is worthwhile to have a set of simple meniscus lenses of varying focal length, such as the Kodak Portra lenses, which can be used in combination, the powers being additive. The Kodak Retina cameras have a fitted set of lenses available in the Retina Close-Up Kits, Type N and Type R.

If the following rather brief description of making the focal-frame device seems inadequate, the reader can obtain Kodak Pamphlet B-10 which contains helpful illustrations and tables to assist in its construction. A commercial device, such as the Cal-Cam Focus Guide,* may be purchased.

Another simplifying factor in focal-frame photography is that **the lens aperture required for proper exposure is unchanged by the addition of the supplementary lens.** The changes of magnification and linear stop values compensate one another.

How to Make a Focal Frame.   To make the device adjustable, both frame size and distance need accurate readjustment and calibration. Hence one focal frame will be needed for each distance and magnification to be used; this usually means one focal frame per supplementary lens. Obtain the following:

1. One hardwood board on which to set the camera and hold the extended frame.

2. Some stiff wire or metal rod such as $\frac{3}{8}$-in. aluminum stock.

The camera must be set up once at the desired distance and a picture taken of a piece of graph paper or crossed yardsticks. This will give the frame size. A rule extending slantwise through the field will check optimum focus and give the depth of field available.

Mark the rod as with a file at all the places to be bent. Clamp the rod in a vise with the mark $\frac{3}{16}$ in. out from the vise jaws. Bend the rod with a wrench. Finish the bend with a hammer.

The diagrams in Figure 4-2, which were taken from Kodak Pamphlet B-10, represent the forms and measurements of focal frames for the three Portra lenses. The angles can be checked against these diagrams. Bend at *A* (both ends) first. Then bend at *B* so that *BC* is at right angles to *OA*. Check this with a square. In fact, check each subsequent operation very carefully, standing the device on the bench to view it.

Some device for positioning the camera is best made. A corner, made with short wooden slats nailed or screwed on, is one good method. Use a $\frac{1}{4} \times 20$ stove bolt to hold the camera on the board.

---

*Cal-Cam, 1564 North Grand Oaks Avenue, Pasadena 7, California.

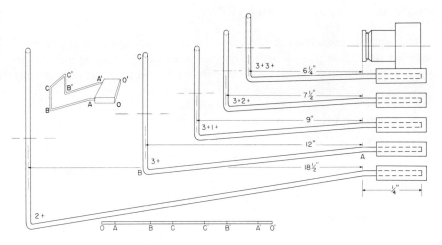

Figure 4-2   Diagram for making focal frame.

Direct View of Image

If time can be allowed for composing the picture, including the illumination, a camera to allow a view of the field on a ground glass has a distinct advantage. In fact, if a great variety of specimens and illuminations is to be encountered, as sometimes is the case in some industrial applications, ability to see the effect of the illumination arrangement on the ground glass may be almost a necessity. The instrument can be a view camera, a technical or press camera, or a miniature camera for roll film, with a device for alternatively removing the camera or sliding it aside to substitute a piece of glass in the plane of the film. If a miniature camera without a camera lens can be readily removed and replaced with respect to the lens system and if it has a focal-plane shutter to protect the unexposed film while it is off, a simple device (i.e., a dummy camera) is applicable. A block of wood is prepared to fit onto or into the camera support, possibly resting there by gravity, and its thickness is made such that its top surface is in the plane otherwise occupied by the film. If it has a large central hole and is covered by a piece of matte film or ground glass, the magnified image can be seen and often focused. This is because of the large depth-of-image focus with appreciable magnification which makes the exact position less critical. Moreover, it also makes the image plane available for a photometer reading for exposure determination. Reflex cameras are very useful for this type of photography; the application becomes rather obvious.

GENERAL OPTICAL CONSIDERATIONS

It is in **this magnification domain** that the exact **optics that are most suitable** (including focal lengths and distances within the system) **change most rapidly** with magnification. The desired magnification may be just that required to fit the image of the specimen into the frame of the camera, but there is a decided advantage in knowing the value of the magnification, at least approximately. Normally it should be known and passed on to the viewer. Very elaborate equipment has been designed for some of the German cameras to cover this range of photography; it may become quite expensive in total, but it certainly provides a great saving in time if much photography in this range is to be done.

As already discussed in Chapter 1, **the camera objective should preferably be an enlarger lens**, properly oriented, or a so-called **microphotographic lens**, but a **standard lens**, especially **with auxiliary (supplementary)** simple **lenses,** may be used. Some of the supplementaries may be useful if the photographer has only a few lenses of different focal length and design. An arrangement for attaching them to the mount of the principal lens is required; even "optical wax" (see Index) can be used with care; centration must not be neglected. Supplementary lenses were discussed in Chapter 1.

Near the magnification of unity, magnification may change faster than the focus of the image in the viewing plane of the camera. Therefore in this same range of about $\frac{1}{5}$ to 5X it is usually wise to fix, by lock or otherwise, the object or image distance (preferably whichever is longer) and to **move the object or the camera and lens** as a unit rather than the lens only within a fixed distance of object to image plane. Thus, above 1X, the correct magnification can be set, and the final touching up of the focus by moving the specimen will not seriously alter it.

This is also the domain in which the **choice**, often required, **between desired depth of field** in satisfactory focus and **desired definition** of important detail can be most agonizing. As already noted in Chapter 2, these two characteristics are affected in the opposite way by manipulation of the objective aperture, either by change of the iris diaphragm or by the choice of the lens. It seems fairly well known that fortunately a somewhat greater depth of field is usually acceptable in photomacrographs, particularly of nature subjects, than corresponds to standard optical formulas for depth of field. Unfortunately, the factor is not usually a large one. This effect and the formulas for calculation of depth of field are given in Appendix 1. The McLachlan method of sheet illumination and multiple exposures to increase the depth of field is described in Chapter 6.

Although we can normally depend on the surface texture illumin- ated by oblique illumination to fill the objective aperture by scatter- ing the light, it is only too easily possible to close down the iris diaphragm of an objective until the resolution becomes inadequate. The practical rules are stated in Chapter 2. A useful table of limiting apertures is given in Chapter 5 which applies equally to reflected illumination. It will be noted from that table that appreciable apertures are required for mag- nifications above 10X.

The apertures of modern camera lenses can rarely be closed to less than $f/16$. This limitation is most frustrating at magnifications close to 1X ($\frac{1}{5}$-5X). The gain in depth of field with three-dimensional fields, such as nature objects, more than offsets the loss in definition of lower apertures; the loss is inconsiderable if the stop size is kept larger than 2 mm. The cause is economic because of the difficulty, and so the cost of providing the thin leaves of the diaphragms with reasonably robust slots to go to these small apertures. Replaceable fixed stops could be used with some old lenses which can be easily and reliably unscrewed to reveal the diaphragm plane. This is either impossible or dangerous with modern lens mounts. A stop in any place but the correct one chosen by the designer will not only produce a defective image but will cause the relation that expresses the image illuminance to fail. (see p. 45) This is needed for exposure determination in flash photomacrography at least. When using a lens at a magnification much outside the lens designer's assumptions and the expected use of the lens, it is easy to cause a situation in which the correct iris diaphragm is no longer the limitation of all light rays reaching all points of the required image area. **At each magnification over the extended range, there is some aperture that must not be exceeded if this transfer of stop is not to occur.** This will vary with the type of lens. When a camera lens normally used at appreciable distances is used in this magnification region, the back of the objective should be scanned, with the eye looking at it from all points of the illuminated field through a small hole. It can then be seen whether some other part of the lens mount than the iris diaphragm sometimes limits the rays.

It is in this magnification region in which the three dimensions are important in the field that there is an advantage in the longer focal- length lenses that provide smaller apertures, together with cameras of larger format than that of 35 mm film, so that no later enlargement of the negatives will be necessary. If the aperture is to be below $f/16$, the modern high aperture corrected lens has little advantage and an old landscape lens or even a suitable process lens may be found.

For a 35 mm single-lens reflex camera with a focal plane shutter several alternatives are available. The Kilfitt Macro Kilars, sold by

Kling Photo Corp. in this country, are available in 40 and 90 mm focal lengths and made to focus through the fractional magnifications through 1X to a low magnification with the extension tubes that are furnished.

The most appropriate lenses should be those made for photomacrography by nearly all the microscope companies. Such lenses include the Zeiss Luminar, The Leitz Micro Summar, and the Bausch & Lomb Micro Tessar. An appropriate flange must be available to take their threads and fasten them to the lens board or camera front. These lenses can be reduced to $f/22$. Actually the diaphragm of the Bausch & Lomb 72 mm Micro Tessar can be reduced appreciably below the marked $f/22$, probably beyond $f/32$. Moreover, it can be disassembled by unscrewing it so that a *thin blackened* sheet-metal aperture of the desired small size can be inserted into the proper plane and the lens reassembled. In fact, this lens can be reassembled to face the opposite direction in the same mount, which is advantageous for magnification.

As already mentioned in Chapter 1 and earlier in this chapter, a good enlarging lens is almost ideal for this purpose if it can be used with the end normally facing the enlarging board facing into the camera toward the film when the magnification is greater the 1X. It is already mounted well for fractional camera magnifications but may require careful machine shop adaptation. Superior results were obtained with a Kodak Ektar than with standard photomicrographic lenses of the Tessar type. Moreover, these lenses can usually be reduced to $f/22$. Unfortunately the number and variety of the enlarging lenses made by American manufacturers have been greatly reduced in recent years. Individual enlarging lenses are listed, however, in the catalogs of some photo-equipment houses.*

Unfortunately the lenses, just mentioned as being so appropriate otherwise, rarely contain a shutter, yet in short exposure this can act as the stop, affecting the very "transfer stop" that we wish to avoid. If there is no focal plane shutter, the exposure can be controlled for the longer periods by capping, which is also applicable for "open flash" exposures for which they are most suitable. Otherwise the shutter must be at the light source or in an illumination beam.

ILLUMINATION TECHNIQUES

Bare Lamps

The general methods and principles of illumination by reflected light were discussed in Chapter 1. It was found that the choice of method was

*See footnote on p. 97.

often dictated by the available working distances in object space and the need for more light [by $(m + 1)^2$] as magnification increased. Relatively powerful sources are needed because of the inefficiency of utilization of the incident light after scattering by surfaces; this condition is aggravated if color pictures are to be taken, since long exposure times, otherwise often the solution to the demands of increasing magnification, may also lead to poor color balance with such films.

One solution is to replace bare incandescent lamps with flash lamps after the illumination pattern is considered satisfactory and possibly after the exposure is determined. This is because it is possible to determine an exposure factor to convert the known exposure with incandescent lamps to that with the substituted flash lamps if a photometric means of determining the former is available. A discussion of the determination of exposure of photomicrographs taken with reflected light by flashlamps is given on p. 597. Resort to directed beams of light (lamp with condenser) is also an alternative to gain more intensity. This method is discussed later in this chapter. It is well to remember, however, that a **lamp plus condenser can give no more light than is obtained by placing the lamp as near the subject as it would be to a condenser** of practical aperture. Thus when circumstances permit the source to be directly nearer the specimen than it would be to an $f/1$ condenser higher illumination will be obtained than from commercially available lenses. In general use of bare lamps as close as this is not convenient, nor does it yield sufficiently uniform illumination if a lamp is used at an angle.

The **quality of the lighting**, including the impression of depth and the contrast in the picture, will be determined principally by the **size** of the source (or the azimuth that it subtends from the object), its **direction, and especially its altitude** from the surface. The ability of a low angle of illumination to disclose surface contour is demonstrated by the familiar appearance of a slightly rough road when illuminated only the the headlights of an automobile. **A relatively flat and very light specimen,** such as a white woven cloth, **can usually be most effectively illuminated by a single source of light from one side,** thus allowing the interstices to be lit by interreflection and transmission. Figure 4-3 is cropped from an 8×10 in. photomacrograph of white handkerchief cloth at 10 diameters. It was made with a single photoflood lamp at a low angle, placed at the apparent upper right of the field. A 3-in. Enlarging Ektar was used at $f/16$, with no color filter and Kodak Royal Pan Film.

An opaque object of light color needs to be lit only in a subordinate fashion from the direction opposite the main beam, possibly only by a reflector. If a photomicrograph in color is being made and the detail of interest shows color contrast, the illumination can be quite flat and

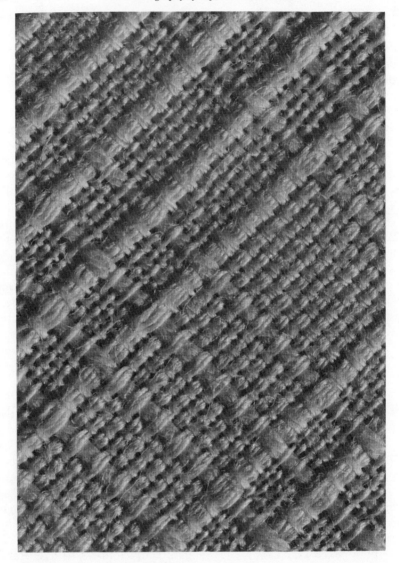

Figure 4-3   White handkerchief cloth X10.

symmetrical and still be pleasing. On the other hand, **the darker the speci-men and the more indented its surface, the more it must be illuminated from all azimuths with nearly equal intensity.** In fact, if the specimen consists of a dark textile with some **interstices**, it is usually best to lay it on a **translucent background**, such as opal glass, which is **illuminated**

from behind. This should normally be subordinate to the top-side incident illumination, so that with a cloth specimen that is quite dark the trans-illumination must be restrained.

The two flash photomacrographs of Figure 4-10 (see color insert) show standard lighting in that a bare bulb is at a 45° angle from the field. The first picture represents the flash bulb on a rather standard camera bracket in the plane of the camera. In the second picture the bulbs are behind the camera on long flexible arms extending from the hand-held camera support, so that the principal lamp is 45° high as well as to the left side. In this case there was a restrained secondary frontal lighting and the background was moved back from 6 to 10 in.

In any case the viewer of the photomicrograph should know the direction from which the principal illumination is coming. He will normally assume that it comes from the top of the picture. If this is so, a specific indication will not normally be needed. If the wrong assumption of the direction of the lighting is made, serious misinterpretation of the picture can occur when the lighting has been quite oblique. The hills and ridges in a specimen will appear as holes and trenches, as can be verified by turning the picture upside down.

Ring Illuminators

As mentioned in Chapter 1, if considerable photomacrography by reflected light is to be done, a ring illuminator is most convenient. Ring illuminators can be used with continuous illumination and with flash. By controlling the arc of the ring, which is lit or uncovered, and adjusting its height quite complete illumination control can be achieved. For this control it is best to hold the illuminating ring completely independent of the microscope or camera objective and support. A weakness of this method is that **an illuminator ring of a given size and diameter is conveniently applicable to only a limited magnification range**, as determined by the focal length and working distance of the objective. Thus several rings are needed if this system alone is to be used in photomacrography.

Several microscope manufacturers have put ring illuminator models on the market in several sizes which were very successful technically but apparently not commercially, for only one or two of the needed range seem to be still available. They have been quite expensive.

A ring illuminator, based on a model described by Preston (3.1931) but easier to make than his, was built in the Kodak Research Laboratories. It is shown, set up for photomacrography in Figure 4-4. As seen more clearly in Figure 4-5, the sockets for automobile headlight lamps (G-E No. 89, 12–16 V, 0.63 A, 6 cp) were soldered to one surface of a flat

Figure 4-4   Ring illuminator for photomacrography.

brass ring, $5\frac{1}{2}$ in. in outside diameter, $\frac{3}{4}$-in. wide, and $\frac{3}{16}$-in. thick, except for a $\frac{1}{8}$-in. wide ledge around its edge which had been turned down to about $\frac{3}{32}$-in. thickness. A Synthane (insulating) tubular ring, $5\frac{1}{2}$ in. in external diameter, $\frac{1}{8}$-in. thick, is supported by this flange and forms a back to the 12 lamp sockets. Instead of the flanged single ring, two rings, $5\frac{1}{4}$ and $5\frac{1}{2}$ in. in external diameter, could have been piled together. A narrow outer brass ring, $5\frac{1}{2}$ in. in internal diameter, forms the other electrode. Through this ring are threaded holes which allow a screw to make an electrical connection to the base of the lamps. Thus it is simple indeed, by a twist of a screw head, to turn the various lamps on or off while viewing the image. If another such illuminator were to be made, the even simpler "banana plug" connectors would be used to connect the outer ring to the lamps. Only simple unthreaded holes would be necessary and the device would be just as easy to use. A small canopy-type rotary switch behind each lamp would also be simple to install. It will be

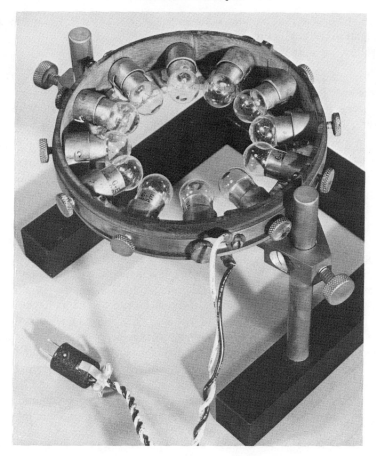

Figure 4-5   Ring illuminator with conical top off.

observed that the live electrodes, including the on-off screws for the lamps, are very much exposed and that a sizeable current of about 7.5 A is drawn by the 12 lamps. Although the 12–16 V across these metal parts are not dangerous, being that of an automobile circuit, provision should be made so that a higher voltage cannot be inadvertently put in. The plug at the end of the lead-in wire should not be the type that can easily be plugged into available house outlets of 115 V. It is not wise to connect a variable autotransformer *directly* to the ring illuminator; such a device can too easily be turned to a high and dangerous voltage.

The power supply for the illuminator shown in Figure 4-4 consists of a variable autotransformer which feeds the input of a step-down transformer, the whole being mounted in a metal cabinet. A warning neon

light to show when the voltage is on was added because it is just possible that all lamps might be turned off.
The following units were used:

1. Variable transformer plugged into 110–115 ac house current: "Powerstat" Model 10B, Superior Electric Co.
2. Step-down transformer: "Directron" Selenium Rectifier Transformer, 115–118 V, 8 A, parallel wiring, center-tapped, Sanford Miller Co.

Small adapters are available to fit into the sockets to allow M-2 flashbulbs to be used in place of the incandescent bulbs. This replacement can, of course, be made after a satisfactory illumination is set up. A flat cone, which rests on projections in the Synthane ring, protects the objective from direct light from the bulbs and forms a reflector. The ring assembly can be suspended from two rods, as illustrated, or be held by one side from a single horizontal (or vertical) rod that can be quickly screwed in to one of the threaded holes. In either case the support for the illuminator is entirely separate from that of the objective, hence the illuminator can be independently adjusted for illumination angle. Figure 4-6 is a photomacrograph of Teredinidae shells *(bankia Gouldi)* 7.5X that was made with this illuminator; only five of the lamps on one side were lighted.

For somewhat higher magnification and smaller object space a much smaller ring illuminator must be used if this method is to be applied at all. The great advantage of this is the ease and speed of setting it up, that is, by merely putting in a ring illuminator, adjusting its height, and deciding whether or not to utilize the whole circle of light. As magnification goes up and the size of the allowable ring becomes smaller, it is difficult to obtain sufficient illumination for a reasonable exposure time, especially if the surface is dark. These small incandescent ring illuminators have been dropped from the market by all microscope manufacturers. It was still possible, at the time of writing, to obtain the lamp for the Silverman Illuminator from the American Optical Co. This almost venerable but valuable lamp consists of a circle of glass tubing of about $1\frac{1}{8}$-in. diameter which contains a single filament that can be operated from a rheostat or autotransformer. It is described in many texts on microscopy. There is also a somewhat larger Silverman lamp. A reflecting holder for these lamps could be made by the user.

A preferable procedure, however, would seem to be that of utilizing **a small ring flashtube**, such as the L432 Ascorlight (American Speedlight Corp.) or the General Electric FT-429, especially if a power supply which allows continuously repetitive low-power flashing is available for visual composition and focusing. This matter is discussed under the title *Flash Photomicrography*. The diameter of the Ascorlight (Figure 4-11) is about

Figure 4-6 Teredinidae (*bankia Gouldi*) Shells X10. Ring Illuminator, lamps lit on one side only. Kodak Wratten B Filter. Micro Tessar at $f/22$. Kodak Panatomic-X Film, DK-50 Developer.

$2\frac{1}{4}$ in.; that of the FT-429 is at least 3 in. and has an input of 400 W-sec. It is the more powerful of the two but does not come equipped with power supply and an appropriate housing ready for use. The power supply of variable power, described in Chapter 13 has been adapted and used with the FT-429 tube (7. Spinell and Loveland, 1956).

It is quite practical and sometimes advantageous to use a **ring illu-**

minator and an independent light source to control highlights, specular or otherwise. For instance, some photomicrographs were made at 15X of a gold-platinum mass (South American Indian, pre-Columbus) of considerable interest as a relic of antiquity (Figure 4-7; see color insert). To make the appearance in the photomicrograph analogous to that observed with a magnifying glass when it was held in the hand, and to give it a truly metallic appearance, there had to be some carefully controlled specular reflections. The best appearance was obtained by laying a flash ring (the FT-429) just above the plane of the specimen and two bare, concentrated flashtubes opposite each other at about a 45° angle above the specimen and relatively close. This would have been too symmetrical a system to be best for black-and-white photomicrography. Obviously the lens must be protected with a tubular or funnel hood so that no light beams may enter it directly form the lamps.

Another illumination arrangement which was very satisfactory and which emphasized surface texture was a **combination of the ring flash and vertical illumination**. The ring flashlamp was kept in the same position, close to the plane of the specimen. A large cover slip was held at 45° inclination just below the objective (a 2-in. Enlarging Ektar) so that a horizontal beam of light from the side would be reflected vertically down onto the specimen. The double-condenser method was used to produce uniform illumination. The two flashlamps could be fired several times and in any ratio to give the desired illumination balance, since the subject was still. Incidentally, it was found that the flashing of the lamps alone caused the very light specimen to jump a bit. The cause of the slight blurring in the picture was not immediately evident but could have been removed by making the specimen stick to its support.

Use of Shadowless Canopy

If a ring illuminator is not available and a specimen has many indentures or otherwise requires flat lighting of the complete horizon, this can be accomplished as follows.

Place over the specimen a hemisphere or even an ordinary box cover painted flat white inside with a highly reflecting but matte pigment. Smoking with burning magnesium wire is a common practice, but a high titanium oxide paint will do if the vehicle is not yellow. There must be a hole in the top of the cover for the imaging beam to reach the objective and one at the side into which a very strong light beam is directed to hit the side of the "integrating dome."

Such a method is exceedingly wasteful of light and may require long exposures. More direct lighting may be combined with it, as with the ring illuminator. In spite of its inefficiency and some clumsiness, this method can be extemporized and some excellent pictures have been made with it.

Directed Illumination from Outside Lamps

As we saw in Chapter 1, when the object space is small, we are forced to **direct the illumination into it from lamps outside the object space by means of mirrors or lenses**. Moreover, this more optically efficient method can be used at all magnifications and relatively little limitation is made on the size of the lamp. If the surface of the specimen is rough or intricate, incandescent coiled filaments can be imaged on it without too much objection, but a uniform incandescent plane like that of the ribbon-filament lamp is a better source for this method. A translucent diffusing surface can be inserted in the illumination beam near the specimen to break up the pattern of coiled filaments. If a piece of ground glass or matte film (such as Kodapak Diffusion Sheets) is too wasteful of light, milder diffusion can be obtained by various means, including the use of a piece of photographic film fixed out with a hardening bath, which has been deliberately reticulated by dipping first into hot and then immediately into cold water. To balance the illumination from such a beam, reflectors may be needed on the opposite side of the specimen, or two or more beams from separate lamps can be used with one predominantly brighter because of its size or distance. The photomicrograph of white crepe acetate cloth 10X, reproduced as Figure 4-8, was made by this method. The beam from a $4\frac{1}{2}$-A carbon arc was directed at a 50° vertical angle onto the cloth but the beam was intercepted *near the specimen* by a ground glass. A ribbon filament was directed on it from almost the oppostie side and at a 45° angle, but its intensity was reduced with a heavy density. It is often convenient in this magnification range (e.g., above 3X) to use lightweight, focusable, and easily manipulated lamps, such as the Universal Microscope Illuminator of the American Optical Co. or the Nicholas Illuminator of Bausch & Lomb (see Figure 4-9). The Daylight Blue Filter of the Nicholas Illuminator makes it less adaptable to photography if much light is needed.

With a larger lamp a simple variation of the technique to **use one or more mirrors reduces the necessity of delicate manipulation of the hot lamps**, yet even increases the ease of the control of the light beam or beams. Set up the lamp with a horizontal beam parallel to the stage, as for vertical illumination, but with the beam passing over the object and under the objective. On the other side of the object let the beam strike a plane mirror held in a laboratory clamp and inclined to direct the beam on the object. In front of the stage and in the beam is an ordinary piece of

Figure 4-8   White crepe acetate cloth X10. Illumination: double-directed beam (see text). Filters: none. Objective: 48 mm Micro Tessar at $f/22$. Plate: Kodak Panchromatic Plate ($\gamma = 1.2$). Exposure: 2 min.

clear glass inclined to throw part of the beam up at a high angle. This beam is then directed back at the object by a mirror, the beam passing between the piece of clear glass and the microscope tube and hitting the object on the opposite side from the main beam (see Figures 4-11 and 12). By making a slight bit of diffusion on the underside of the clear-glass plate or partially silvering it the relative intensities of the two beams can be controlled. If the back mirror is cut to a narrow oblong and held by a rod half flattened at one end, this method can be used to include the 10X (16 mm) objective, although the auxiliary reflectors cannot be used.

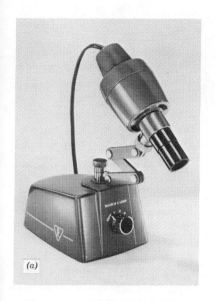 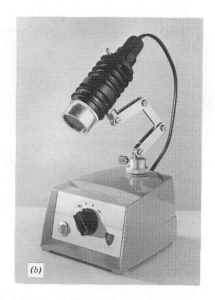

Figure 4-9 (*a*) B&L Nicholas Illuminator; (*b*) AO Universal Microscope Illuminator.

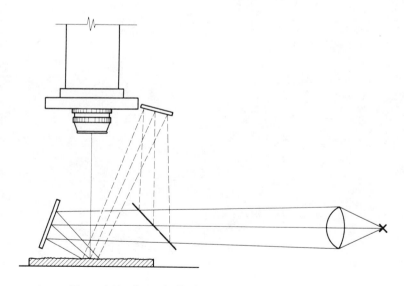

Figure 4-11 Control of incident illumination with mirrors.

Figure 4-12 Use of mirrors with incident illumination: (*a*) with simple microscope; (*b*) with compound microscope.

Complete Diffuser as a Light Source

In practice, whenever the photographic contrast is high or when the specimen is flat and very light in color, the use of some diffusion to make the illumination uniform will prove unsatisfactory. Moreover, with high contrast it will be quickly noticed that even the image of a ribbon filament in the field is not uniform is brightness. This trouble can be overcome by utilizing a piece of flashed or pot opal glass as the light source. It should be illuminated from behind by the primary lamp to cover any desired area. There must be a diaphragm practically in contact with it, preferably on the side of the specimen. Whereas it is convenient if this is an iris diaphragm, a sharp hole of correct size and shape cut from black paper or sheet metal will do. This is then imaged by a simple lens of adequate focal length onto the field. By means of the focal scale (page 32) it is simple to make the illuminated area of the size to just cover the field with appropriate reduction of magnification. This method is wasteful of light but this may be unimportant for the case. On the other hand, we will see this method is most useful for transmitted illumination.

Double-Condenser Method

Another method, which is slightly more trouble originally, will almost invariably be satisfactory if correctly set up and will always furnish: 1) a uniform field of illumination and 2) one of the desired size. This method, which requires a light source and two condensers, is utilized by most lantern-slide projectors and is sometimes designated by this name. Some projection lamps use it, such as the Nicholas Illuminator, but unless rather elaborate mechanical means are incorporated in the lamp, its range of adjustment will be strictly limited. By using this method *ex tempore* with extra condenser lenses, which can be purchased unmounted quite inexpensively, it can be utilized over a very wide range of field sizes.

Almost any lamp can be used, since its filament will be completely out of focus unless the lenses are of unusually short focal length. A lamp-condenser lens (Lens I, Figure 4-13) now focuses the light source in the plane of a second lens (II of Figure 4-13), which is used to focus the first onto the field. It is wise to magnify the source S to make its image as large as possible in the second lens, which should have a reasonable aperture to avoid annoying depth-of-focus effects of small apertures that will prevent the source from being completely out of focus and the field from being entirely uniform in brightness. The source and its first condenser may consist of a commercial photomicrographic lamp, but one of the two condensers must be chosen to be suitable for the specific applica-

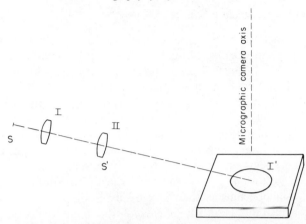

Figure 4-13   Double-condenser method of incident illumination.

tion if the distances of the elements involved are to be reasonable.

If the source size, the size of the two condensers, and the desired diameter of illuminated field are known (and usually they can be), it is quite **simple to determine the required positions of the elements of the system by successive application of our focal rule (p. 32) to the imaging of the primary source into the second lens by the first and then to the imaging of the first lens at the desired magnification in the field by the second.**

Example. The lamp is the 100-W, 120-V projection coil filament (CDS) bulb used by the American Optical Co. in their #735C lamp. It has four filaments in parallel lines with the small dimension of the rectangle formed equal to about 5 mm. A lamp condenser of 50-mm clear aperture is in the commercial lamp. An illuminated field of 1-in. (25-mm) diameter is wanted. This corresponds to a magnification of Lens I of $\frac{1}{2}$X as projected on the field. The *total distance* between the lamp condenser (Lens I) and the field is: $(m + 1/m + 2)f_2$, where f_2 is the focal length of Lens II. In this case the distance is $4\frac{1}{2}f_2$. A 60-mm $f/1$ condenser is available. This corresponds to a total distance of 27 cm (about $10\frac{1}{2}$ in.) which is reasonable. Lens II will be $(m + 1)$ or $1\frac{1}{2}f_2$ units from the field or 90 mm and $(1/m + 1)$ or $3f_2$ from Lens I, that is, 180 mm. This distance is $3.6f_1$ so that the filament is magnified 2.6X to 13 mm in its narrowest dimension but about 24 mm on the diagonal. This is about $f/4.5$–$f/2.5$ and proves enough to give a uniformly illuminated field. Obviously, a lens of smaller aperture for Lens II would have been satisfactory if the $f/1$ lens were not handy.

With the general choice of the best lens and the approximate distances being chosen by these simple calculations, the final focusing of the components is quickly done. It is advantageous to have Lens I of comparatively short focal length to allow adequate magnification of the filament within reasonable distances. For very low-power work, especially, it is sometimes advantageous to assemble a "double-projection lamp." This merely consists of the double-condenser system assembled to a unit, for instance by utilizing a rod as a common support for the lamp and the two condensers. This assembly can then be moved about and directed as a lamp. Moreover, the mirror system just described can be used with the compound-condenser system so that the lamp assembly need swivel only a bit to allow quite versatile lighting. Good and stable centration of the "double-projection lamp" is a necessity.

The photograph shown as Figure 4-14 was made with this system and the lamp combination just described and also shown in Figure 8-3. In this case, however, a larger illuminated field was necessary, that is, one with a diameter of 40 mm (about $1\frac{5}{8}$ in.). Thus the magnification of the 50 mm Lens I was 0.8X. Lens II was only 9 cm (about $3\frac{1}{2}$ in.) from the field and the magnification of the filament into Lens I was only 0.85X. Ideally, Lens I should have been of shorter focal length, but the spot was uniformly illuminated to the eye when projected on a white card. There is usually bad fringing at the edges of the illuminated field because of the nature of the relatively simple lenses used. This is even more evident when the circular section of the beam is stretched to an extended ellipse or hyperbola. In this case the elevation was 20° and the azimuth from the vertical was 14°. With the tilt, there is some decrease in brightness of the spot with distance from the lamp. For presentation other than such an illustration the picture would normally be cropped or the illumination area decreased. The details of the subject showed better in this illumination than with a higher elevation or any other way of making the illumination absolutely uniform. Often photomicrographs of flat metal surfaces can be shown by this method with satisfactory uniformity of illumination.

Oblique Incident Light at Higher Magnifications

We have already seen in Chapter 1 that, with increasing magnification, when the cramped space below the objective makes it impractical to illuminate the specimen surface with *direct* beams from a lamp either with or without a lamp condenser, the same type of reflected illumination can still be obtained by means of special objectives with a condenser built annularly around them. Such objective-condenser combinations are available from several microscope manufacturing companies, (e.g., the Tri-Vert Illuminator of Bausch & Lomb, the Ultropak of Leitz, the Epi-Condensers of Zeiss, and Wild). At the time of this writing the Ultropak

Figure 4-14   Fossil fern (Alethopteris), X2.5.
Double condenser illumination. Green filter (Wratten 58).
Panatomic-X, DK-50, $\gamma = 1.2$.

objectives form the most complete line available for use with most microscopes. They are illustrated in Figure 4-15. A picture made on Kodak Ektachrome film with an 11X Ultropak objective (NA 0.25) and 7.5X Hyperplane ocular is shown in Figure 1-28. In this case the specimen, made as a mounted slide by standard technique, might have been taken in routine as a photomicrograph by transmitted light. The incident illumination gives a much more natural appearance.

A built-in light source is usually supplied with all of these objective-condenser combinations for reflected light. With the Ultropak equipment,

Leitz

ULTROPAK

after Heine

Illuminating attachment for intense lightsources

Sector diaphragm

Slot for light filters

Ultropak with lightsource attached

U-O-Objectives with ring condenser

Reflecting condenser

Illumination penetrating below surface with reflecting condenser

Figure 4-15 Leitz Ultropak.

183

however, a separate condenser (code name AGNEE) is supplied for use with an external light source and is very helpful in photomicrography. This is because a very bright source may be needed, especially if the surface to be examined is not very light in color. Too prolonged exposure times may lead to troubles from the failure of the reciprocity law (see Chapter 15) especially for colored materials. Since the admitting aperture is a ring around a central opaque stop, the circular source of light, provided by arc lamps, including the tungsten arc, is an advantage. Provision is made in the equipment to include both the small build-in light source for visual use and the special condenser for the external lamp with a flip-flop mirror to allow choice between them. This is convenient for setting up the external light source.

These special objectives are corrected for use without a cover slip. They can, however, be used with a cover slip within the limitations discussed in Chapter 2. Special caps and cones are provided for the Ultropak objectives for use when the specimen is covered with water, thus obviating the effect of a moist surface on image quality or a tendency to dry the sample. The black glass slides that are available are most convenient when the specimen is small.

A special wide-angle annular condenser which gives much more grazing incidence to the illuminating beam is supplied for the dry Ultropak objectives. It is appropriately called a "relief condenser" and has found more use in our laboratory than the standard condenser. Moreover, it seems to be rare that the entire ring illumination of these illuminators is used without some sector being blocked out to give directional lighting. **The limitation of this type of illumination with an annular condenser is that it is impractical to make an objective of greater NA than 1.0.** Therefore, as the specimen becomes even smaller, this aperture may be inadequate for resolution. Again mechanical considerations have been the limit to the illumination with increasing magnification.

This limitation can be obviated. If a very small particle is to be examined by oblique reflected light, it can be done by **illuminating it with a thin ring of annular light by the technique of vertical illumination**, using **conical illumination** and **a metallographic objective of NA 1.40**. Arrangement of the apparatus is the same as that for vertical illumination, which is described on p. 204 and illustrated by Figures 4-25 and 26. In this case a small round opaque patch must be held near the plane of the aperture diaphragm which allows only a thin ring of light to pass through. A low-power micromanipulator is convenient but, of course, not necessary.

A radioactive particle under investigation by this system is shown in Figure 4-19 (see color insert). It was found worthwhile to be able first to locate and focus the particle by use of transmitted light. It then became possible to illuminate the particle with a combination of reflected and transmitted light, keeping the latter subordinated. It will be found that the focus for transmitted and reflected light at this high aperture is different and the depth of focus apparently much less by reflected light, which may prove disappointing, especially photographically because it is the periphery of a particle on which we focus by transmitted light and any loss of definition in the opaque center is unseen. At 1.40 NA the angle of illumination is about 23° from the horizontal plane of the field.

Control of Specular Highlights

In illumination by oblique reflected light the presence of *too much* specular highlight may obscure some detail or reduce image contrast. One solution, already discussed, is to resort to diffusion of the illumination. More often a preferable way is to **use polarized light for the necessary control**. Sometimes one polarizer, such as a Kodak Pola-Screen or a Polaroid filter, inserted in the image beam is all that is required. It can be laid on the rear of a simple microscope lens, which is pointing downward, and rotated to the azimuth that is most effective in reducing the unwanted highlights. This depends on the fact, discovered by Malus, that when light is reflected from all materials except metals there is a certain angle of reflection at which the reflected ray is **plane-polarized**, that is, the light consists of transverse vibrations in a single plane. Brewster found that this angle is that at which the reflected and refracted rays are perpendicular to each other; hence it is dependent on the refractive index. Therefore for most common transparent materials the angle of incidence for most effective polarization lies between 53° (water) and 58° (flint glass). A polarizer, acting as an **analyzer**, can then be made to absorb the plane-polarized beam completely by orienting it properly.

A more reliable method **places a polarizer**, such as a Polaroid filter, **in the illumination beam with an analyzer in the image beam**. In both cases the analyzer should be of excellent quality, since it is in the image beam; for example, the Polaroid filter should be mounted in "A" quality glass. This method depends on the phenomenon that scattered light is depolarized. Since this method is wasteful of light, especially strong light sources may be required. It is especially useful for eliminating the excessive highlights of moist specimens. This method is illustrated by Figure 4-16 which shows mold on leather at 4 diameters. The only difference between the two is that two Polaroid filters were used for the second photomicrograph (Figure 4-16*b*) whose planes of polarization were crossed. One Polaroid was in the illumination beam, the other resting on the lens mount above the specimen. The way in which specular

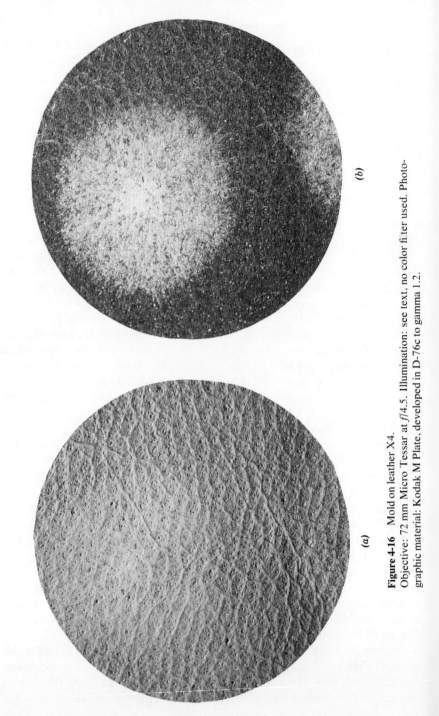

(a)

(b)

Figure 4-16 Mold on leather X4.
Objective: 72 mm Micro Tessar at f/4.5. Illumination: see text, no color filter used. Photographic material: Kodak M Plate, developed in D-76c to gamma 1.2.

surface illumination hides tone and color difference is shown in Figure 4-16a, and the effectiveness of removing this light, at the expense of greatly prolonged exposure requirements, is shown by the other. Two lamps whose beams passed horizontally over the specimen and were directed down onto it at 60° by mirrors were used for these pictures. With black specimens much light was required; the principal lamp was a 4½-A carbon arc, the other, a ribbon filament at right angles (in the horizontal plane).

The fact that **specularly reflected light is the component by which we judge surface contour** can be seen by comparing the obvious surface texture shown in Figure 4-16a with its lack in Figure 4-16b. This effect is illustrated much more dramatically by comparing field a with field b in Figure 4-17; both show some artificial-leather at 10X magnification. In the latter case crossed Polaroid filters have not only eliminated nearly all scattered light but all surface contour, which is the significant detail. In metallography the surface contour is also the significant detail and is exhibited by directing the illumination beam down the axis of observation. This use of zero angle, however, is not the best technique for exhibiting surface contour when the latter is rough on a gross scale, but some specular illumination should still be used. Figure 4-18 shows another example of this same important effect of specular light, to show surface detail only.

SPECULAR (VERTICAL) ILLUMINATION, METALLOGRAPHY

This method of illumination finds its chief application in the photography of metal specimens and minerals. Anyone doing a considerable amount of metallography should become familiar with the *Metals Handbook* of the American Society for Metals and with the Standard Methods for Metallography specified by the American Society for Testing and Materials (see Category 6 of the Bibliography).

In vertical illumination the illuminating beam is directed to the specimen as nearly as possible along the same axis as it is observed by the objective. Since the light source itself must be at one side, this usually involves directing a horizontal focused beam to the optical axis and then deflecting it to the specimen by a reflector inclined at 45°.

Satisfactory specular illumination consists in imaging on a field of the microscope a uniform source of light of sufficient area that its image is as large or larger than the desired field, which is acting as a mirror bearing a pattern of variable reflectance into the objective. From the preceding discussion involving the sample of leather it can be seen that specular

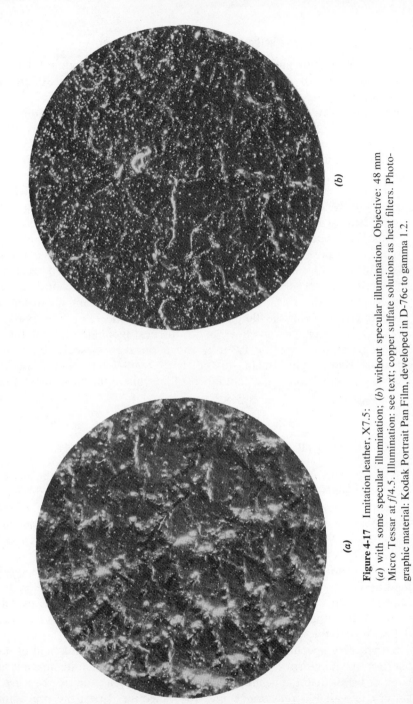

(b)

(a)

Figure 4-17 Imitation leather, X7.5:
(a) with some specular illumination; (b) without specular illumination. Objective: 48 mm
Micro Tessar at $f/4.5$. Illumination: see text; copper sulfate solutions as heat filters. Photographic material: Kodak Portrait Pan Film, developed in D-76c to gamma 1.2.

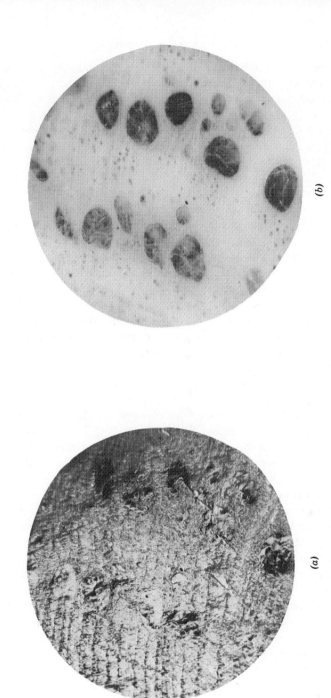

(a)

(b)

Figure 4-18 Opalized wood, X35.
Use of polarizers with oblique incident light to reduce specular reflections: *(a)* unfiltered
tungsten *(b)* crossed Kodak Pola-Screens. Optics: Ultropak, 0.15 NA. Kodak M Plate.

illumination is most useful when it is the topography of a specularly reflecting surface that is desired; color is usually desaturated.

For very low-power work with the simple microscope such illumination is most easily set up by erecting **a vertical piece of ground or opal glass** which is illuminated uniformly from behind, possibly by a bare lamp, and **reflected specularly onto the field by a thin sheet of clear glass held below the objective** at 45°. Only about 4% of the light will be reflected by the sheet of glass down onto the specimen. When specular illumination is desired with a compound microscope and the object space is adequate (as with a 4X (32 mm) objective), the best method is to hold a microscope cover slip, selected for flatness, at 45° *below* the objective. Then a horizontal beam of light is aligned and focused to comprise the double-condenser method of illumination, already described in this chapter and illustrated by Figure 4-13. The second condenser most conveniently is about the same focal length as the objective, or a little bit longer, and at least of equal aperture. As first noted in Chapter 1, with specular illumination **it is now necessary to arrange that the illumination beam has an aperture adequate** for resolution at the magnification that is used. With objectives of somewhat higher power and shorter focal length, it is preferable to focus the light source in the rear focal plane of the second condenser rather than into the lens itself. This image plane of the light source is then acting as an aperture plane for the illumination beam and should be imaged into the aperture plane of the objective if the system is correctly set up. A diaphragm, therefore, can well be held in this plane of the light-source image. The second condenser is now functioning as a microscope condenser. (See Index, *vertical illumination*.)

With most 10X (16 mm) objectives and nearly all those that are higher powered, it is no longer possible to put a 45° plane-glass mirror below the objective. It is placed directly above the objective to form the conventional **vertical illuminator**. All microscopes for general metallography are equipped with this device, which reflects the illumination beam to the specimen through the objective. The objective first acts as a condenser and then passes the image beam, it forms from the light reflected from the specimen on to the eyepiece. The vertical illuminator may be a simple holder of the plane-glass mirror or narrow prism, with or without an auxiliary lens and diaphragm, as illustrated in Figure 4-20a, and one that can be screwed into any microscope which will in turn hold a microscope objective. Much more frequently it will consist of the same device optically but built irretrievably into the casting for a commercial specialized metallographic microscope.

The 45° reflecting surface within a vertical illuminator may consist of a fully reflecting mirror or prism that covers about one-half the total

(a)

(b)

Figure 4-20 Simple vertical illuminators for tube lengths of 215 mm or 160 mm: (a) B & L (31-34-90) with single nosepiece showing auxiliary corrective lens for 160 mm tube length; (b) B & L (31-34-82) with triple nosepiece, and alternative cover glass (full aperture) and mirror (half aperture) reflectors.

objective aperture, that is, half the light beam transmitted by the objective, or it may be a thin glass plate that only partially transmits or reflects the beam but covers all of the objective. The completely reflecting mirror allows considerably more contrast, since at least half the objective aperture is clear as usual and also because a reduction in objective aperture provides greater contrast. However, as discussed in Chapter 2, this will be at the expense of resolution of detail. In this case, with a semicircular aperture, there will be more resolution of the specimen in one direction than in the other. Often the extent to which the prism or mirror extends into the aperture toward the optical axis can be controlled. More or less oblique illumination is then obtained.

The plane reflector covering the aperture allows complete resolution up to the limit of the objective in use. Here the sacrifice is the introduction of flare and lower contrast. If an ordinary glass cover-slip reflector is used, only 4% of the illuminating light will be reflected. Thus, with longer exposures, a dimmer image is obtained.

In spite of this the completely reflecting, half-aperture mirrors should be used only for low magnification, that is, with the 16 mm (10X) objective or the 32 mm (4X) objective (an outside plane glass is best in this case). However, with the advent of the specially coated reflectors (see p. 201), the prism type may become obsolete except as a device to obtain oblique illumination.

In Figure 4-20a, which illustrates the Bausch & Lomb "Simple Vertical Illuminator 31-34-90," the "box" holding the plane glass illuminator is shown at the left. It contains no optics save the plane glass reflector for which there is rotation and some translatory adjustment. The plane thin glass reflector has been coated to act as a single reflecting surface of 25% reflection, although there are probably many units in laboratories with the simple cover-slip reflector of 4% double reflection. With the empty ring adaptor to fit the single RMS nosepiece of the microscope, the tubelength should be 215 mm for Bausch & Lomb metallographic objectives. An alternative adaptor, shown below and at the lower right, contains a lens with which the mechanical tube-length may be 160 mm with the same objectives. The component at the upper right consists of the optical illuminating system with lenses and diaphragm. A small lamp can be screwed onto the end of this assembly and carried by it. Alternatively, an independent light source may be used.

In this optical illuminating unit the outer lever at the right operates a field iris diaphragm; that is, it is imaged by the combination of a lens inside the unit and the microscope objective onto the surface of the specimen. The rear lever of the unit lies just behind the auxiliary condenser lens and also operates an iris diaphragm. Although called an aperture diaphragm, it is not accurately in a conjugate aperture plane. The small lamp unit (not shown) carried on the illuminator uses a microscope ocular in front of a 6.5-V lamp which has the outer planoface so frosted that it acts as the light source. When an *outside* light source is used, an auxiliary in a cap, designated 31-34-85, should be screwed onto the coarse outer thread. The outside source should be focused 13 mm in front of the face of the supplementary lens. The lamp should then be 6 in. away. This prescribes the focal length of the lamp condenser if the image of a particular source, filaments or arc, is to fill the opening of the field diaphragm; with an arc lamp a lens of 45 mm e.f. is satisfactory.

The vertical illuminator shown in Figure 4-20b is somewhat different; it is designated by Bausch & Lomb as 31-34-82. It carries a nosepiece for three microscope objectives and also an alternative half-aperture complete reflector to the full aperture cover-glass type used by both illuminators. This semicircular mirror reflector is pulled into the optical axis by the rotating knurled knob. Since the whole aperture cover-glass-type reflector is optically coated, there is little use for the mirror reflector except to obtain oblique illumination.

The optical system of the illuminating unit of this vertical illuminator is preferable to the first one. The advantage of the first is the ability to use only the reflector block (separately obtainable) for a completely external optical system described later in the chapter. The same auxiliary small lamp is available with the frosted ocular for visual work. The illumination from it is not critical, but an external light source can be used with it, also with the aid of the component 31-34-85 lens screwed onto the unit. The external lamp, 6 in. away, is focused onto this external lens and the outer diaphragm directly behind it. This is now a true aperture diaphragm in a plane conjugate to the objective aperture. Behind it is a field diaphragm onto which the lamp condenser is imaged by the auxiliary outer lens and which is, in turn, imaged onto the specimen by another lens plus the objective. Another way of stating the same condition is that the field diaphragm is imaged (virtually) by the second lens at a point 210 mm above the objective. Again, the auxiliary lens for a 160 mm tube length should be employed when that type of microscope is in use.

The American Optical Co. has two vertical illuminator units that fit below standard upright microscopes; one, for the Polarstar, is somewhat simpler and the other is for the Metal Star microscope. Each has a 6.5-V lamp attachment. The Polarstar vertical illuminator, unlike the Bausch & Lomb-90 unit, does not break down to a simple semireflecting box for cases in which a completely outside condenser system is to be used. Both have Köhler illumination; the image of the illuminant is placed in the aperture of objective.

A Polarizing Vertical Illuminator. A vertical illuminator of novel design* has been incorporated in the Research Metallographic Stand made by Bausch & Lomb. The illumination beam enters a specially cut and recemented calcite prism before entering the objective (used first as a condenser as usual). The beam is split into two components polarized mutually perpendicularly in the normal way for calcite, and one of them (the ordinary ray) is reflected at the dividing surface between *A* and *B* (see Figure 4-21) and absorbed. The other (extraordinary ray) passes out of the prism through the objective to the surface of the reflecting specimen. However, between the prism and the objective is a quarter-wave plate through which the beam must pass twice, going and coming from the specimen. This rotates the plane of polarization 90° so that on its return the beam is totally reflected by the dividing surface and thence, after a second total reflection at a silvered prism surface, it passes into the eyepiece of the microscope. The calcite wastes 50% and more is lost at the quarter-wave plate, but about 40% of the initial illumination beam is still available at the image if all is reflected by the specimen. Moreover, by removing the quarter-wave plate the illuminator functions as a polarizing microscope, for now only that part of the light depolarized by the specimen (or that whose plane of polarization is rotated by it) will be reflected by the prism to form the image.

Illumination Types

Because of the inefficient use of the illumination beam in vertical illumination, especially before the advent of coated optics, most commercial metallographic equipment was originally designed to image the light

*See paper by L. V. Foster (6. 1938).

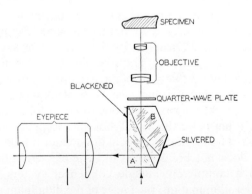

Figure 4-21   Polarizing vertical illuminator, B & L Research Metallographic Camera.

source in the field, that is, on the surface of the metallic specimen, since this method is the most efficient in utilizing the illumination. A great many metallographs so designed must still be in use. **In more recent years most equipment has been designed to image the lamp condenser in the field.** These two methods, the Abbe-Nelson and the Köhler, respectively, are compared in more detail in the chapter on transmitted illumination, since a mirror-type specimen causes the illumination method to be *more like* the Köhler than the other incident-light methods.

Incandescent ribbon-filament lamps are used quite widely in metallography, in spite of the fact that exposure times of fractions of a minute are often required. **An arc lamp, xenon, zirconium, or carbon, is usually considered more suitable** and often reduces the required exposure time about 15-fold (for the 10-A arc), that is, from minutes to seconds. Automatic carbon-arc lamps are now available. If the carbon-rod electrodes have been baked at 400°C and then kept in a desiccator, these lamps become relatively simple to use and are in a different class from a clockwork arc served with carbon electrodes that have lain around to absorb water from the air. Carbon rods cored with amorphous carbon should be used and are normally supplied for this purpose. However, the xenon arcs and the 100-W zirconium arc are definitely more convenient. A more complete discussion of light sources is included in Chapter 7 and is applicable here.

All metallographic stands have both **field and aperture iris diaphragms** and are **especially important** in this type of photomicrography; both must be closed down correctly. Their function has already been discussed thoroughly for transmitted-light photomicrography in Chapter 2 and their characteristics and use are the same; they are best demonstrated in the more extended setup of the vertical microscope described later in this chapter (see Figure 4-25). It is only necessary to emphasize at this point that the field diaphragm should be kept centered, in focus in the field, and closed down just to clear the photographed area. If the diaphragm is not in the best focus, the last cannot be done critically. The locations of the field and aperture diaphragms will differ somewhat in the different models of metallographic stands but will be pointed out in the manufacturers' direction pamphlets.

Although the field diaphragm may be so built in and prealigned that it cannot be adjusted, provision is usually made for decentering the aperture iris diaphragm. The light source is normally focused on this diaphragm. For brightfield vertical illumination the diaphragm should be centered and the image of the light source should be somewhat larger than the largest opening that will be used. This, in turn, will be that opening that just subtends the back lens of the objective in use, as determined by the light beam projected on it and as seen by looking at the back of the objective through the pinhole ocular in the eyepiece tube. In most commercial outfits the distance of the light source, hence the size of the light-source image, is not changed as the magnification and objectives are changed.

The metallographic stand should be set up for centered brightfield illumination and returned to this type after some specialized illumination is used. The simplest variation is that of **oblique illumination** obtained by decentering the iris diaphragm in the aperture plane of this system. This is not only possible in most commercial metallographs but it is also possible to rotate the "aperture diaphragm" so that the oblique illumination may come from any azimuth. As already stated, the direction of the illumination should be noted and the information furnished with the photomicrographs, since this is the so-called "relief illumination" in which shadows are present. A "click stop" is normally provided for bringing the system back into centration.

Oblique illumination over all or most of the azimuth simultaneously (i.e., conical illumination) is advocated by some (3. Hughes, Syers, and Thomas, 1962). Although not frequently built in, it can be obtained by opening the aperture diaphragm to the limit *for the objective* and placing an opaque disk just in front of the aperture diaphragm that is somewhat smaller than the hole. In Figure 4-1*b* it can be seen that at the extreme a completely darkfield effect may be obtained.

The extension of this technique leads to **darkfield illumination**. In this case, which is analogous to that of transmitted light, the method calls for all the angles in the cone of the illuminating beam to be greater (from the axis) than the maximum angle that will enter the objective. In this case the illuminating rays are reflected by the mirror surface of a completely plane specimen at such a shallow angle that they miss entering the objective. Only irregularity of the surface will then scatter light into the microscope to show against a dark field. This, we have seen, requires special objectives that carry the light down the outside of the objective proper, such as the Bausch & Lomb Tri-Vert. To date the maximum aperture of these special objectives is NA 1.0. But commercial metallographic stands do have built-in darkfield illumination, with which the field and aperture diaphragms are opened completely and one of these special objectives is used. The effect and advantage of darkfield may be observed in Figure 4-1. The usual characteristic of darkfield, the detection of very small particles, is also present with reflected light and specular objects.

Another type of illumination available on commercial metallographic stands is **phase illumination.** Its general principles were discussed in Chapter 2. Since metallographic samples normally act as mirrors to give specular brightfield illumination, these same principles apply except that it is impractical to put the **phase-retardation annulus** in the aperture plane in the objective itself. The light beam traverses this plane twice and surface reflection from the illumination beam would introduce bad flare. Therefore it is **put in an image of this aperture plane**, that is, a conjugate plane. The exact method differs somewhat with the manufacturer. In the Bausch & Lomb Balphot, an extra unit is substituted for the usual eyepiece and its tube. At any rate, some auxiliary lens is required, in practice, to allow easy visual examination of the aperture plane during the process of exactly superposing the annular ring of illumination onto the annulus of the phase plate. However, the transfer of the position of the phase plate makes it possible **to use ordinary metallographic objectives for phase microscopy**. Somewhat greater care may be required in preparing the specimen, since any diffusion or scattered reflection will spoil the efficiency of the phase illumination. It is especially nice to have conversion to phase illumination made relatively easy because the micrographer must usually decide with each specimen whether it is advantageous. In brightfield "vertical illumination" all metallic surfaces perpendicular to the optical axis will reflect with nearly equal brightness. When the surface slopes from this plane, which is chiefly at grain boundaries, or when it is not flat, the darker tones appear to fill in detail. With phase illumination, planes at different levels will be imaged in different tones of gray, which may mean a uniform but distinct gray for each grain face rather than only a line to outline the grain.

Illumination with **polarized light** is yet another special illumination that is built into most of the modern metallographic stands. With the exception of the Bausch & Lomb Research Stand, already described, Polaroid polarizing filters are used. There is also a sensitive tint plate (first-order red). With this plate in the beam between crossed polarizers, the field will be magenta but slight differences in anisotropy may become more evident. The theory and use of polarized light in reflected light work can be quite complex. This is especially true in the study of the anisotropy of the metal crystals themselves, in which exacting technique is required, but some workers state that the results justify it. It is more often utilized for location and identification of nonmetallic inclusions. Cameron (6. 1957, 1959) has organized the study and identification of minerals with polarized light and a vertical illuminator into a reliable and valuable method.

There would be little advantage in listing the specific steps in setting up a commercial metallographic stand for taking pictures. These steps differ somewhat with the apparatus and are listed and discussed by the manufacturer for each specific instrument. It can be more adequately treated here for the case in which the operator sets up his own stand, as discussed later in this chapter. **Once a good image is on the ground glass**, the **procedure is relatively standard for all forms of illumination**, except as the latter effects the contrast and except for the general level of illumination. This is discussed in the section "Photography."

Commercial Metallographic Stands

Nearly all large commercial metallographs now use the inverted design, due to LeChatelier, so that only one surface of a specimen need be prepared for specular reflection because it can lie face down on a plane table stage with the microscope objective looking up at it.

Normally special short-mount **metallographic objectives** (see Chapter 3) are used so that the rear glass lens will be closer to the reflector than it would be in the conventional compound-microscope objective designed for work by transmitted light. These objectives are corrected for use without cover slip. They are also corrected for much longer tube lengths than the standard objectives for biology. This is primarily to allow for the insertion of the vertical illuminator and must be taken into account when ordering them for or using them with other than the specific equipment for which they are sold.

Some metallographic stands utilize microscope objectives corrected and focused so that the image-bearing light is a collimated beam. These objectives are said to be corrected for infinity, that is, an infinite tube length. They require an auxiliary lens between them and the ocular and form the intermediate microscope image at their focal plane which must also be the proper plane for the eyepiece to pick it up. This design has several advantages. It liberates the designer of the whole stand by allowing him to include more optical path between the objective and eyepiece for his folded image beam without optical sacrifice. A collimated image beam is also optically less degraded than a converging one by the interposed slanting sheet of plane glass that constitutes the vertical illuminator. Therefore, with objectives corrected for infinity, a thicker, more rigidly plane sheet of glass can be used as the inclined reflector; distortions of this glass reflector can seriously degrade the image. It should be remembered, however, that **placing the reflector above the objective is only a necessary expedient and that when possible it should be used below it**. This can be done only with low-power objectives with which the coverglass error is unimportant.

The optical system of the Balphot Model I of Bausch & Lomb, which is typical of the metallographic stands in most general use, is shown in Figure 4-22.

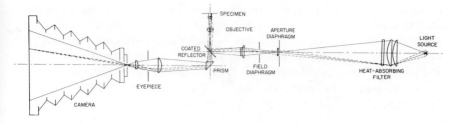

Figure 4-22   Optical system, Balphot Metallographic Camera.

The tendency lately is to design metallographic stands that are highly specialized. Bausch & Lomb now sell three general models of metallographs, all three unique by older criteria.

The Dynazoom Bench Metallograph has an optical system that is folded compactly into a rectangular box with a microscope stage on top near one end and the observing binocular oculars on top at the other end. The optical path from the enclosed 100-W tungsten lamp (in the middle) goes up and down one end, then along the bottom and up the other end to the oculars. It takes standard metallographic objectives, but now the real image is allowed to form at exactly 215 mm from the objective (with a reticle turret in that plane); a relay lens system then picks up the beam to image both field and aperture planes in the same relative position farther along. Thus the standard zoom system, already described (Chapter 3), can pick it up for binocular viewing or photography with the vertical tube of a trinocular. Photomicrography is therefore the same as using a biological microscope with a camera placed above it.

There are two "console models," Balphot II and the Research II Metallograph. Both have the same console, which resembles a metal desk in general appearance, with the microscope stage in front of the operator and the oculars, projection screen and photographic plane behind it and part of a low rectangular tower. Both models use flat-field objectives designed by the principles already discussed (Chapter 3), *except that* they are designed for use without cover glass and, if the back component fixed to the stand is included, for a tube length of 215 mm.

Research Metallograph II utilizes the Foster vertical illuminator which incorporates polarization. Balphot II has a simple beam splitter but in the same form as the Foster prism for mechanical and optical compatibility in the console. Therefore a Polaroid polarizer and analyzer are used in this model. A sensitive tint plate to make polarization colors critical is included in each case, as has been standard practice. Since the lateral chromatic aberration of all the objectives has been made equally bad, a single lens in a field plane corrects this defect, and therefore an ocular uncorrected for color is used. Only one ocular is considered necessary, since a zoom system of 1:6 ratio is included. This unusual range is obtained with a somewhat different design than the one described for the other microscopes (Chapter 3). A collimator is used before the mirror system (flip-flop plus partial) by which all or part of the image beam can be viewed directly or projected and also sent to a cadmium sulfide photocell. Therefore a telescope, which may be binocular, is used for viewing the image, and a good lens system, focused for infinity, for the projection screen.

The Balphot II has a 150-W xenon lamp. In the Research II Metallograph a 100-W tungsten lamp or a 450-W xenon lamp, each with its own condenser, can be slid into the optical axis.

The great advantage of this design, of course, is convenience; it does seem to be very stable. All decisions are made for the operator, although most types of illumination are available with simple manipulation. An excellent book of instructions is furnished with all large commercial metallographs; in fact, they are usually more complete than the corresponding ones for other types of microscope. **Once a good image is on the ground glass, the procedure is relatively standard for all forms of illumination**, except as the latter affects the contrast, and for the general level of illumination which affects exposure. These problems are discussed in the sections on photography.

Metallography in Color

Considerable metallography in color has been done quite satisfactorily, but most of it seems to have been of specimen images primarily colored by polarized light or interference patterns and with the use of monopack color films. Although the broad sensitizing and color characteristics of such films normally do not allow accurate color reproduction of this kind of subject, fortunately it is not usually required. Another type of very common metallography is more of a problem; for example the specimen is a neutral metal with some areas colored by staining or inclusions (see Figure 19-1 in color insert). A pinkish or even greenish steel, for instance, may not be acceptable. Yet commercial processing is usually balanced for good flesh tones and excellent neutrals may be sacrificed. They can be obtained by compensation. Poor or variable ultraviolet reflection in the mirrors of the metallograph can be a problem, as can variable or excessive ultraviolet in the arc source. Although expecially critical in some metallography, these problems do occur elsewhere and are best discussed in a special section (Chapter 19).

Flare From Glass Surfaces: Surface Coatings

Flare from light from internal surfaces of the optical system not used in producing the image may be very troublesome in metallography, especially when the contrast within the specimen is low and its total reflectance is also low. If a plane-glass reflector is used above the objective, as is usually done, most of the light is transmitted by the reflector and, in the case of the first reflection of the illumination beam, must be absorbed within the illuminator.

General flare is most effective in decreasing the brightness *range* of the image. This is easily demonstrated.

Let I_{min} be the minimum illuminance in the image (no flare) and I_{max} the maximum illuminance in the image (no flare). The contrast range of the image (brightness scale) = I_{max}/I_{min}.

If the flare light added at all points is I_f, let $I_{min} = 0.02$ ft-c and $I_{max} = 0.4$ ft-c. Let $I_f = 0.05$ ft-c. With no flare, contrast = 0.4/0.02 = 20. With this flare, contrast = $(0.4 + 0.05)/(0.02 + 0.05) = 6.5$.

The brightness range of brightfield metallographic images may be lower than 6.5. Therefore it is natural that materials of relatively high potential contrast are chosen for metallography, although this tendency is sometimes overdone, as discussed later.

Use of a solid-mirror reflector increases the contrast. This is sometimes done for low and intermediate magnifications and requires that half the aperture be used for the illumination beam and the other unobscured half for the image beam. Obviously, at high magnifications a mirror reflector would provide inadequate aperture for resolution.

The discovery and advent of **antireflection films** that could be coated on glass surfaces to reduce the reflection at these surfaces was a sufficiently important advance in the technique of vertical illumination and metallography to warrant special discussion. When light strikes a surface of transparent material, a fraction of it, R, is reflected by the surface. Fresnel's famous formula for reflection at normal incidence is

$$R = \frac{(n-1)^2}{(n+1)^2},\tag{4-1}$$

where n is the refractive index of the material. For crown glass (about $n = 1.525$) $R = 4.3\%$. It is true that the second surface reflects almost the same amount and that the reflector utilizes about 8% of the illumination beam, but each successive air-glass surface in the lenses, etc., will deduct a similar percentage of the transmitted beam and may distribute much of it as flare. If there is a surface film of different refractive index on the glass and its optical thickness is $\frac{1}{4}$ wavelength, there not only will be a reflection at both surfaces but the two beams will be **exactly out of phase as they turn back, hence tend to destroy each other.** This they can do completely if they are of equal strength. This, in turn, depends on the selection of the refractive index of the antireflection coating. For a surface film $\frac{1}{4}$-λ thick reflection is completely eliminated when $n_2 = \sqrt{n_1}$ (where n_1 is the index of the glass and n_2 that of the film). Since refractive index will vary with wavelength, it is obvious that **for only one wavelength can the reflection be entirely eliminated.** It will, however, be reduced for other wavelengths also. See Figure 4-23, which is due to Mooney (2. 1945). It is difficult to obtain a material of the correct refractive index that will also produce a stable hard coating. Magnesium fluoride ($n_D = 1.38$) has been much used. By using two coatings, with the first one (against the surface) of high index, another variable is obtained which allows greater choice of materials and so makes the complete elimination of reflection more practical *at one wavelength.* The formula for optimum refractive index is now $n_2{}^2 = n_1 n_3{}^2$, where once again n_1 is the index of the glass and n_2 and n_3, that of the successive layers. The penalty for use of two or more coats is the fact that the wavelength effect is steeper.

Such coated microscope objectives have been commercially available only since World War II. The back surface of the lens will, of course, have a color complementary to that of the wavelength whose back reflection is most completely eliminated. In the years since the war the

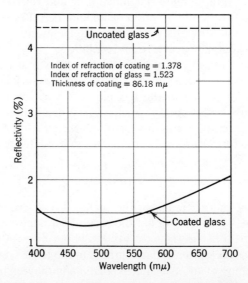

Figure 4-23 Reflectivity of glass as affected by a monolayer coating one-quarter wavelength thick.

technique of producing these coatings has advanced greatly, and they are now hard and stable. However, since their exact thickness and refractive index is so vital to their functioning, they should not be needlessly rubbed or contaminated, especially with a silicone lens-cleaning paper. Even more than most microscope objectives, the rule should be to **keep them clean in order to avoid having to clean them frequently.** A further discussion of the applications of coated objectives is given by Benford (6. 1946).

Just as a glass surface can be coated with a film to decrease its reflectance, **it can be coated to increase its reflectance.** By Fresnel's formula the index of the coatings should now be higher, a great advantage if applied to our plane cover-glass reflector. Zinc sulfide ($n_D = 2.37$) has been used for this purpose. With it a reflection of 16.3% is made possible by (4-1). The total transmission is then about 14%. If an antireflection coating is then applied to the opposite side, **the reflector becomes, in effect, a single reflecting surface.** By making the cover glass reflect 40 to 50%, transmitting the balance, the final image beam would be 24 to 25% of the original illumination beam (assuming no absorption), which is the maximum possible by this system. Unfortunately, a metallic film, with which such reflectance could be achieved, also has appreciable absorption that would annul the gain in reflectance as far as gain in final illumination is concerned. Most plane-glass reflectors of new commercial vertical illuminators are now coated, which makes the use of a mirror or prism illuminator above the objective less attractive. At the time of this writing the majority of metallographic equipment in use still has uncoated optics. Often some coated objectives may be obtained for a metallographic bench, whereas the rest of the optics, including the reflectors, remain uncoated.

A direct experimental determination of the effect of completely uncoated optics, as on any older metallographic stand versus completely coated optics, as on any new stand, yielded results most significant for metallographs. The same metallographic specimens were set up on the same equipment, except for the antireflection coatings, and the same magnification. The specimens, such as a stainless steel of low brightness range, which is very common, were most interesting; for example, one had a brightness range *in the image alone* of $1:3$. *With coated optics*, the brightness level was increased about 4X, but much more significant for metallography the *brightness range* increased by a factor of 2X. A somewhat similar effect was produced by an etched specimen of wrought iron whose brightness scale was $1:15$ with coated optics. As explained in Chapter 15, when modern equipment with coated optics is used, **the illuminance range of the image may be so increased over that with uncoated optics that the reproduction contrast of negatives suitable for older** uncoated optics may result in negatives of excessive density range that cannot be reproduced on any printing paper.

USE OF UPRIGHT STAND FOR METALLOGRAPHY

Although most metallography is by far most conveniently done with a specialized commercial stand that has an inverted stage and all illumination optics prealigned and prefocused, **equally good** (and some claim superior) **work can be done with an upright microscope** and with a vertical illuminator such as is shown in Figure 4-20. **A microscope with a focusing stage is almost a prerequisite** unless the light source is attached to the vertical illuminator, which is somewhat undesirable for photography. The author has, on occasion, improvised a focusing stage by placing the specimen on a focusing substage condenser mount.

Several microscope companies offer upright microscopes with focusing stages and otherwise simple design especially suitable for metallography. Although the large commercial inverted stands usually have excellent optical arrangements, as well as being expensive, the author has found the optical elements and arrangements attached to the simple vertical illuminator for the upright microscope to be a less desirable compromise, presumably with expectation that they will be used only provisionally and visually. The lens and diaphragm can usually be unscrewed from the box like device that holds the reflector, leaving only the latter with the proper screw flanges and holes for the light beams. This stripped-down illuminator can even be purchased directly and is quite inexpensive (see Figure 4-20b). It is this device that is assumed in the description below of the assembly of elements for an optical arrangement that will provide excellent quality and complete control. The original vertical illuminator assembly, with its lens or lenses and diaphragm, will provide the most convenient way, short of a standard inverted stand, of making metallographs. If this assembly is used, care should be taken to use all extraneous light sources and lamp condensers at the distances suggested by the manufacturer. These distances are usually quite short, perhaps 6 in., in order to make the whole assembly quite compact. It is normally more convenient to extend the arrangement when setting the system up from its elements on a bench.

Setting Up a Metallographic Microscope with an Upright Stand

Prerequisites. In order to set up a system for metallography that is both convenient and satisfactory for the quality of its product, certain prerequisites must be provided. Some will be commercially available or already at hand; the others can be made or obtained at a fraction of the cost of the commercial equipment, although some shop facilities should be available.

Assume that a microscope with a focusing stage is set under a camera, both having a vertical axis, with the stage and the photographic plate horizontal. A horizontal axis must be established from the light source to the vertical illuminator above the objective (see Figure 4-24). Usually the height of the light source, such as a carbon arc, is the limitation for the height of this axis. There must be a reasonably convenient capacity for holding lenses and iris diaphragms in this axis in the correct plane and in alignment.

The author has found a low wooden bench to be very convenient (e.g. 22 x 60 in. although it could be somewhat shorter). If the microscope is blocked up stably to include the right height for the optical axis of the illumination beam and sits on a thick plate of brass or wood, a vertical plate screwed onto the end of the horizontal one forms an excellent support for both the aperture condenser lens and the diaphragm described later (see Figure 4-26). The other elements of the illumination system are held on an optical bed running along the middle of the bench. Although such a commercial optical bed with its clamps is most convenient, the practicability of a home- or laboratory-made track should not be overlooked. The simple edge board described in Chapter 3 is an example. It should be possible to move all elements up, down, or sidewise or to rotate them. All of this is simple if they are held on a single rod of adequate diameter and length and on a support that centers sidewise. It is worthwhile to use stop rings (collars). These useful simple rings, usually of metal with a set screw, are made to fit well but to slide on a rod (one can be seen on the rod in Figure 4-25). When set, they allow the rod to be loosened in its holder for adjustments without disturbing the height. A lens with a thin edge is best held in a grooved V-support such as the

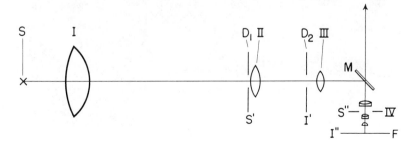

Figure 4-24 Optical system for metallography.

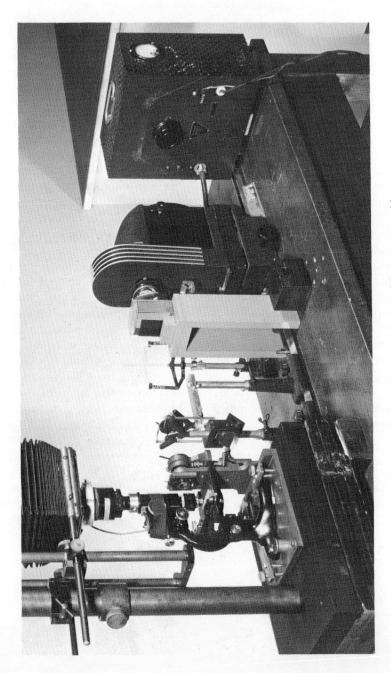

Figure 4-25   Metallographic camera as laboratory assembly.

Figure 4-26 Metallographic camera as laboratory assembly. Closeup of illumination system.

S1900 models sold by the Gaertner Scientific Company of Chicago (Figure 4-26). A thick lens is best held in a V-groove made from a piece of 90°-angle brass or iron, cut to about $\frac{1}{2}$-in. width and fastened V-edge down into the top of the vertical support rod of wood, Synthane, or metal. A piece of electrical tape will hold the lens onto its V-mount very well, but the mount must be securely fastened to the rod.

The following **directions for setting up the system**, which uses Köhler illumination (see Chapter 8), assume that the elements are separately held, except as commented on, so that the vertical illuminator consists only of the plane-glass mirror over the objective. They also assume the possession of a **pinhole ocular** (see Chapter 3) and a piece of **matte focusing film or plate**, which may consist merely of a piece of matte film or ground glass with a penciled cross drawn on it. The author has found a pair of black crossed scales, photographically printed onto Kodak Translucent Plate, Type 3, to be so generally convenient that one is always kept available. A **centering slide** can be made by marking a cross on a **first-surface mirror** with a sharpened wax pencil. If such a cross is drawn, it has been found worthwhile to make it with a rule, rather than freehand, and to keep it small.

Quite a number of light sources can be assumed. For photomicrography of this type the author much prefers to have **the light source and its lamp condenser independently mounted.** This arrangement permits the use of a field diaphragm behind the lamp condenser, which greatly reduces flare. However, a complete lamp with condenser is practical and may be the only unit available.

If a high-intensity light source is used, as really should be the case for extensive metallography, **a cooling cell should be available** (see Chapter 3). For a 10-A arc and for black-and-white photography a 3-cm thickness of a 5% acidified solution of copper sulfate is appropriate. With this light source, **a rather high density** (e.g. 2.0) **is needed** during the alignment process. Although smoked glass could be used, glass densities are more convenient.

Optics of Vertical Bench

The optical arrangement for the system described here is fundamentally no different from that of the horizontal bench for transparent illumination, described in Chapters 3 and 7, but certainly does differ in practical details. Figure 4-24 is a diagram of the arrangement.

The fundamental new factor is that the specimen in the field F is a plane mirror, or can be considered one for a uniform brightfield in this setup. The pattern of the specimen is made by departures from this plane mirror and brightfield, but if most of the field consists of such departures the normally best illumination would be oblique or ring, that is, reflected darkfield illumination.

The first consequence is that the objective first serves as the subfield illumination condenser for the incident beam and then as the objectve for the reflected beam.

There is still a *single, linear, optical axis* from light source, through the center of the field, to the center of the final image plane, but now it is *bent at right angles* by the partially reflecting mirror M; *this axis is then folded back on itself* at the field and finally passes through the glass M to the image plane above.

The components, including the various reimagings of the relay optical system, can be followed from the diagram of Figure 4-24 by noting the successive primed letters or Roman numbers. Note also that theoretically the diaphragm D_1 should occupy the same plane as Lens II and that if a lens, such as Lens II, did exactly lie in the plane of an image it would not alter the positions of any subsequent images of this plane. In practice, this cannot be precisely true. However, the source S is imaged on the diaphragm D_1 by the lamp condenser I and is reimaged by Lens III in the focal plane of the objective (i.e., S'' in Lens IV). Lens II, which should not affect this process, is used to image the lamp condenser I in the field F. The diaphragm D_1 will also be imaged in the focal plane of the objective and serves as an aperture diaphragm.

Finally the diaphragm D_2 is inserted somewhat in front of Lens III at such a position along the axis that it is sharply imaged in the field. Obviously it is also the point at which Lens II images the lamp condenser I. This is the field diaphragm.

The position of the rear focal plane of a microscope objective varies greatly with the type of objective. In a simple 16 mm achromatic objective it just above the rear lens. In well-corrected, high-power objectives it may lie far down in the objective. In commercial equipment, when it is not expected that the user will readjust the optical components, a compromise is made, usually in favor of the complex high-power objective, and the source is imaged about 20 mm below the base of the shoulder for the objective. In this system *this compromise need not be made*. It is a simple matter to adjust the position of the diaphragm D_1/Lens II pair to place the image S_1, correct for the objective in use; only a little readjustment of Lens III and diaphragm D_2 is needed.

The positions of the various elements along the axis will depend chiefly on the focal lengths of the lenses of the illuminating system. If longer focal lengths were used, the system would become even more extended along the bench. **Commercial metallographic apparatus generally uses shorter focal lengths than are suitable for this type of setup in order to be more compact.** This is not only no advantage for apparatus that is set up on an optical bench, but great compactness in this case is very inconvenient. The optical clamps themselves occupy much length along the bench. The distances and focal lengths mentioned below, used for this example, are almost a minimum for convenience.

The focal lengths of the various lenses must be somewhat compatible with one another because of the relay characteristics of the system. In the example, illustrated in Figure 4-25, the lamp condenser Lens I, was 60 mm equivalent focus and $f/1$, aspheric; Lens II, which images Lens I in the field, was a simple spectacle lens, 57 mm equivalent focus, 25 mm diameter, obtainable from any optician; Lens III, then, should be about 50 mm focal length. Although it, too, is usually only a simple lens, in this case an old 48 mm Micro Tessar, $f/4.5$, was used, since for metallography in color it is an advantage to have Lens III an achromat.

It does pay to have the first image of the light source at S' appreciably larger than it will be when imaged in the objective aperture at S''. This allows accessories to be used easily, as in the darkfield lighting mentioned on p. 184. In the example (Figure 4-25) the magnification of the first image of the source was about 5X and was placed about 7 in. in front of the vertical axis and six focal lengths from Lens I.

Alternate Optics

The particular vertical illuminator used in the apparatus shown in Figures 4-25 and 26 is a once-common one, now discontinued; that in Figure 4-20 is essentially the same. The latter is the Bausch & Lomb Cat. No. 31-34-90-01, which also allows a tube bearing illumination optics to be pulled off, leaving only a transmitting partial reflector at 45° in the body, coated so that the reflection is essentially from a single surface. Therefore, with this model all discussion is pertinent.

Of course, the total illuminator with its side arm bearing the optics, may be used. As already mentioned, with such a commercial unit, the focal lengths of the lenses have been made so short that the whole system is short. A lamp that attaches directly to the unit also may be used, but it will be inadequate for much photomicrography. When this lamp is in use, the image of the source is placed on the outside iris of the tube and made of sufficient size to cover it; it is a field diaphragm. This optical system is not the one described in this chapter; rather it employs the Abbe-Nelson method to place the illuminant image in the field. There is a lens close in front of the aperture diaphragm in the tube.

Bausch & Lomb has a separate vertical illuminator based on the system described here, that is, their Cat. No. 31-34-82-01. This illuminator is placed on top of a triple nosepiece, and therefore a rhombic prism is inserted between Lens III and Diaphragm D_2 (Figure 4-24) to raise the optical axis above the nosepiece. The axis resumes its horizontal direction with some compaction gained. The same lamp used with the other illuminator can be attached to the illuminator tube; there is considerable diffusion on the rear surface of the lamp condenser. The same remarks apply to it. For appreciable metallography an outside source can, and should, be used. In this case an auxiliary lens (Cat. No. 31-34-85), which is Lens II, is screwed onto the end of the side tube. With Lens III, which has a focal length of 27 mm, and Lens II, an equivalent 26 mm, this system can be much more compact than the system illustrated in Figure 4-25. With these focal lengths the lamp with its condenser should be about 6 in. from the vertical illuminator; its exact distance must be determined by trial until the iris diaphragm on the lamp, or an arrow against the lamp condenser, is in focus in the field and the light source is focused on the outer iris diaphragm (D_1). With an arc lamp, a lamp condenser whose focal length is 45 mm is about right to make the light-source image just fill the iris aperture D_1.

An upright, relatively simple microscope with a focusing stage, obtainable from Bausch & Lomb, utilizes both the first (-90-) vertical illuminator described (Microscope Model AM) and the second (-82-) with Köhler illumination (Microscope Model DM).

These seem to be the only commercially available vertical illuminators that are not incorporated into the microscope itself. Unitron offers two relatively simple models, therefore relatively inexpensive; one, the MMU, is an upright model with a focusing stage and a quadruple nose-piece; the second, the MEC, is a model with an inverted focusing stage (LeChatelier type) and a 35 mm camera for photomicrography.

For routine metallography the metallurgist will want one of the more elaborate inverted stands in which all illumination and accessories are contained. These stands have already been discussed briefly. Setting up the image is best done according to the manufacturer's direction booklet. For its photography the second section of this book should prove useful.

Directions. The first step is to establish the optical axis. This involves the tilt and rotation of the mirror in the vertical illuminator and its alignment into the axis. **Put the coarse centering slide** (cross on a first-surface mirror) **on the stage and center it in the field with a low-power objective** (e.g., 10X) **and eyepiece**, with any convenient illumination. **Then remove all lenses from the system,** including the lamp condenser, if that is simple to do. **Insert the pinhole eyepiece. Move the lamp,** which may be a carbon or metal arc, **to its farthest distance along the bench. Looking down the microscope, tilt and rotate the mirror of the vertical illuminator so that the light source itself can be seen, apparently superposed on the cross of the slide.** This is merely a provisional setting.

Next place the matte glass, with its cross, on a bench clamp on the horizontal optical bed and bring it as close as possible to the microscope.

Looking down the microscope through the pinhole, **shift the matte glass laterally until its cross is visually superposed on that of the slide. Then shift the clamp, bearing the matte glass, to the other end of the bench** (near the lamp). **Rotate the vertical illuminator**, if necessary, **or tilt the plane mirror a bit to keep the two crosses visually in line.** The adjustment of the vertical and horizontal angles of the mirror (tilt and rotation) are best done independently. This will require **repeated sliding of the vertically held matte glass back and forth**, close to the microscope, where the matte glass is shifted to align the crosses, and to the other end, where the mirror angle is changed to align the crosses. Of course, the position of the bed clamps must be reproducible when clamped, for which some care is often necessary. After this procedure the angles of the mirror of the illuminator are established with respect to the horizontal optical bed; the rotation of **the vertical illuminator is locked** (usually with a set screw) and **not disturbed thereafter.** It may be possible to slide the support of the plane-glass mirror, within the vertical illuminator (see Figure 4-20), without affecting its tilt, in order to center it optimally to the aperture. This affects the reflected image the same as a slight height adjustment.

Now the light source should be set back along the bench to its correct position, either finally, according to directions or previous experience, or approximately. The principles governing this are discussed a bit later. In this example a 10-A carbon arc was placed $24\frac{1}{2}$ in. from the vertical axis of the microscope. **The position of the source with respect to the axis is again observed through the pinhole ocular**, still with no objective or other optics in the system. **This time it is visually superposed on the cross of the slide by laterally translating either the source or the microscope** with respect to the optical bed **and** for vertical alignment **by raising or lowering the body tube** of the microscope. This should be untouched thereafter, except for fine focusing; the coarse focusing is done by movement of the stage. It is well either to lock the coarse focus or to indicate, by tape or other cover, that it is not to be used to forestall an unthinking reversion to habit.

With the optical axis established, the various lenses and diaphragms can be individually added and centered in. Some care must be taken that the plane of the added element is orthogonal to the axis.

Before adding any lenses, or even before using a high-density filter in the focused beam to shield the eyes, **a cooling filter should be added into the axis.** These filters were discussed in Chapter 9. If a thick water cell is used (one containing 5% copper sulfate is used for black-and-white photography with an arc), this is an opportunity to note how well it is placed perpendicularly to the axis to avoid decentering the light source. The author found it to be advantageous to set the cooling filter on a wooden support straddling the optical bed rather than on a table support held by a bench clamp that would use up more valuable length of the optical bed.

At this point the matte glass bearing the cross should be reinstated on the bench in front of and close to the vertical illuminator **and centered into the axis.** An optional check step can be made at this point by temporarily reinserting the 10X objective and eyepiece and focusing on the cross. The scale of the matte glass should also be in focus, with its center on the cross of the slide. With the objective removed (if it had been used), **set the centered matte glass in the position later to be occupied by diaphragm** D_1, which in turn is dependent on the position Lens II will occupy. In our example D_1 and, of course, the image of the source S' were $7\frac{1}{4}$ in. from the vertical axis.

The lamp condenser is the first lens to be added, if it has been taken out. If the source and its condenser constitute a lamp in which the source cannot be separately supported, these two elements must first be mutually centered by the methods described in Chapter 6. The alignment of such a lamp into the axis then procedes quite similarly to the method just described. The source itself can be seen through the microscope if the lamp condenser can be focused in until it is very close to the source, possibly, in the case of an arc, without its being turned on. In either case **the lamp condenser is now focused to give an image of the source on the matte glass. By centering the light at once on the cross of the matte glass** any later deviations can be assumed to be due to the slight decentration of the carbon arc in burning, which is touched up at the arc itself. Some other sources can be counted on to remain steadily in axis.

Now the illuminator condenser (Lens III) is inserted, focused, and centered. This new lens should form an image of the cross, or scale, of the matte glass in a plane 20 mm below the vertical illuminator, as marked by a piece of white paper temporarily held there. The function of this lens is to reimage the light source in the aperture plane of the objective. A lens should be chosen to form an image of the source large enough to fill the largest linear aperture of an objective in the plane S'' of Figure 4-24.

The low-power (10X) objective (Lens IV) and ocular are now reinserted and focused on the cross. Another lens (Lens II) is added in place of the matte glass and in the plane of the image of the source **to focus an image of the lamp condenser (Lens I) in the field of the microscope.** This focus should be done carefully by holding the matte glass with its cross or scale temporarily against the lamp condenser, either in front of or behind it, while looking through the microscope. Lens II, of course, should also be centerable and the illuminated disk of the condenser image should be left well centered in the field with an eyepiece of very low power. Sometimes a little refocusing of Lens III, Lens II, and the lamp distance is necessary when available lenses are not optimum. If Lens II were added in the plane of the image of the source, it would not affect the reimage of the source at S'' when it is added to the system; actually it draws up the image of the source slightly.

Now a diaphragm D_1 is placed just in front of Lens II. It is convenient to have it, rather than Lens II, in the plane of the image of the source and so is reimaged at S'' in the objective aperture. It should be separately held, well centered, and as close to Lens II as possible. Then, if the system is stable, it can be used as the criterion for recentering a carbon-arc image (S') if the crater of the arc tends to wander. This diaphragm should appear sharp and well centered when viewed down the microscope through the pinhole ocular. It should be partly closed to be seen within the objective aperture.

Finally, the field diaphragm D_2 is placed in the plane between Lens II and Lens III, where it is sharply imaged in the field and on which Lens I (or the focusing screen) is sharply imaged. In our example this was about $3\frac{3}{4}$ in. from the vertical axis. It should be well centered and closed within the limits of the visual field. Now it will be found that all objectives from 10 to 100X, NA 1.4, can be substituted in turn with only a little touching up of these diaphragms.

HANDLING THE SPECIMEN

Since oblique incident illumination is the type most usual for human observation and, as we have seen, applicable at almost any magnification, a wide variety of specimens can be encountered. Preparing them to give a suitable photomicrograph may be the major problem, but specimen preparation for microscopy has always been of predominant importance. With this type of illumination, it is usually a surface with which we will be concerned. A slightly concave surface will normally give us our biggest field; more often our surface may be convex. A simple **tilting stage** is very convenient. This may be merely a hemisphere held by a ring so that its flat surface can be tilted at any angle and in any direction. The Micro Lab-Jack sold by Central Scientific Company is also convenient for raising the stage to any height. Cloth and paper form a common type of specimen. For such specimens it is convenient to have a flat lead disk with a central hole to form a ring, onto the bottom of which is cemented a piece of cloth such as flannel. Such a disk has a very flat bottom if cast on a flat surface. It is most convenient for quickly holding samples flat and in position. A segment of the ring should be removed to allow very low glancing illumination, required by some specimens.

The orientation of the specimen in relation to the direction of the illumination and the orientation of the final photomicrograph as it is viewed by the observer is very important. Normally the illumination should seem to be coming from the top and one side. Capacity to rotate the specimen easily with respect to the lighting is desirable when examining it before photography.

Photomicrography of bullets has become so important in crime laboratories that special microscope assemblies are sold just for this purpose, in particular to allow comparison of bullets. A bullet is usually easily held from its back end, with wax or an adhesive, as an extension of a rod of slightly less diameter.

A word might be spoken to remind light microscopists of the **use of replicas of a surface for microscopical examination**. This method has been highly developed by electron microscopists, but light microscopists consider that they can examine the original surface by reflected light. This may not be easy or indeed it may be impossible without a forbidden injury to the subject. If a portion of a huge surface is to be examined, use of an intermediate replica may be the simplest procedure. Many resin solutions or polymers have been used, including simple collodion or gelatin solutions. A single replica skin produces a negative of the surface. The negative may be satisfactory or a positive replication may be obtained by "pulling" a second replica from the first, using another material that will not stick to the first or by dissolving the first from the second. An opaque coating or some density can be given the replica chemically (e.g., silvering) or by vacuum evaporation. Shadowing from an angle can also be done. Some references to the details of these procedures are given in the bibliography.

METAL SPECIMENS

The preparation of metal specimens for microscopical examination has become a highly developed technique, which is often associated with the kind or type of metal, particularly its hardness. An uninitiated individual might easily, yet unwittingly, change the character of the structure by his method of preparation, especially in cutting or grinding. Excellent directions, however, for the preparation of metal specimens are given in *Metals Handbook*, published by the American Society for Metals, Cleveland, Ohio, and by the American Society for Testing and Materials in their ASTM Standard, E 3-62. The professional metallographer will certainly have available an inverted metalloscope of the LeChatelier type so that only one face of his specimen, including any mount for it, will need to be plane and well polished. If his specimen is small, it will probably have been rapidly mounted in Bakelite, or a similar resin, in a heated press before one surface is ground and polished, although some specimens do not lend themselves to this method.

This book, however, considers the preparation of metal samples from the viewpoint of the individual who has metal specimens only infrequently and more generalized microscopical equipment. As mentioned before, the specialized equipment is more important in this case for speed and convenience that for the quality of the results except as noted below.

The specimen is preferably not less than $\frac{1}{2}$ in. nor more than 1 in. square or in diameter and only thick enough to hold while grinding and polishing. When cutting the sample from a larger piece, great care must be taken to avoid local surface heating, usually by lubrication of the saw or other cutting tool, often with water. The specimen may be much smaller, in which case it must be mounted to achieve a size that can be finished without being rounded from the edges.

If the specimen is to be held on the stage of an upright microscope and below an objective, it must have the top and bottom sides plane and parallel or be held with its top side parallel to the stage. This can be accomplished fairly easily. If one surface has already been prepared by polishing, ready for examination, the specimen can be placed on a piece of Plasticine in a container with a top edge that is parallel to its bottom and then pressed into the Plasticine with a sheet of glass, such as a microscope slide, until the slide touches the rim of the container at all points. Special metal holders just for this purpose can be purchased from dealers in metallographic supplies.* If simple machine-shop facilities are available, they can be made by facing short metal rings on both ends or by facing the tops and bottoms of pipe caps (e.g., gas caps) parallel to one another.

The pipe caps can be used also to hold a specimen during grinding, polishing, and etching as well as examination. In this case the container material should be at least as hard as the specimen that is to be ground and polished. The specimen may be pushed, with the flat surface up, into cooling sealing wax or a low-melting alloy, such as Wood's metal. Molten yellow sulfur has been recommended. In fact, it is convenient to cement the face of the metal specimen temporarily to a glass slide which is then laid across the top of the cap. In any case the final specimen in its mount should consist of a cylinder whose height is appreciably shorter than its diameter, for otherwise it would be difficult to maintain a flat surface during grinding and polishing.

*Such as Buehler Ltd.

A more elegant and a very satisfactory method is to cast the specimen into a block of an epoxy casting resin, such as Araldite 502 (Ciba). The following procedure was found to be most satisfactory for small metal specimens in the Kodak Laboratories.

First, molds are made as follows: cylinders are cut from a rod of inert material, for example, Lucite or polystyrene, which are of adequate diameter, say 1 to $1\frac{1}{4}$ in. and $\frac{3}{4}$ in. high. A disk of polyethylene of the same diameter is placed on top and a dish formed by wrapping the whole with aluminum foil at least $1\frac{1}{2}$ in. wide.

The following formulation is used:

Araldite 502	100 parts by weight
Hardener HN-951	14 parts by weight

This is a higher concentration of hardener than suggested by the manufacturer, but the resulting resin is very hard and grinds well on abrasive paper. The hardener must be thoroughly and uniformly mixed into the resin (taking at least 5 min.). In spite of the viscosity– and because of it– great care must be exercised to avoid forming many bubbles in the mixture. In fact, one advantageous step is to subject the mixture to a vacuum, as from a water aspirator, to break any bubbles. An alternative to the vacuum treatment, or an added procedure, is to flow the liquid resin mixture into the mold over a sheet of glass when many bubbles will burst and others can be pricked. One precaution, necessary if bubbles are seen in the original resin, is to heat the Araldite *before* the hardener is added to at least 140°F and then allow it to cool before adding hardener. The metal sample is immersed in the resin after wetting it with benzene. If the metal specimen contains crevices, it would be well to warm it before putting it in the resin mixture; the viscosity is much reduced with temperature. A paper label bearing the identification of the sample in pencil or carbon ink can also be immersed in the clear resin if care is taken to rid it first of incorporated air. This is quickly and easily done by soaking it in benzene, especially if it can be "boiled" in it with an aspirator and given a preliminary dip in resin before it is put into the casting. The sample block can be cured either at room temperature for 24 hr or, after 1 hr at room temperature, in an oven at 140°F for 3 hr. The aluminum foil is peeled off to remove the sample.

An alternative to the molds just described is the use of a ring mold, avoiding brass, which is placed on a sheet of polyethylene covering a flat plate. The specimen is placed face down on the polyethylene, inside the the ring, and the liquid monomer mixture poured over it. The ring is expendable, remaining on the specimen, and insuring that the face of the specimen is parallel to the stage.

Grinding and Polishing Metal Samples

The sample may be assumed to be cut and, if necessary, mounted at this point. It may need to have a roughcut surface "faced" to a flat surface, which is done by laying a medium mill file on the bench surface and rubbing the sample over it *in one direction only*. Care must always be taken to avoid overheating the surface and deep scratches. Usually a stone is poor for facing.

Next, the specimen is ground on successively finer flat abrasive surfaces. This is done by rubbing the sample back and forth, but in one direction only, on an emery or carborundum paper, starting with a No.2 grade, which is laid on a very flat hard surface. Sometimes, however, this grade may be omitted. Grinding with each grade, No. 1, No. 0, No. 00, and No. 000, successively, is done with the sample turned 90° from that used with the preceding grade and until the pattern of the new finer scratches obliterates the last of those caused by the preceding coarser abrasive. Since it takes only one coarser grit particle to make a sorry pattern of a few coarse scratches in a surface that is otherwise marred only by uniform fine ones, the necessity of strict neatness and cleanliness will soon be emphasized if not observed *a priori*. The grades of paper must be strictly segregated. Even the paper itself, as purchased, may be suspect, since some commercial abrasive papers are not free from a few coarser particles in a surface of finer ones. With such papers excellent samples could never be obtained. If aluminum or other soft metal is being ground, the surface should be wet with kerosene to prevent particles of abrasive from being imbedded in it.

The above steps can be done also on a horizontal wheel with abrasive powders of the successive grades imbedded in the paraffin. The paraffin may be poured as a melt onto a disk that has an edge, or a billiard cloth may be impregnated with the paraffin and stretched while warm over the disk. Grooves should be cut as a spiral in the disk and a solution of green soap used as a lubricant.

In any case, the sample is now ready for polishing, which is done under water on a revolving horizontal disk. The disk should be covered with a "kitten's ear" cloth and the abrasive suspended in water shaken or poured on. The abrasive should have been selected by "levigation." For that, a quantity of the abrasive, such as gamma alumina,* is stirred in a glass cylinder and drawn from the top after some time. The specimen

---

*The polishing powder is best obtained directly from a laboratory supply house. Alpha and gamma alumina (the latter being finer) are good. Air-levigated magnesium oxide can be excellent but its tendency to form hard coarse particles of magnesium carbonate on standing wet in air makes it a poor choice for the novice.

should be rotated contrary to the rotation of the disk. The speed of importance is the linear speed of the disk under the sample and not the rpm of the disk. Villela recommends a speed of 280 rpm when the sample is $4\frac{1}{2}$ in. from the center.

The routine sample in many laboratories is likely to have been polished entirely electrolytically and even etched merely by reducing the voltage. For this technique the reader should refer to the ASTM Standard E 3-62T.

Etching the Specimen

Sometimes the sample is ready for microscopical examination and photomicrography directly after polishing. Often this is the case when the nonmetallic inclusions are the items of interest. More often, however, a chemical **etching agent** is applied. This is because a skin of amorphous metal has been formed and smeared over the surface of the sample, thus hiding the crystal structure of the body of the metal. With poor sample preparation this layer of damaged crystals may extend comparatively deeply. Normally, however, the etching by the corrosive agent should be as light as possible. The best modern technique is to etch lightly, examine under the microscope, repolish very lightly, re-etch, and then re-examine until a constant appearance is obtained.

A multitude of etching agents for the different metals and for the different components of the sample are of special interest. For a survey of those available, their functions and formulas, the reader should consult ASTM E 3-62. An agent that is in common use for most metals is a solution of nitric acid. It is added to water, about 25 parts in 75 parts of water for copper, nickel, and aluminum and their alloys, and stronger (50:50) for lead and the precious metals. For iron, steels, zinc, tin, and their alloys nitric acid is dissolved in methyl alcohol (1–5%) and goes by the name of **nital**.* As high as 10 cm³ of nitric acid per 100 cm³ of alcohol is used for stainless steels.

---

*This should be made as needed and not stored in closed containers.

Chapter 5

Low-Power Photomicrography with Transmitted Light

When the magnification needed is low, a comparatively large area of the specimen is usually wanted in the final image. Moreover, the image has a familiar aspect; for example, when the object is seen directly or with a hand magnifying glass. These considerations influence the technique; the photographer is often willing to sacrifice some central definition in order to obtain greater depth of field or a field of sufficient area. As we saw in Chapter 1, the lens designer assumes that a field of appreciable size is wanted when he designs a camera lens; such lenses usually include an iris diaphragm to control the aperture, whereas objectives for a compound microscope are computed to give the best possible central definition at the required aperture or the highest practical aperture for the grade of the objective. In the magnification range in which either the simple or compound microscope can be used (20–50X), the choice should depend on whether field, in extent or depth, or central definition, within a more limited field, is the more important. There may be no detail within the specimen fine or important enough to be worth sacrificing depth or extent of field for it.

The employment of the proper lens for the magnification was discussed systematically in Chapter 1 and further for low-power work by reflected light (see Chapter 4); the principles are the same for transmitted light. Special lenses are made for photomacrography (defined in Chapter 1), the Micro Tessars, Microtars, Micro Summars, etc. They tend to be corrected for 10X magnification. Some modern lenses sold for photographic enlargers, such as the Kodak Enlarging Ektars, were found to be especially good for this purpose. At about 35X and above reversed camera lenses, especially cine lenses, are preferable to either of the preceding types and also can provide the required apertures.

Relationship Between f/α and NA

The reader has probably already noticed that on lenses for photo-macrography the series of available apertures is marked according to the f-number system; for example, $f/4.5$ and $f/5.6$ (in general, denoted as f/α), whereas the apertures of the objectives for the compound micro-scope, also engraved on the barrel, are given as NA 0.1, NA 0.2, etc. Yet if we are to select the aperture deliberately, according to our require-ments, the needed aperture value must be calculable. How are the systems intercompared?

We met both systems in Chapter 1 (p. 37). Let α of the f/α system be defined as follows:

$$\alpha = \frac{\text{focal length}}{\text{linear diameter of the aperture}}.$$

On the other hand, $NA = n \sin u$, where u is the angle subtended by the semiaperture (radius) from the object point on the axis. The object point is at a distance L from the lens, where $L = (m+1)/m$ focal lengths. Considering the difference between radius and diameter in the two definitions, we find

$$\alpha = \frac{m}{2(m+1) \cdot NA}. \tag{5-1}$$

$$NA = \frac{m}{2(m+1)\alpha}. \tag{5-2}$$

For large magnifications of the compound microscope,

$$\alpha = \frac{1}{2NA}. \tag{5-2a}$$

In the development of these formulas the difference between sine u and tan u is neglected or, more logically, the principal plane is assumed to be concentric with the object point [2. Kingslake, 1963 p. 109].

Determination of Requisite Aperture

The aperture requirements for adequate resolution in low-power work are not considered so frequently as they should be. As the magnification increases, the aperture requirements become stricter — faster than does the realization of them — and, depth-of-field considerations offer a special temptation in this range to keep the aperture as low as possible. The criteria for the determination of the required aperture were presented in Chapter 2, but there it was more convenient to consider aperture in terms of NA, whereas photomacrographic objectives are usually designated in *f*-numbers. However, we can use the formulas on that page in which a wavelength of $0\cdot5$ μ is assumed for the illumination (which is good enough to represent white light) and put the requirements for low-power work into a simple table. This also utilizes the fact that antipoints smaller than 75 μ *in the final image that is viewed in the hand* cannot improve definition and that antipoints larger than $250\,\mu$ (0.01 in.) are unacceptable. The definition required in the photograph, when the latter is to be viewed by an audience by projection on a screen, is discussed in detail in Chapter 21. In general, the subsequent magnification must be considered in angular terms; it may average 4X for a lantern slide.

$$\text{NA (required)} = \frac{\text{magnification}}{4D}$$

where D = diameter of acceptable antipoint.

$$\alpha \text{ (required)} = \frac{2D}{m+1}$$

$$\text{If } D = 75\,\mu, \text{NA} = M/300; \alpha = \frac{150}{m+1}$$

$$\text{If } D = 250\,\mu, \text{NA} = M/1000; \alpha = \frac{500}{m+1}$$

Table 5-1 Aperatures Required in Photomacrography at Various Magnifications

Magnification	D in μ	NA	f/α	Magnification	D in μ	NA	f/α
1.0X	75	0.003	75	15X	75	0.050	9.4
	250	0.001	250		250	0.015	31.2
2.0X	75	0.007	50	20X	75	0.067	7.1
	250	0.002	167		250	0.020	23.8
3.0X	75	0.010	37.5	25X	75	0.083	5.8
	250	0.003	125		250	0.025	19.2
4.0X	75	0.013	30	30X	75	0.100	4.8
	250	0.004	100		250	0.030	16.1
5.0X	75	0.017	25	35X	75	0.117	4.2
	250	0.005	83.3		250	0.035	13.9
7.5X	75	0.025	17.6	50X	75	0.167	2.9
	250	0.008	58.8		250	0.050	9.8
10.0X	75	0.033	13.6	75X	75	0.250	2.0
	250	0.010	45.5		250	0.075	6.6

Illumination

It is the methods of illumination and the optics involved that constitute the chief difference from photomacrography by reflected light, for now **the aperture requirements and the uniformity of the illuminated field must be achieved by deliberate arrangement**. As we shall see, the illumination optics become tied to the camera lens, that is, the microscope and the optimum characteristics of both change fastest near 1X magnification. This has led many workers to believe that excellent low-power photomicrography by transmitted light is much more difficult than the other types. If the range of magnification is considerable, more individual lenses are required, but, on the whole, they are inexpensive ones.

It is worthwhile to know the principles of illumination optics for low-power photomicrography. These principles are discussed in a series of *Cases*, the purpose being to segregate the useful ones. In this discussion it is important to know that **when a thin lens is placed in the plane of an object or image**, it becomes a field condenser and **will not affect the position of subsequent images of this plane nor the magnification**. With a real object, the field condenser will usually have to lie immediately behind it (since it cannot occupy the same space), hence the lens may slightly affect the two factors mentioned.

Case 1. *Diffusion in the Field Plane*. This method, the simplest case, may involve no substage optics at all. It consists merely **in mounting the specimen on a piece of opal glass**, illuminated uniformly from behind, and with **an extensive opaque mask**, such as black paper, **cutting out all but the field** immediately surrounding the specimen. The opal glass usually must be selected for uniformity if the field size is appreciably large. This case has already been mentioned. Figure 1-29*b* shows an example made by this method. It is definitely preferable to one that gives inadequate aperture, field size, or uniformity because it can provide all three. With it, however, there is poor control of flare and, therefore, with some subjects inadequate contrast. It is most valuable for large fields at magnifications between $\frac{1}{4}$ and 4X. Many photographic enlargers, however, are designed to utilize this illumination.

Case 2. *Objective — Single Condenser — Light Source*. This case may be called "the specular-enlarger or lantern-slide type of illumination." However, the illumination provided by Case 2 can be better than that of most specular-type enlargers if **Case 2 illumination is specifically set up, with its elements and arrangement, to suit the particular camera objective and magnification**.

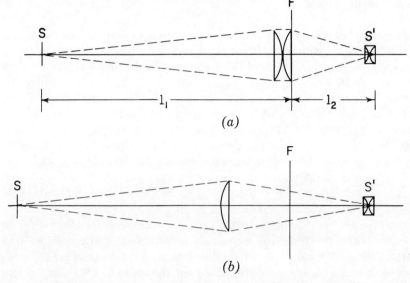

Figure 5-1   Optical system for low-power photomicrography, Case 2.

Only two lenses are required (although both may be composed of several elements), the camera lens which is the simple microscope and **a condenser which is both the object and the lamp condenser**. The lamp may be a monoplane-filament lamp or even a diffusing surface, such as opal glass or good matte film, illuminated from behind and *diaphragmed to a definite size*. It is advantageous, as discussed later, to have the area of the light source large enough for easy filling of aperture and field. This may require projection lamps of large filament area or a large secondary source. The latter, such as an illuminated and diaphragmed opal glass, is much used for this case. **The light source is moved back and forth to allow its image to be focused into the objective**. This is important. The image of the source should lie in the exit pupil of the lens. Its proper focus is ascertained most easily by viewing it from the image plane of the camera through a small hole and scanning with the eye through it across the plane. *The image of the light source should not move with the eye.*

The image of the source should fill that aperture of the objective that is required by the magnification, as discussed in the first part of this chapter. This value will normally be known before the apparatus is set up if the desired magnification is known.

The ideal case is represented by Figure 5-1*a*, which shows the condenser directly behind the field. This requires some selection of the focal length of the condenser and the size of the source if a given aperture is to be filled. Some leeway is available, however, the exact conditions are described in the equations in Appendix 3.

Moving the condenser back from the field plane F and readjusting the distance of the light source from the field is often recommended and practiced. It is also often acceptable, but it is now obvious from Figure 5-1*b* that the field plane F lies in the noncritical space between the uniformly illuminated lamp condenser and the image of the light source S', formed in the objective. If the light source consists of the multiple filaments of a projection lamp, their out-of-focus image will produce unevenness of illumination in the field in the resulting picture at relatively short distances and at all but low photographic contrast. (This can also be obscured by much pattern in the picture.) On the other hand, if the light source is itself a uniformly illuminated area, as it can be for an illuminated and diaphragmed piece of opal glass, then the tolerance is much greater.

As can be seen from the equations in Appendix 3A, the relations within **the system become quite simple when the focal length of the condenser is the same as that of the objective** and the arrangement is that of Figure 5-1. For this subcase the ideal or "correct" size of the light source is easily calculated. For instance, if a 100 mm objective is to be used for a magnification of 3X with $f/22$ directly filled by the lamp image, the light source should be about 14 mm in diameter (or square). This is close to the size of some monoplane-filament lamps. Moreover, **for this subcase the lamp should be m times as far from the field as is the objective**, where m is the magnification of the camera image. This rule (Appendix 3 A) at least provides a guide to the selection of the elements for a setup with this type of illumination. It also indicates the useful rule that best results, with good uniformity of illumination, are obtained with condensers of relatively short focal length. **They are even advantageously of shorter focal length than the objective.** The difficulty then is in obtaining one of sufficient diameter to cover the field. **The size of the illuminated field is that of the condenser;** the field is somewhat smaller if the condenser is some distance back from the field.

A simple plano-convex lens, with its plano side toward the objective and field, is often used. This will produce an image of the light source in the lens which consists of the direct image surrounded by a halo of light, caused by spherical and other lens aberrations. This direct image should at least fill the minimum aperture required for the magnification and is usually easy to obtain. However, **unless the objective aperture is left open adequately to admit the illumination halo also, the field illumination in the image plane of the camera will be quite uneven.**

A corrected lens, such as an achromat, produces a good filament image and the phenomenon just described is not encountered. Aside from the expense, it may be difficult to obtain achromatic lenses of sufficient diameter for this purpose. The most satisfactory method is probably to **set up a lens composed of two plano-convex elements with the plano sides outward**. Ideally, the two components should have focal lengths, as follows, when F is the focal length of the combination and m is the camera magnification: $1/F = 1/f_a + 1/f_b$, where F is also the focal length of the camera lens and $f_a = mf_b$. Since this last condition is not so important, it may be possible to pick up a compound condenser of approximately the correct focal length with both components already mounted together. Separate elements are available from various optical supply houses.

The equipment for low-power photomicrography by transmitted light furnished by Bausch & Lomb utilizes single plano-convex condensers for the various objectives.

In any case, **if the condenser is larger than the desired field for any specimen, a diaphragm should be placed in the field plane or just behind the condenser**. Although an iris diaphragm is convenient, an opaque sheet of any kind is just as efficient optically.

As discussed under Flash Photomicrography, in Chapter 13, the illumination system of Case 2 is especially satisfactory for use when flashbulbs are the primary source of illumination. They should be placed directly behind a diffusing surface, such as several pieces of matte acetate film, with a diaphragm of proper size for the optimum light source directly in front of the matte-film surface. If a continuously burning control lamp is substituted behind this diaphragmed diffuser, the rest of the system is unaltered by the use of the control lamp.

On the whole, **Case 2 illumination is the most satisfactory of all types for very low powers** (below 5X) and for transmitted light. Case 7, discussed later, is preferable for photomacrography above 5X. Both cases can be used, however, over a greater magnification range than this.

For greater usefulness when photomacrographs by transmitted light are to be made with a range of magnifications, the operator is better off if he has a number of projection bulbs with different filament grids of different sizes. There lamps are discussed in Chapter 8. Then a candelabra-size bayonet socket and a medium-size socket on an arrangement which allows them to be moved along the optic axis for focusing should be available. This arrangement may be merely a flat-based type of socket, screwed onto a square board, which is carried at the end of a rod extended along the optical axis by a support clamp. A simple cylinder bearing a hole serves as a lamp house.

If a means of focusing a substage lens of adequate diameter is available, a convenient arrangement divides the condenser and places the strongest plano-convex element just behind the field. A considerably weaker component just behind it is used to vary the focal length of the combination in order to focus the lamp filaments accurately in the objective. Thus, with some choice of lamp size and position, the filament image is easily brought to the correct size *and* focus in the objective.

Case 3. *Lamp with Condenser; No Object Condenser.* Case 3 is most instructive, although it cannot be recommended for practical photomacrography. However, it can be recommended that a student set up the cases described here, in each of which **a camera fitted with a lens** at a magnification of one or higher **is focussed on a stage in the field plane *F*.** The variations in the succeeding case have to do only with factors behind the field. In this simplest case there is **but one condenser** — that of the lamp — which conveniently can be a ribbon filament *s*. The lamp should be focused along the optical axis of the camera so that the **image of the filament is in the field plane.** The object on the axis can be considered as lying at the vertex of an X. The characteristics of this system are obvious and are illustrated in Figure 5-2.

1. *The uniformly illuminated field consists of the image of the ribbon filament* whose magnification by the lamp condenser determines the object-field size relayed to the camera-image size by the objective.

2. The aperture (angle) of illumination of the objective *from* the field is the same as that *to* the field and is determined by the angle subtended from the field by the edges of the lamp condenser. These relationships are illustrated in Figure 5-2. Observations of the objective aperture by inspection from the back of the camera will show it to be a round luminous disk in the total aperture of the objective if it is not filled.

If the lamp is quite close to the field, the size of the filament image will be relatively small but the subtended angle from the condenser (i.e., the aperture) will be relatively large, possibly larger than the objective can use. If the lamp and its condenser are withdrawn, the size of the illuminated field will become larger but the aperture subtended will

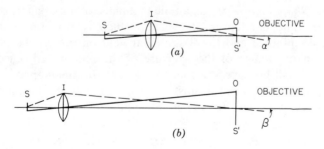

Figure 5-2   Case 3: (*a*) lamp close to object field; (*b*) lamp far from object field.

become smaller. Therefore there is a reciprocal relationship between these two important characteristics. In fact,

$$\frac{\text{size of uniformly illuminated field}}{\text{size of aperture illuminated}} = F/\alpha = \text{constant.}^* \qquad (5\text{-}3)$$

(When the lamp is withdrawn a considerable distance, a secondary limitation, discussed in the next section, occurs.)

The aperture size required depends on the magnification, as just discussed, and is shown in Table 5-1. The size of the illuminated field desired is limited only by the purpose of the photography and the nature of the sample. Both factors can be increased only by obtaining a lamp with a larger incandescent surface or a lamp condenser of larger aperture and refocusing at the appropriate distance. In practice, an $f/1$ lamp condenser is about the largest commercially available. The ribbon filament has a width of about 2 mm. Larger lamps can be obtained and are discussed in Chapter 8.

Independent Pinhole Effect.   In this case a secondary effect intervenes to limit the size of the illuminated field as seen on the ground glass of the camera and is independent of the size of the original illuminated field. As **the distance between the lamp and the field is increased** while the filament is kept in focus in the object plane, what appears on the ground glass to be **the image of a round vignetting diaphragm closes down to limit the apparent field size**, making it smaller **just when the increasing size of the filament image would have made it larger**. The cause is as follows:

Behind the objective the image cone of light becomes very small in cross section at a point that is the **exit pupil of the system**. The small disk of light formed by the cross section at this point is **the image of the lamp condenser formed by the objective**. This is easily tested by placing a paper arrow against the lamp condenser and observing its image on a piece of white paper held at the exit pupil. Moreover, its position and size can easily be calculated by use of the focal scale (see Chapter 1).

If a small diaphragm, such as a hole in a piece of dark paper, is placed against the lamp condenser, its image, which is now the exit pupil, behaves as a pinhole and the illuminated field on the ground glass becomes a pinhole image of the aperture of the illuminated objective in front of it. It will be observed that the instant the small aperture in the paper diaphragm is placed against the lamp condenser the illuminated field becomes (a) small, (b) absolutely uniform in brightness, and (c)

---

*This is, of course, a form of expression of the famous sine condition, h sine $u =$ constant, where u is the angle of the extreme rays to the axis and h is the half-diameter of the beam.

sharply edged. Its shape is, of course, that of the aperture of the *objective*, which may be polygonal from the iris and is reduced or enlarged as the iris is changed. The diameter of this illuminated field is easily calculated and is simply proportional to the distances x and y in Figure 5-3a, with the objective aperture and ground glass at the two ends.

When the exit pupil is the image of the whole $f/1$ lamp condenser, the relationship is only slightly more complicated, as shown in Figure 5-3b. With a disk instead of a pinhole, the overlapping images cause a vignetted zone to lie outside a central circle, c-d, on the ground glass from which the whole aperture is seen.

This vignetting effect **depends on the size and position of the exit pupil** and is therefore influenced by the following factors:

1. The *size* of the exit pupil depends on **the absolute diameter of the objective**. Therefore it changes with the focal length of the objective for the same relative aperture; a 32 mm Micro Tessar lens might reduce the field when a 72 mm $f/4.5$ would not.

2. The **size** of the exit pupil obviously depends on the **distance of the lamp condenser.** This is the reason why moving the lamp back suddenly causes the illuminated field on the ground glass to grow smaller instead of continuing to become larger. The lamp can be brought near enough to eliminate this effect.

3. The **position** of the exit pupil is also changed by **change in the lamp distance**; in this case the size effect is the more important.

4. The **position** of the exit pupil is changed, normally, by interposing a **lens in front of the objective. A positive lens brings the exit pupil nearer to the objective.** Of course, this is the usual effect of a combination of positive lenses but it is important here.

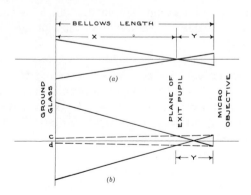

Figure 5-3   Pinhole effect in simple system. (Case 3).
(a) very small effective objective aperture. (b) larger effective objective aperture.

Case 4. *Single Object Condenser. Filament Focused in Object Plane.*
A. As noted in principle before, **a lens can be placed in the object plane**
with no change in (a) the size of the illuminated field or (b) the course of
the rays through the central point which determine the aperture of the
cone of illumination entering the objective, hence the fraction of the latter
that is filled. However, more of the total beam may be included in the
objective so that a better image of the ribbon filament may be formed on
the ground glass. Also, as already noted, the **exit pupil of the objective
will be drawn closer to the lens potentially to clear a larger cone**, the shift
being proportional to the power of the condenser. **Some power of a condenser will be just sufficient to clear the vignetting effect for any prescribed
angle of the beam.** The field will be as uniform as the light source itself,
for example, the ribbon filament, and is any desired size. **The inverse
relationship between field and aperture size remains the same as before.**
B. **If this second condenser lens is moved behind the object (field) plane**,
with the lamp distance unchanged and the filament constantly focused in
the field, the aperture of the illumination cone will increase but the size
of the filament image will decrease. If the lamp distance is readjusted
to make either one of them (aperture or illuminated-field size) the same as
before, the other will also become the same as it would be without the
second condenser. Therefore **the relationship of (5-3) still holds**, as it
should. An intermediate diaphragm or diffusion disk may seem to alter
the relation. The equation, however can be considered as setting the
limiting field and aperture without intermediate diffusion in the beam.

The **function of the object condenser** is twofold: (a) to remove the
secondary pinhole (vignetting) effect, that is, **to control the exit pupil of the
objective**, and (b) **to act as a field lens**, thus causing more of the outside
rays of the illumination beam to enter the objective aperture.

However, Case 4 requires an illuminant with an area of absolutely
uniform brightness. Usually Case 4 has no advantage until refocusing
converts it to Case 6.

Case 5. If the **filament is focused near or in the objective** instead of the
field, an image of the filament will be seen in the objective, looking from
behind it, rather than a round disk; it will be found that the fraction of
the aperture that is filled is much smaller for a given lamp distance.
Moreover, **the illuminated field will consist of an out-of-focus image of
the filament** that looks particularly bad if the condenser is comparatively
weak. As the power of the condenser increases, the aperture of the
objective illuminated decreases for a given lamp distance. There is danger
here that the field may be so bright that its nonuniformity will not be
evident. Indeed, it may be sufficiently uniform for a photomicrograph

made at low photographic contrast, whereas, if one is taken with a high contrast, the nonuniformity of the illumination may be revealed.

When the focal length of the object condenser is equal to or shorter than the working distance of the objective, the field becomes (a) absolutely uniform, and (b) **equal in size to the condenser lens**, but it is impossible to fill anything but a very small aperture with a lamp filament of practical size.

Although this method has been recommended and is used in commercial equipment, it should not be considered really satisfactory. A practical compromise is a series of object condensers for various focal lengths and apertures of the objective, which can be as strong as possible, yet fill the required aperture.

Although the field is no longer a sharp image of an element of the system, it will be observed by trial that our invariant still applies so that the angle of the aperture that is filled and the size of the illuminated field will still have a rigid reciprocal relation and the ratio will be closely the same as before.

Case 6. *Object Condenser Behind Field Plane, Imaging Lamp Condenser.* If a single object condenser is used and is located behind the field plane (toward the lamp), it will soon be found by trial that the only logical position for it is that which makes it **image the lamp condenser in the object plane.** This is done most easily by placing a paper pointer against the lamp condenser. Moreover, if the lamp is focused to and fro, it will be found that only when **the filament is focused in the rear focal plane of the second condenser** will the field be evenly illuminated.

With this arrangement, the height of the filament on the second condenser determines the aperture of the objective that is filled. This is larger, the farther the lamp, so now the effect of lamp distance is reversed from that of the preceding cases. As should be expected, the objective aperture/field size relation is the same as in those cases.

Since the object condenser must still subtend the objective aperture, a larger lens is needed than when it is in the field, the nearest possible distance to the objective. A new defect now becomes important.

Since the lamp condenser usually has absolutely no chromatic correction, the blue, green, and red images of the filament will be strung along the axis to be picked up by the object condenser. This will increase the discrepancy unless the condenser is achromatized. Moreover, **the field is now the image of the lamp condenser.** A corrected lens will reproduce its uniformity better. Therefore a **simple achromatic lens is now advantageous for this condenser.** There should be a diaphragm in the plane of the filament image and also at the lamp condenser.

This method is satisfactory; it is certainly preferable to Cases 3 to 5. However, Case 7, introduced by Köhler, is the most satisfactory.

Case 7. *Use of Auxiliary Condensers. Recommended Method.* Case 7 might be considered a combination of Cases 5 and 6 or a **splitting of the object condenser** into a **field lens** in the object plane and a rear lens which is now strictly a **collimator** (see Chapter 1). This excellent system allows any objective aperture to be filled and a perfectly uniformly illuminated field to be obtained with the greatest ease of adjustment. Flare can be most easily controlled by this method, although an extra lens is required. Its **greatest disadvantage** may be the **extent of the system** along an axis. However, **a mirror can be put in the collimated beam, to allow a horizontal stage and vertical camera axis** but a horizontal beam from the lamp, with no fundamental disadvantage. The photomicrograph of Figure 5-4 was made at a magnification of 6X by this system. Its superiority over the pictures of the same specimen made with simpler systems and shown in Figure 1-29*b* was obvious on the original, especially in a transparency. It was made on a Kodak "M" plate developed to a gamma of 1.2. A 72 mm Micro Tessar objective was used at *f*/11. Lens II (of

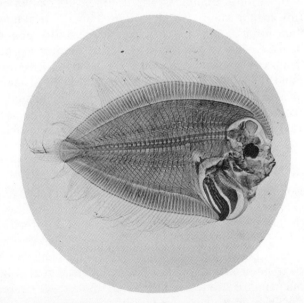

Figure 5-4   Larva of Flat Fish, X6.
Stain: Acid carmine. Optical system: Case 7. Objective: 72 mm Micro Tessar, *f*/4.5. Ribbon filament. No filter. Plate: Kodak "M" Plate. Developer: Kodak Developer D-41 (cf. Figure 1-29).

Figure 5-5) was a simple lens of 20 cm focal length. A ribbon-filament lamp was used in which Lens I had about 38 mm focal length.

The optical system recommended for low-power photomicrography with a simple microscope, especially above 4X, and with transmitted light is illustrated in Figure 5-5. Note that the filament *s* and the lamp condenser Lens I are imaged and reimaged independently along the optical axis. To control the illumination of the aperture a diaphragm, *D*, is in the plane on which the filament is first imaged. This plane is, of course, also imaged in the objective, which may have its own diaphragm. The first aperture **diaphragm should still be used; it is a good principle to have diaphragms as early in an illumination system as possible**. A single field diaphragm, however, is preferably placed against the lamp condenser on the opposite side from the lamp. Two diaphragms, each of them against one face of the lamp condenser, fore and aft, are most effective in removing flare light. Most commercial lamps either have no field diaphragm or have it some distance from the lens. An efficient field diaphragm can be made by merely pushing against the lens a round sheet of black paper or metal, which is held vertically at the end of a stick and which bears a hole of the desired size. The field diaphragm, which can be considered part of Lens I, determines its effective diameter.

The aperture diaphragm *D* is placed at the rear focal plane of Lens II. It is convenient to join this diaphragm and Lens II to a rod so that they can be moved as a unit. **Lens II is focused to place the image of Lens I in the field.** A simple plano-convex lens (III) in the field (or as close behind it as possible) will focus the filament from the parallel beam into its own focal plane. This is the objective aperture itself, since **the focal length of the field condenser is chosen equal to the working distance of the objective.** The photomicrographic objective IV should be observed from behind to ensure that the specified aperture is filled. If the opening of the diaphragm *D* is set at the specified aperture, as viewed from behind the objective, **the lamp is merely set at such a distance that the image of the source will fill the opening of *D*.** Then Lens II is

Figure 5-5 Optical system for lower power photomicrography, Case 7, triple condenser.

refocused to image Lens I in the field as determined by placing a piece of paper or a diaphragm against it. This may make a slight readjustment of the lamp distance necessary and again a slighter adjustment of Lens II.

Instead of this empirical method of setting up the system, it is both convenient and faster, especially if one is handy with a slide rule, to set it up from the equations of this system given in Appendix 3B. In any event, it is important to remember that $A_L \cdot s =$ constant, where $A_L(= 1/\alpha_L)$ is the aperture of the lamp condenser and s is the size of the light source. **If an illuminated field of insufficient size is obtained for a required aperture, either a lamp condenser of larger relative aperture or a larger light source is needed.** An $f/1$ lamp condenser is about the highest aperture obtainable in such lenses. Larger sources are usually obtainable. On the other hand, **an excessively large source is conducive to flare and lower image contrast.** As may be surmised, it is more important that the lamp condenser be a corrected lens for low-power work than for photomicrography at high magnifications. This lens should be at least partly corrected for spherical aberration, and lamps containing aspheric or compound condensers are sold by the microscope manufacturers.

A simple achromatic lens, which can be obtained from any optical supply house,* is an advantage, but not a necessity, for the collimator (Lens II) for the reasons discussed in Case 6. It is especially advantageous for maintaining illumination quality for color photography. Its aperture must be larger than that of the objective, also as discussed in Cases 5 and 6. For instance, a lens of approximately $f/3.5$ is needed for $f/4.5$ at the objective. Its focal length is in general not critical; a 6- or 8-in. (15 or 20 cm) lens is usually convenient. A longer focal length makes a spread too extensive along the axis. The field condenser (Lens III) need be only a simple plano-convex spectacle lens. Such a lens, cut to a specified diameter to allow it to fit a standard holder, should be obtainable at little expense from any optician. There should be a small supply at hand if considerable low-power photomicrography is to be done at various magnifications and with several objectives, since the focal length of Lens III should approximate the distance $f(1 + 1/m)$, where f is the focal length of the objective and m is the image magnification. Slightly greater

---

*Two concerns that operate a mail-order business and frequently list surplus lenses, mounted and unmounted, are Edmund Scientific Co., 101 E. Gloucester Pike, Barrington, New Jersey, and A. Jaegers, 691 Merrick Rd., Lynbrook, New York. An excellent reference, especially for all parts of the United States, is *The Optical Industry Directory,* Optical Publishing Co., Lenox, Mass. Simple thin lenses can most easily be obtained from one of the many concerns that finish them for spectacles; sometimes they are not prepared to make them of small diameter.

distance is needed because of the displacement of the field condenser behind the field and also the thickness of the objective.

An optical bench for such a system is convenient and will save time, but it is not essential and its absence should not deter anyone from making photomicrographs at low power by transmitted light. Usually the objective is mounted directly on the camera. The objective and a cross on the ground glass of the camera back should each be so centered that together they establish an optical axis which is parallel to the camera support; for example, an optical bench. On the other hand, if considerable work of this type is to be done, certain innovations become worthwhile. The author has mounted both the objective and the stage directly, and independently, on a horizontal optical bed. For this purpose a simple focusing rack for a compound microscope was purchased and the body tube replaced with one considerably larger that could carry the flanges for camera and enlarger lenses. Moreover, it is worthwhile to obtain a **pinhole objective**. One can be prepared in any machine shop from a sheet of brass; it allows a pinhole of at least $\frac{1}{32}$-in. diameter to be screwed in place of the objective lens to be used. The front surface should be flat to show the image of the light source that will eventually be focused on it.

On the other hand, a vertical camera with a horizontal stage can be used instead. Some means of centering the field condenser III must be provided, probably by moving the whole stage bearing it. The rest of the illumination system can then be horizontal. If small magnets are cemented to the holder of Lens III, it can be held against a steel plate under the stage and centered easily by sliding.

The light source may be one of two types: (A) with the lamp and its lens (condenser), each **supported independently** on a horizontal optical bench, or (B) with the lamp and its condenser **built together** in a singly supported housing. The first case is decidedly preferable, but the second is much more common, especially with commercial equipment.

For Case A setting up and centering the lamp and subsequent optical system are relatively straightforward. The problem and procedure are almost the same as those for the horizontal bench that bears a compound microscope. The lamp is set up alone, preferably with the current turned low if it is an incandescent lamp. If the system is to occupy one long horizontal bench, with the camera at the other end, the lamp must be brought to the same height and center alignment as the cross on the ground glass of the camera back. This is most easily done with an auxiliary pointer (see Figure 6-5). The lamp filament and camera-back cross then determine an optical axis parallel to the bench. The lamp condenser, with its plane carefully made perpendicular to the axis, is centered into position by seeing that the image of the lamp filament is imaged centrally on the camera back when properly focused.

When the camera axis is vertical and Case A is used with a mirror below the stage, the best procedure is to slide the lamp back and forth on the horizontal component track of the illumination system and, by adjustment of both the mirror and the lamp, ensure that it is lined up in the projection of the axis established by the central cross of the camera back and the objective aperture. A pinhole objective (Chapter 3) is very helpful to use in place of an objective lens or the large hole left by its absence.

If Case B is used with a commercial lamp unit, the first step must be to ensure that the lamp filament is on the axis of the condenser. One or other is usually adjustable for centration. Three methods for doing this are described in Chapter 6 (p. 245), since the best alternative among them will depend on the specific lamp at hand. When the lamp is well centered as a unit, it is set up at one end of the bench or at the appropriate distance on a table top and pointed to center its image on a mirror or directly on the cross of the camera back. If the mirror is used, it is then tilted to center the beam.

The other elements of the final system are added and are themselves centered when the illuminated bright spot is again centered on the cross of the camera back. A convenient order of insertion is, successively, the objective, or preferably, a pinhole objective, the diaphragm, D, closed down, the field condenser, and finally the collimator Lens II. Each one, of course, is centered before the next element is inserted, except that the correct positions should be established approximately before final centering. This is another advantage of using the equations in Appendix 3B. The centration of the field condenser is very critical, since it is at the end of the short arm of the optical lever, with the objective as fulcrum. Focusing the pinhole objective to and fro makes the small disk of light on the ground glass a convenient size. Care must be taken that the stage (probably bearing the field condenser) and the collimator are not tilted with respect to the optical axis. If the pinhole objective is used, the objective is screwed in last. If the objective has a focal length of 32 mm or longer, it is not very important if its center is not *precisely* that of the pinhole. The precision is needed in the alignment of the whole system. Appreciable decentration of the objective is undesirable, of course.

Low Power with the Compound Microscope

Photomicrography at magnifications appreciably below 100X can be done with a compound microscope, usually for one of two reasons: (a) because such work is only occasional and the more appropriate equipment is not easily available, that is, for expedience; and (b) because the required field is small and the highest central definition is desirable.

Use of the compound microscope for photomicrography within this magnification range will normally be done with a 2X (48 mm), a 4X (32 mm), or a 10X (16 mm) achromatic objective or a special objective within this focal range. Use of an apochromatic objective would restrict the size of the acceptable field still further, under conditions for which the improvement in definition is not usually worth it entirely aside from its financial cost. However, the remarks in Chapter 2 p. 51 should be remembered here. The achromatic objectives mentioned, especially those of lower power, give superior images when used as a simple microscope but with somewhat smaller fields (as measured in the object plane).

The special simple microscope lenses for photomicrography, such as the Micro Tessars, and Micro Summars, can also be used in the objective carrier of the compound microscope, since those small enough for the purpose (48 mm and shorter focal lengths) have the standard R.M.S. thread. With them, the limiting condition may be the size of the body tube of the microscope. With a microscope which has interchangeable body tubes for a monocular or a binocular, a tube of adequate diameter can be improvised.

The problem may be one of utilizing a suitable substage condenser. The NA and also the focal length of the object condenser should be comensurate with that of the objective. With the better substage condensers, the condenser for high-power objectives can be converted to one of suitably lower NA and longer focal length by merely unscrewing one or more top elements. When the microscope is used with a vertical axis, it is sometimes even possible to remove the top element from an Abbe condenser, which is not meant to be convertible, and let the bottom element remain in place by gravity.

An alternative procedure substitutes a lens of suitable focal length for the usual condenser, holding it on the condenser mount by any expedient, such as an adhesive wax* if the condenser mount is separately focusable. If it is not, and an Abbe condenser is merely held in place against the stage by a ring, it is usually preferable to abandon it and turn to the optics of Case 7, described in this chapter. This is preferable, anyway, for magnifications below 50X. In this case a simple spectacle lens is held in

---

Note. An adhesive wax is sold as Tackiwax by the Central Scientific Co., Chicago, Illinois 60613. A wax often used for lenses on optical benches can be made by mixing beeswax and turpentine to the desired consistency (approximately 1 : 1) and adding finely powdered Venetian Red.

the ring against the stage to serve as Lens III; Lens II is supported independently and on the other side of the microscope mirror.

Control of Exposure

The image in the plane of the film and ground glass of the camera may be very bright in low-power photomicrography by transmitted illumination, thus requiring a very short exposure time and allowing the use of slower photographic films, all of which are highly desirable. The disadvantages are that the **shutters** must be accurate for small fractions of a second and **should be accurately located in a field or an aperture plane** or a conjugate plane to one of them. This cannot always be done. If the shutter used is the common type with leaves that close in from the periphery and the opening *and* closing time is an appreciable fraction of the total exposue time, the greater open time of the central portion will cause a corresponding greater exposure of the center of the field **unless the shutter is accurately in an aperture plane**; that is, **there will be vignetting**. Sometimes it is satisfactory to mount the photomacrographic lens directly against the face of the shutter. Whether there is vignetting for the time of exposure involved, with consequent apparent nonuniformity of the illumination in the field, can be tested by making two camera exposures. Two films should be flashed to produce a rather light, uniform density over the film, after removal of any specimen from the field. One exposure should be made with a rather long exposure time, after reducing the illumination brightness by any mechanism that does not also introduce nonuniformity, such as a neutral density filter or possibly a diffuser in the substage aperture or at the lamp. The second film is then exposed at the high brightness and short exposure time.

The problem of providing satisfactory shutters for short exposure times, or truly neutral filters to extend exposure times for shutters available for photomicrography, is discussed in Chapters 3 and 9, respectively. In Case 7 the shutter should be placed against diaphragm *D*. Protection for the photographic film is then required, which may be the dark slide of a film holder or a shutter mounted near the camera lens.

Chapter 6

Use of the Compound Microscope in Photomicrography

Before the author, and finally the reader, undertakes such a chapter as this, it should be re-emphasized that an even higher priority should be given to reading the instruction booklet written by the manufacturer for the microscope and the assembled photomicrographic outfit, if such is to be used. Not only does the manufacturer know his own mechanism, and its optics well, but it is very much to his interest that it be used to produce the best possible results. Besides an instruction booklet for the specific instrument, most manufacturers distribute or sell a more general booklet on "Use and Care of the Microscope" with which the microscopist should be familiar. This instruction book should be kept handy, yet safe, for reference in interrupted periods of use or for new personnel. If none is available because a microscope or photomicrographic outfit was inherited from a predecessor, new instructions for the same or a similar model can usually be obtained from the manufacturer.

TYPES OF APPARATUS

Although the apparatus for photomicrography has already been covered in Chapter 3, it is necessary to classify it here according to the discussion of its use. There necessarily is some overlap or repetition in such a classification but it should not be confusing.

239

IA. Large "universal" photomicrographic stands.
IB. Microscopes with built-in light sources.
II. Bench with vertical axis of microscope and camera.
  A. With view camera for final picture.
  B. With small camera (e.g., 35 mm) requiring later enlargement.
III. Horizontal photomicrographic bench (see p. 251)
IV. Specialty types; for example, the metallographic bench, which constitutes a large fraction of the photomicroscopes in use (see Chapter 4).

IA. Large "Universal" Photomicrographic Stands

As discussed in Chapter 3, the large "universal" photomicrographic stands that are sold by some companies are already set up for use; most of the decisions of optical arrangement have already been made, from light source to image. The nature of the optics in these stands may not even be disclosed or may change with the model. The operator is completely dependent on the manufacturer's instructions and this part of the chapter will not be of direct use. On the other hand, such outfits must employ the same principles of optics. Study of the directions for the more general case should reveal the decisions that have been made for the operator. The other chapters, such as those on selecting the contrast of the image and photographing it, should prove directly useful.

IB. Microscopes with Built-In Light Sources

Most of the microscopes now sold provide for incorporating the light source into the stand as a unit. Most frequently there is provision for its removal and the use of an outside lamp, which is fortunate, since microscope stands made primarily for visual use lack lamps of sufficient intensity for convenient and sometimes even for practical photomicrography at the higher magnifications. On the other hand, some incorporated lamp units definitely are sufficiently bright for most photomicrography. With such outfits the manufacturer's instructions can best be used for adjusting the lamp. The instructions in this book should be useful for subsequent procedures. Some of the criteria for good illumination discussed later may be useful.

In a number of cases of the author's acquaintance an external lamp with mirror has been substituted for the built-in lamp in order to be able to use an external shutter to time the actual exposure. This shutter (e.g., an Ilex No. 3) is mounted in the path of the illuminating beam from the lamp. A shutter on the camera or in the photomicroscope is used to protect the film but is opened earlier and closed later than the true exposure time. This is because it is so difficult to avoid vibration with such an impulse-device mechanically connected with the microscope. It is recommended that a test with an external shutter be made as an alternative to improve definition if a shutter in the camera or microscope body is in use.

II. Setting Up a Photomicrographic Stand with Vertical Microscope Axis

A stand for photomicrography, which utilizes the microscope in its usual upright position, may be set up (a) in a fundamental way or (b) in an expedient manner. The former is preferable if considerable photomicrography of good quality is desired with the least time involved between periods of use. On the other hand, the situation does arise when a particular specimen is at hand and an acceptable photomicrograph is needed soon. This case is best discussed from the perspective of the more fundamental method.

Normally the best procedure is to **establish the axis of the microscope first and then align the illumination axis with it. The camera is then placed above the microscope** in such a way that the microscope axis coincides with the camera axis and passes through the center of the field (cross on the camera ground glass, if there is one).

There are two optical arrangements that can be used for photomicrography with an upright microscope:

A. With the illumination system on a horizontal optical bed and the illumination beam horizontal and parallel to it. The substage mirror or prism is then at an accurate 45° to bend the axis to the vertical. See Figure 6-1.
B. With both the lamp and the microscope supported on the same flat surface. (For this skip the discussion of Case A.)

Case A is not used so frequently as formerly, especially if the microscope is also to be used visually, but is an excellent arrangement when considerable photomicrography is to be done. For this arrangement the microscope must be raised above the level of the optical bed to correspond to the height of the light source above the bed. Here it is advantageous to have the light source and lamp condenser independently supported. In this case the procedure given in Chapter 4 for adjustment of the mirror of the vertical illuminator can be followed, except that the slide with a cross on it will be the ordinary transparent centering slide and the mirror, is now substage.

Case B is the more common one and includes the situation in which the same microscope is frequently used both visually and photographically.

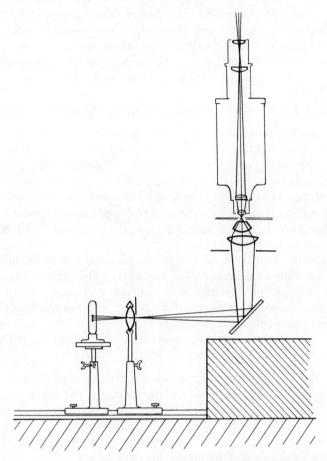

Figure 6-1   Case A: illumination system on optical bed with vertical microscope.

The following assumes that the illumination used for photography is *not* built in. The other condition has already been discussed. This important case deserves more complete coverage.

Mounting.   In this arrangement the microscope, its lamp, and its accessories are supported on the same flat surface (e.g., a table). It is very advantageous to mount the microscope and its illumination system stably and permanently on a separate board. An arrangement in use in the author's laboratory is shown in Figure 6-2. With this board the microscope can be removed and replaced without altering its fundamental centration. Moreover, several microscopes, all equipped with horseshoe bases, it is true, but of different shape and size, have been used on the

Figure 6-2   Microscope system on board with independent support for the substage mirror.

board with only minor adjustment of the location stops and clamps. It is unfortunately true, in the writer's opinion, that microscopes with the horseshoe base have almost disappeared from the market. However, even a round base with a mirror and a lamp can be clamped to a board and located by stops, if desirable, for temporary removal. This board can well be supported by the small rubber feet used on chairs, two at the microscope end of the board and one in the middle of the lamp end, to form a tripod.

The wide variety of cameras that can be employed may or may not be supported independently of the board. At any rate, this assembly can be moved about without disturbing it optically. Thus several outfits could be moved under a large vertical camera alternately as fields were found desirable to have as photomicrographs.

In Figure 6-2 the vertical support for a small camera is shown on the board, as discussed in Chapter 11. A double Polaroid light-modulating filter is also shown in the beam. It can be flipped out of the way when not wanted.

Substage Mirror and Support.   One of the advantages of the separate board support for the microscope is the ease with which the substage mirror can be supported below and independent of the microscope. This is true for the open type of base such as the horseshoe in which the mirror fork usually has a central post support that can be stuck vertically into a hole below the microscope. Since it is *very* desirable to get the mirror very low below the microscope, it is desirable to have a hole or slot cut in the plate supporting the microscope. With the board shown in Figure 6-2 and again in Figure 6-3, a long slot was cut in the board itself, with a slight inverted V of a "dovetail" to take an independent piece to fit it and with a depression cut in it to carry the mirror. The post of the mirror fork actually fitted into a hole in a ¼-in. brass plate screwed to this tongue. This allowed longitudinal centration of the mirror. Any lateral centration was easily accomplished by sliding the whole microscope sidewise before screwing down the location stops.

This independent support of the mirror has three advantages:

1. It allows optimum alignment of the mirror.
2. Substage condensers can be inserted and withdrawn without disturbing the mirror.
3. The mirror itself is less likely to be disturbed in use.

Considering the second point first, it will be found that it is a great convenience in centering the illumination, especially if alternative condensers are used—brightfield, darkfield, phase, etc.

With respect to the first point, it can be stated that unfortunately substage mirrors are frequently *not* in good optical alignment as mounted on the microscope (see Chapter 3).

Figure 6-3   Board for microscope system with mirror support.

Centration of the Lamp. Since the axis of the illumination beam is tilted, it is usually considered most convenient to use a lamp in which the source itself and its condenser lens are mounted as a unit. This does not preclude holding a tilted lens separately. In the latter case a light source can be used that does not have to be tilted also. When the lamp and its condenser are a unit, **the source must first be centered to the condenser or the converse.** One of the following three methods can be used:

1. If the back of the lamp house can be lifted or temporarily removed, the lamp can be optically centered to the lamp condenser by utilizing the reflections from the lens surfaces. When available, this method is normally the best because it is optically unequivocal and is independent of how well the relatively cheap condenser lenses may be centered in their mounts. In fact, if the condenser lens is made up of a series of spectacle lenses, a compromise may have to be made to align the images. Looking through a heavy density into the lamp from the rear, past the filament, the latter is first centered in a horizontal and then a vertical plane. This is, of course, because we cannot look along the axis through the filament.

2. If the back cannot be opened, another method must be devised. If the lamp is very badly off axis, or a new one is being inserted, the simplest and fastest way is to look directly into it from the front with one eye; thus the lamp, with the condenser appears round and not elliptical. Obviously the lamp must be dimmed, either by a variable transformer or rheostat or with a high visual density. Some almost mutually exclusive color filters are sometimes advantageous but care must be taken that high-strength ultraviolet (with mercury lamps) or infrared (heat) radiation is not allowed to pass through to the observer's eye. Then, with the lamp quite well centered, it can be still more accurately aligned by looking at it, still with one eye, first from one side so that the condenser tube *just* occludes the filament from direct view. The images formed by the lens surfaces can be seen and aligned. The process is then repeated by looking down the vertical plane and across one edge at the condenser lens or lenses. *If* the centration of the lamp and its condenser is easily effected by centering screws and *if* it is fairly well aligned already, its final accurate alignment can be accomplished after looking through the microscope.

3. The above methods are sometimes impractical or inconvenient. The following may then be preferable and usually is for the horizontal bench. To use it, it must be possible to focus the image of the source with the condenser lens from a near to a far distance.

a. Throw a horizontal beam from the lamp onto a white screen a short distance away. This may be a pad of standard-size white paper, but its reflectance is needed later to see the unilluminated portion of the lamp condenser.

b. Interpose a relatively long focal length lens (e.g., a hand reading glass). Obviously its focal length must be less than one-fourth of the distance from the screen to the lamp.

c. Focus the lamp filament on the auxiliary condenser.

d. With a paper pointer against the lamp condenser, focus its image onto the screen. The lamp, auxiliary condenser, and screen should be aligned as well as can be done quickly. This is the double-condenser method of Chapter 4 and a round, uniformly illuminated disk of light should appear on the screen.

e. Now rack in the lamp condenser toward the lamp to show the source. At this point it may be advantageous to move the auxiliary condenser in somewhat, and possibly the screen as well, depending on the range of focus of the lamp condenser. At any rate, it should be possible to see the image of the source and also the rim of the condenser lens on the screen *and to center the two to each other.* The whole provisional system, set up for this centering operation, must be in reasonable alignment; but if it is not obviously misaligned, there will be no appreciable noncentration of the images.

Now **the internally well-centered lamp** on its clamped base in front of the microscope **is directed at the substage mirror so that its beam is centered about the point at which the axis of the microscope intercepts the mirror surface.** This is further discussed below. **The lamp should be focused** so that an image of the source will lie in the rear focal plane of the substage condenser, which is normally at the substage iris diaphragm. The exceptions to this statement, which assumes that Köhler illumination is desired, are discussed in the appropriate places.

Establishment of the Optical Axis. The next step is the establishment of the axis of the microscope. For this **all lenses are removed from the microscope.** Almost invariably the top point of the axis is best established by **dropping a pinhole ocular** (see Chapter 3) **into the top of the draw tube.** Rotation of this mechanical ocular will test its centration.

The procedure for the establishment of a second lower point to locate the axis will depend on the microscope involved. Consider three cases.

1. If the substage iris diaphragm can be returned into position unequivocably, possibly with a click stop, after the condenser lens is removed, it is definitely the best agent. Looking down through the top pinhole through the diaphragm, closed or opened appropriately, easily establishes a vertical axis. Sometimes, when the substage condenser is centerably held by a ring but its diaphragm is not separate and so cannot

be used as a fixed axial point, it may be worth the effort to prepare a sub-stitute mechanical centerable condenser in the machine shop. This is merely a ring of equal diameter to that of the condenser with one end closed to an opening just large enough to view the whole lamp condenser.

2. The objectives for the microscope are precentered in place.

3. The objective sockets are centerable.

In all three cases the next step is to **put a centering slide (Figure 6-4) on the stage.** If one is not available, a cross can be marked on a micro-scope slide; it is important to have its lines straight and perpendicular.

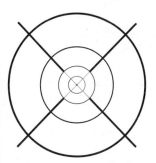

Figure 6-4 Microscope center-ing slide.

The target on the slide is **then centered into the axis** to act as a temporary fiducial mark to establish it visually. In doing this it is wise to **place a diffusing sheet**, such as matte film or even thin paper, **directly under the microscope** to obliterate direc-tional illumination. In case 1 above, this step is obvious and simple. In case 2 **a low-power objective is first used with an ocular** to center the test slide. After that the pinhole ocular is returned and one space on the objective carrier is left empty to look down without lenses. In case 3 a 48 mm objective is screwed into the lower part of the draw tube and, together with an ocular, is used to position the center-ing slide. **All lenses are then removed from the microscope.**

Now that the microscope axis is well established, **look down it through the pinhole ocular at the substage mirror.** Lean a small sheet of paper against its surface. With a pencil point it is relatively simple to **determine where the axis of the microscope intercepts the mirror surface.** If the mirror is centerable, it can be centered at this point or a mark can be made on the white paper. In any case, **the lamp is now directed to center its beam about this axial point on the mirror surface. The mirror is then tilted** while looking down the axis through the pinhole **to center its beam** around the target of the center slide and possibly with the substage diaphragm. It is at this stage that the centration of the incandescent lamp to its lamp condenser can be perfected, *if it is easily done with centering screws.* Slight refocusings of the lamp condenser to show the filaments somewhat out of focus, by making the color fringes symmetrical, is the most sensitive method. Unless easily done at this time, however, it is best to start with a precentered lamp. It may help to tilt the lamp at this point *very slightly,* since the appearance is very sensitive to centration.

This step should be finished with the well-centered lamp having its beam aligned with the microscope axis, that is, by interception by the mirror at the locus of the vertical axis and by proper tilt of the mirror, and with the light source focused on the substage iris diaphragm or where the diaphragm is expected to be.

Insertion, Focusing, and Alignment of Microscope Lenses.   At this point it is probably wisest to **put into position the lowest objective and eyepiece available,** usually a 4 or a 10X objective. In doing so, with no condenser in place, **hold a piece of matte film** or ground glass just below the stage **and focus on the centering slide. If the objective holder is centerable, the objective can be centered on the slide** at this point. Now the **substage condenser is put into place and focused.** If this can be done without touching the mirror, it will be quite worthwhile; normally it can if the mirror is independently supported below the microscope. It can be often done anyway with a little care by first racking the condenser ring to the top and then down onto the condenser. Assuming that this can be done, the condensor is brought up into focus as the field is viewed with a low-power objective. This will occur when a uniformly bright round disk, the illuminated lamp condenser, appears. **A cardboard arrow placed against the surface of the lamp condenser should be sharply in focus.** If the substage condenser is centerable, it should be adjusted so that the bright disk will be centered to the cross on the slide. This is Köhler illumination.

If the condenser cannot be inserted without tilting the mirror, the latter should be tilted flat **on one set of pinions only.** With the condenser in place, it can be tilted back in order to center the beam. There will be no question if the condenser is not centerable but factory centered. If the condenser is centerable and not necessarily correctly centered, it is necessary to rely on the preceding steps in which the precentered lamp was properly aligned. Successive approximations of mirror tilt and condenser centration will establish the disk of light, which represents the illuminated field, centered to the slide and uniformly illuminated or centrally illuminated when the lamp is slightly defocused. In this case, however, it is especially advantageous to have a mirror low enough to be untouched during insertion of the substage condenser.

When the substage iris diaphragm is part of the substage condenser and we can assume that it is well centered to it, the following technique is more convenient if it was necessary to tilt the mirror to insert the condenser:

1. **Place a piece of matte film just below the substage condenser; close the diaphragm.**

2. Using a low-power objective (32 mm, 4X, or 16 mm, 10X) and low-power ocular, **focus on the small opening of the iris diaphragm by lowering the substage condenser** or raising the objective. It may pay to manipulate the opening of the diaphragm at this stage. **Center the substage condenser.**

Notice that it is best to **set up the microscope with an objective of lower power than that to be used** for work, so that the size of the lamp condenser will be small enough in the field to be seen as a whole. This is not necessary for focusing the substage condenser, since an arrow or the edge of a piece of paper against the face of the lamp condenser can be used for focusing the latter in the field. Sometimes the lamp is equipped with an iris diaphragm, that is, a field diaphragm, which can be closed to center its image into the field around the cross of the centering slide. However, the author has found many such diaphragms not very well centered on the axis of the lamp and prefers to center the lamp condenser directly. Then the centration of the important field diaphragm can be checked. Incidentally, the ideal field diaphragm would be very close to the lamp condenser.

At this stage the whole illuminated field should be obvious as a centered round bright disk, assuming that a higher power objective is to be used. **The higher power objective is now put into place and the test pattern brought into focus.** Either the cross of the test slide will appear well centered in the field, thus showing the two objectives to be well centered, or it will not. **If the cross is decentered, the centration of the higher power objective should be corrected** if the latter is held on a centerable mount. If not, **the cross is centered to the field by moving the slide.** Then the low-power objective is replaced and the illuminated disk of the field, now decentered a bit, can be seen. If the condenser is centerable, it is re-centered to bring the illuminated field to centration with the present position of the slide (i.e., with the high-power objective as the standard).

If, as is so often the case, neither condenser nor objective is centerable, the fundamental misalignment of the two must be accepted—but the **illuminated field is centered to the cross for the objective to be used by touching up the substage mirror.**

With the objective to be used in place and the field in focus, **the pinhole ocular is placed in the draw tube** and the **rear lens of the objective is viewed.** This viewing of the objective aperture should become a habit; the appearance of both the field (seen through the eyepiece) and the aperture (seen through the pinhole ocular) should be known and satisfactory for uniformity and centration of illumination and absence of flare. The diaphragm of the substage condenser should be seen through the pinhole ocular as a reasonable black peripheral limit which closes down symmetrically. Rarely is there occasion to have it opened to more than 0.9 or even

0.8 of the diameter of the objective. The optical reasons for this were discussed on p. 77.

Focusing the Objective. There is often little space indeed between the bottom of a high-power objective and the top of the cover slip, or the specimen, when an objective for use without cover slip is employed. In spite of this, there should be little excuse for hitting the slide or specimen with the objective. If there is the slightest doubt that a proper field is in the optical axis, **the slide should first be viewed with a lower power objective** at adequate working distance. When a high dry objective (e.g., a 40X or 4 mm) is to be focused, the eye should be looking from a point almost in the plane of the stage as the objective is lowered, preferably with a white background on the opposite side. The reflection of the objective and the objective itself can then be seen to approach one another and they can be brought almost into contact without touching. Then the proper focus will be found by looking through the microscope as the draw tube is raised by the fine-focus adjustment. One trick is to place the edge of a light card temporarily against the edge of the lamp condenser or the plane of the field diaphragm while examining it with the low-power objective so that its edge is in the center of the field and (by touching up the focus of the substage condenser) in very excellent focus. This edge is then used to focus the high-power objective when there is little detail of high contrast in the specimen.

The following technique is both safe and time saving if there is some detail of good contrast in the field; it may even be dirt. **The medium or high dry objective is brought close to the slide** as before. **The aperture is viewed through the pinhole ocular,** preferably with the iris reduced to at least $\frac{2}{3}$ aperture. Now the objective is being used as a simple lens; look for specks that represent object points. They may be noticed **first by moving the objective up (or the stage down)** with the focusing adjustment. Any specks that represent field should apparently move outward toward the periphery of the objective. If they move to the center, reverse the vertical movement of the objective, whether up or down. **As the objective moves toward the proper focus, all detail moves outward,** vanishes, and then, with continued movement, returns from the periphery and goes to the center. The correct focus has been passed. If the ocular is substituted for the pinhole when the pattern is outside and the aperture clear, the image will be seen at once and its focus will need only slight touching up.

When an immersion objective is to be used, a small drop of immersion oil should be placed on the slide, in the optical axis, and another on the front of the objective. Now the objective is *slowly* lowered toward the slide. The two drops will apparently reach for each other, leaping to final contact in a thin thread. In this manner an air bubble will never be caught below the objective. High-power immersion objectives will have to be brought still nearer the slide after oil contact is made to focus the image. The technique described above is especially useful here. It works well with the cross-test slide. It may be worthwhile with certain specimens to include high-contrast marker material deliberately. Sometimes lines can be drawn on the slide, before or after the specimen is mounted on it. In work in far ultraviolet bits of gold leaf are sometimes added in the field.

Further discussion of the adjustment of the microscope for high-quality imagery is resumed on p. 258 under the subtitle *Aperture Examination and Adjustment.* With the microscope set up, we can turn attention to the camera. This is discussed under that subtitle on p. 272 and in Chapter 11.

III. Setting Up a Horizontal Photomicrographic Bench

It is best to **level the basic bench** first, both longitudinally and laterally. This is only a convenience for later procedure but a considerable one.

The two ways of the bench can be considered to form two parallel lines and a plane (which has just been leveled). *The optical axis, then, is an invisible line that will lie above this plane and parallel to it.*

The following accessories will save much time and make the procedure easier.

1. Two centerable pointers, each on a bed clamp. The two pointers should face each other and point from below to clear the line of sight in establishing the axis (see Figure 6-5). These pointers can be made from a rod and a daub of optical wax or modeling clay.

2. A piece of matte (ground) glass to be held on the bench in a vertical plane and bearing a graduated cross which can be made with a pencil. The author found this device so generally useful that some were printed on Kodak Translucent Plate Type 3. Kodalith Ortho Matte Film Type 3 (Estar Base) is also suitable when film can be used.

3. A pinhole ocular (see Chapter 3).

4. A pinhole substage condenser, unless the substage iris diaphragm can be separately used without its condenser lens. This is usually just a tube of a size to fit the ring for the condenser and with a closed bottom bearing an accurately centered hole about $\frac{1}{32}$-in. diameter. It can be made with quite limited shop facilities or quite cheaply by an independent machine shop. The bottom can be a plate cemented on the tube with an epoxy cement.

5. A centering slide. The pattern on the slide is shown in Figure 6-4.

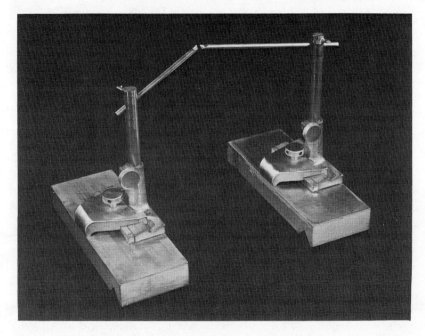

Figure 6-5   Pointers for centration of optical bench.

Establishment of the Optical Axis   1. Every component on the optical bench may be centerable. In this case **first establish the vertical plane in which the optical axis will lie.** Any point in such a plane establishes an axis parallel to the bench. For this purpose use the matte glass plate, set perpendicular to the bench, and one or two pointers. If the optical bed is such that a bed clamp can be unequivocably set and also reversed on the bed, by far the best method is to use it to establish the midplane. **With a pointer touching the scale of the matte glass at its center (by first guess), reverse the pointer with its clamp on the bench to touch the glass on its other side.** The true central vertical plane passes halfway between the two pointer positions. This method can be used at any time to re-establish this plane. The height of the axis above the bench may be a matter of convenience.

If the clamps cannot be reversed on the bench, the central plane will be more arbitrarily established and a good record of the decision should be kept.

2. If one component on the bench is already fixed in position (it may be the cross on a camera back), it will establish the axis. The pointers are needed to help establish it along the bench.

The optical axis should now be established at its two extremes on the bench. One end point is the light source, the other is often a cross on a camera back. However, if by previous work the microscope is already on the bench and aligned with it, it will be simpler to establish the general optical axis by projection of the axis of the microscope, with no lenses in it, and then to align the light source into this established axis.

Setting Up the Light Source. The nature of the light source establishes two subcases for further alignment.

1. The light source and its condenser are each separately supported from the bench on centerable supports. This is preferable with the horizontal bench. **A screen**, either reflecting or translucent, **should be set up near one extreme of the bench** (it may be the camera back) and **should bear a cross** representing the optical-axis point of intersection. The light source as a point, or its center as a point at the other extreme of the bench, established the axis. Therefore **center the light source by placing a centered pointer as close to the source as possible**. Note the shadow of this pointer and the alignment of the shadows of the two pointers on the bench with the cross on the screen at the other end. The lateral centration of the light source is quickly established. The height of the source can be judged quite well from the side if a pointer can be brought very close to it; otherwise it is best to hold a pinhole in an opaque screen near the center of the bench and center it with a pointer before using the projection of its aperture on the rear screen as a fiducial mark. The pinhole aperture becomes the fulcrum of the lever of the light beam.

With the end points established, the rest of the procedure is merely a matter of introducing the other elements, one by one, and ensuring by careful positioning that the light beam will again become centered on the cross at the other end, usually the camera back. However, elements may be out of alignment (a) by a rotation or tilt and (b) by displacement. Although (a) and (b) can apparently compensate each other, this is not a good arrangement, since movement along the bench would destroy it. The two effects can be separated by first centering an element being introduced by *displacement* with the help of a pointer or cross on the matte glass close to the element and then by rotating the element, using projection of the beam to a distance on the cross at the back of the bench. A yardstick held against the face of some components will allow them to be rotated very closely perpendicular to the bench.

First add the lamp condenser. Focus the image of the source on the rear screen and center it on the cross or a pointer. Discussion of the selection and final focusing of the lamp condenser is given on p. 256. For this case (1), now skip to the section on *Introduction of the Microscope*.

2. The lamp contains the light source and lamp condenser as a unit. In this case the lamp must first be internally aligned by the procedure already discussed under Case II.

Mount the self-aligned lamp at one end of the bench and focus an image of the source centrally on the cross on the rear screen (e.g., camera back). When the camera is used, it is an advantage, in this case, to be able to open up the front of the camera.

Now the lamp, itself, is centered on the bench by ensuring that its beam is centered near the lamp as well as at the screen at the other end of the bench. There are some lamps in which the act of focusing will disalign the condenser. In this case it is best to focus the source on the rear screen provisionally, then touch up the centration after refocusing the lamp near the middle, where ultimately it will be focused.

Introduction of the Microscope. Usually the microscope is put on a "sole plate," which is a flat surface resting on the bench, with the microscope tilted back to make its axis parallel to that of the bench. **Level the sole plate** to make it parallel with the bench.

Establishing the Optical Axis of the Microscope. The first step is taken, with no lenses in the microscope, by establishing two pinhole apertures on axis at its two ends. However, for aligning the entire microscope on the bench it is important that these two small holes be easily removable, either by pulling out the element or by opening up a diaphragm. **Insert a pinhole ocular** to establish the axial point at the extreme upper end.

The best method for establishing the axial point at the other end of the microscope depends on its design. There are four cases.

1. A substage iris diaphragm is mounted, independent of the condenser, directly on the microscope. This is the simplest case. **Close down the iris diaphragm to a small hole**. With the pinhole ocular, this establishes the axis.

2. The iris diaphragm may be mounted on the substage condenser and the whole unit fitted into a ring under the stage. There are two subcases here.

a. The condenser lenses can be removed, leaving the iris diaphragm still mountable in the ring. This case is essentially the same as Case 1.

b. A separate iris, independent of the lenses, is unavailable. **Insert the pinhole condenser**. There is almost always a tubular sleeve into which it will fit. It is pulled out of the ring to open the aperture.

3. The whole substage is centerable but the objectives are precentered. Here, as in Case 2, the iris diaphragm can be used without the condenser

lenses or a special pinhole condenser should be obtained. **Using an objective of very long working distance with a low-power ocular, focus on the small substage aperture and center it** in the field. An ocular with cross hair is desirable for this step. It is also wise to put a diffusing sheet (matte glass, film, or paper) below the substage.

A small hole in a plate could be put on the stage and centered into the axis in the same way for use as the lower centering guide. This is not so desirable as having the aperture near the lower end of the microscope, but is a practical alternative that is always available.

4. Both objectives and substage are centerable. **Screw a long working distance objective**, for example, the 2X (48 mm), **into the lower part of the draw tube** which is not centerable. Use it to **focus directly on the small substage aperture and center it**, as in Case 3. If it is impossible to rack the objective far enough to focus on the substage, accomplish the same thing in two steps. First focus on the centering slide on the stage, center it, and then use the slide to center an objective in its normal position, which in turn is focused on the substage to center it.

An alternative procedure is to center an objective directly into its normal mount, after having centered in the centering slide, with no optics but the pinhole ocular and a diffusing screen below the substage. Here the round mechanical aperture of the body tube is used as a guide.

Alignment of the Microscope on Bench. The alignment procedure is simple in principle *if* the axial beam of light, from lamp to rear cross on the screen at the back of the bench, has already been established. **The spot of light transmitted through the small apertures at the two ends of the finally aligned microscope should be concentric with the cross on the rear screen with both holes in place and also with either one removed.** Some precautions should be taken, however, to prevent the alignment procedure at this stage from becoming unduly tedious. An assistant to observe the centration on the screen accurately does speed the process; the spots of light on the rear screen, with the three different aperture combinations, have quite different diameters.

It is a great help if the centering movements of the microscope in space, with its axis, can be made separately according to the individual components, that is, translation, up or down and sidewise, and rotation in a vertical or a horizontal plane. Although the microscope can be slid about, freehand, on the leveled sole plate, being finally clamped firmly in position after alignment, this procedure can be quite tedious. It is preferable to have the base fit into the corner of an L-shaped piece on the sole plate and to be able both to rotate this L-stop and slide it sidewise when screws are loosened. Vertical movements of the microscope, both of translation and rotation, are normally effected with the help of the leveling screws of the sole plate (which should have lock nuts). With a little planning, only minor vertical adjustments will have to be made.

By far the best technique is to center the light spot on the screen by using only one pinhole at one end of the microscope at a time; when both, used together, do not alter the centration, the job is finished. The closer the axis of rotation to the substage pinhole the better. This aperture is used as the criterion for the translation of the microscope that may be needed for centration. Since the center of rotation is not really at the pinhole, it is best to make the movement in the right direction but only to the extent of bringing the spot about halfway to center before checking and centering with the pinhole at the other end of the microscope.

With only the substage pinhole in place, check the light spot on the rear screen for centration. Slide the microscope in a horizontal plane on the sole plate to bring the spot to center. Remove the substage pinhole and insert the pinhole ocular. Rotate the microscope horizontally on the sole plate to bring the spot to center. Repeat alternately. Then confine the adjustment to the vertical plane with the supports of the sole plate and again to the horizontal plane by sliding the microscope on the sole plate. For these later adjustments, however, bring the spot only halfway to center each time. From time to time insert both pinholes as a check but use only one during the adjustments. With everything under control this method of successive approximations succeeds most efficiently without consuming excessive time. The microscope is then clamped into position.

Selection and Focusing of the Lamp Condenser. A more thorough discussion of the function and the effects of the quality of the lamp condenser must be deferred to the chapter on illumination. The conclusion that may be stated at once is that the quality of the lamp condenser is important, especially for color photomicrography, and that with the old type of high-aperture, spherical, "bull's-eye" condenser it is difficult to obtain a field in the microscope and camera that is uniformly illuminated in both brightness and color quality. A great deal of quite satisfactory work has been done with aspheric (normally pressed) condenser lenses and condensers of multiple components of simple lenses. Here the gain is the removal of the single highly curved surface explained in Chapter 1.

An advantage of the horizontal bench with a separately supported condenser is the ease with which it can be changed and its focal length chosen for the specific case. This choice is often a matter of convenience and availability in the aperture needed. With Köhler illumination (see Chapter 7), the diameter of the lamp condenser actually used is limited by the size of the field in the microscope and camera. This is generally very small at high magnifications and a lens of low aperture is all that is needed as a lamp condenser.

Sometimes, on a horizontal bench it is desirable to use a condenser of quite long focal length, such as the 60 mm $f/1$ aspheric condenser of Bausch & Lomb, with a stable, incandescent light source. Then other light sources, such as a mercury arc, can be set up from time to time in the space between the original lamp and the microscope by using a condenser of short focal length without disturbing the primary lamp. Such a condenser is the 34 mm sold by Bausch & Lomb.

Assuming that Köhler illumination is to be used, **focus the image of the light source by movement of the lamp condenser onto the diaphragm of the substage condenser.**

Addition and Adjustment of the Microscope Optics. Essentially the process of alignment is continued by the same principles for using the established optical axis and is also the same as that for the vertical microscope at this stage, except that the advantages of the horizontal bench can be utilized: the absence of a complicating substage mirror and the ease of projecting the centering pattern on the camera back.

A centering slide should be placed on the stage, if it is not already there, and its cross brought to center in the axis of the microscope and bench. Insert a low-power objective, for example, the 4X (32 mm), and a low-power eyepiece. If the objective is precentered on the microscope, the order of this action may be reversed. **Center the slide and the objective on each other** by moving the slide if the objective is not centerable.

Insert the substage condenser into place. The mechanism for doing this, inserting into a ring or sliding on a dovetail, will vary with the microscope. Place a pointed marker* against the front surface of the lamp condenser. **Focus the substage condenser so that the edge of the marker against the lens can be seen in the field simultaneously with the specimen** — in this case, the cross of the centering slide. The combined image can also be projected on the camera back. The lamp condenser should be clean; if there is a slight pattern imaged in the field, it is preferable to focus the substage condenser slightly in too far rather than out too far. **The image of the lamp condenser will be seen as a bright round disk of light.** If not of uniform brightness, the focus of the lamp condenser may need touching up a bit at this point. Even the centration of the lamp condenser may be touched up if color bands in the disk are not symmetrical when the lamp condenser is not quite in proper focus (a very sensitive test). For use on a professional horizontal bench either the substage

*Such a marker can be quickly cut from paper or an index card.

condenser or the objectives should be centerable. **Center the substage condenser to bring this disk central with the cross.** If the objectives rather than the substage condenser are centerable, the two should already be centered but might need very slight touching up by movement of both the slide and the objective to correct a slight inherent lack of centration of the condenser.

Now substitute the higher power objectives that are to be used. A note on focusing those of very short working distance is given on p. 250. If the objectives can be centered, **each should be centered to make the image of the cross centered in the field.** This is the center of the illuminated field by previous adjustment.

With each objective, **substitute the pinhole ocular for the eyepiece and look at the back of the objective.** The partly closed iris diaphragm should appear sharply in view and centrally located. An image of the light source will also be in view, but this is discussed later. However, the technique habit of looking at *both* the field (through the eyepiece) and the aperture (through the pinhole ocular) is important at this point of setting up the microscope. The quality of both field and aperture can be judged and kept under control.

When an oil immersion objective is to be used, we learned in Chapter 2 that oil should also be added between the top of the substage condenser and the bottom of the slide. The phase objectives are an exception, since the directly transmitted light has an NA of less than 1.0. Oil between condenser and slide is advantageous even with high dry objectives if highest image contrast is important and if all other sources of glare have been eliminated.

APERTURE EXAMINATION AND ADJUSTMENT

In Chapter 2 the importance of adequate aperture was discussed, as was the loss in image contrast with increasing illumination aperture. This discussion should be kept in mind at this stage. As the substage diaphragm is opened, the source of light should apparently increase to fill the objective aperture. If it does, the diaphragm should be closed down, usually to 0.8–0.9 diameter, according to the principles of Chapter 2. If the direct illumination consists of a small disk in the center of the objective or is markedly nonuniform over the area of the aperture, the reasons should be investigated. As shown in Chapter 7, nonuniformity may be caused by the imaging of coils of a lamp in the aperture. Lamps with several parallel coils may be a problem. The image of the coil or coils must extend over a reasonable diameter of the aperture and be symmetrically located unless the unsymmetrical position is deliberate in order to produce oblique illumination effects. Nonuniformity of illumination in the aperture, with a central spot, rings, etc., may be due to several causes.

1. The lamp condenser could be improperly focused.
2. The quality of the lamp condenser could be inadequate.
3. The substage condenser may be improperly focused.
4. The focal length of the substage condenser may be inappropriate and its aperture inadequate.
5. The quality of the substage condenser may be inadequate.
6. The optical system may have become decentered or never have been correctly centered.

The last named condition may be due to the unintentional movement of the substage mirror with a vertical microscope. With a horizontal microscope it is simpler to remove the substage condenser and even other optics, if necessary, to test alignment, if fundamental decentration is suspected. With a vertical microscope, it is probably best at this stage to check the alignment with a piece of paper along the axis before repeating the centration procedure. Possibly only a retilting of the mirror is involved to gain both a central field and aperture.

The remaining points are discussed in order.

1. A quick check will tell **if the light source is focused in the rear focal plane of the condenser.** This can be done with a piece of white paper. With some sources, such as a vapor arc of mercury, the exact focus is a little difficult to ascertain. The focusing of the lamp condenser should be done with the color filter to be used in place, since lamp condensers are rarely achromatic.

With a complete substage condenser of high numerical aperture, it can usually be assumed that the rear focal plane is the plane of the iris diaphragm. This may not be true of some achromatic condensers with their diaphragms in the principal plane of the lens. A somewhat different method of illumination is involved when the light source is projected into the condenser, almost at the field (see Appendix 3C).

The iris diaphragm is no longer even approximately at the rear focal plane when the top is unscrewed from a substage condenser, as is often done to increase its focal length and decrease its aperture. In this case its rear focal plane is appreciably in front of the iris diaphragm (toward the lamp). The usual excuse for still focusing on the iris is that with the condenser of weaker power this error is less important. It is true that it is preferable to focus the lamp too far toward the microscope rather than too far toward the lamp. The correct focus can be obtained as follows: With a provisional focus of the lamp, focus the substage condenser with a pointer against the lamp condenser and an ocular in the microscope to bring the image of the lamp condenser in the field. Then **with a pinhole** eyepiece and looking at the back lens of the objective, **use the edge of a piece of paper below the substage condenser to determine where this edge is in best focus in the objective aperture.** If the lamp can be focused smoothly without decentering it (not always possible with all lamps), refocus the lamp to obtain a sharp image of the source in the aperture or to fill the latter uniformly with light.

The image of the source should, of course, be just large enough at the rear focal plane to fill the hole (i.e., the aperture) of the substage iris. Any greater size of the image of the source will unnecessarily reduce the size of the illuminated field (see Appendix 3C). The size of this lamp image on the substage is determined for a given lamp by its distance from the microscope.

2. If a simple spherical lamp condenser is used, such as a single lens of large aperture with a substage condenser of reasonably good quality, the complete lack of any correction of the lamp condenser may give a focused image of such poor quality that it will preclude uniform filling of the aperture. **High-quality photomicrographs usually require some correction of the lamp condenser**, if only the splitting up into a series of weaker components to reduce the lens curvatures. Some useful ones are aspherical.

A poor lamp condenser in particular calls for a field diaphragm. This is best almost against the lamp condenser, which is rarely the case in a commercial lamp (see Chapter 3). On a horizontal bench with a separately supported condenser it is possible to place a diaphragm against the condenser (see Chapter 8).

When a secondary light source is used, for example, the matte surface discussed in Chapter 8, the diaphragm is efficiently at the source.

3. The focusing of the substage condenser in order to place an image of the lamp condenser in the field has already been described. It is an important step that should not be omitted. A field diaphragm in front of the lamp is often focused on, but in the author's opinion this is not quite so good if the lamp diaphragm is far in front.

One of the reasons for incorrect focusing of the substage condenser occurs when it hits the slide or substage stop before a correct focus can be obtained. This means that the microscope slide is too thick for the substage condenser. The simplest procedure is to use thinner slides. If the lamp is far away, bringing it closer will somewhat increase the thickness of the slide that can be employed. If the microscopist has a slide that is too thick but must be used, a negative lens at the rear of the sub-stage condenser will also increase the allowable slide thickness. This is not the best solution if a well-corrected condenser is being used, unless this lens is otherwise prescribed.

Selection and Adjustment of the Substage Condenser

The remaining points listed under Aperture Examination merit dis-cussion. Some of this material has already been given in Chapter 3.

If the objective aperture is still inadequately filled with direct light, as observed through the pinhold ocular, then either (a) the light source image is too small in the aperture of the substage condenser or (b) the aperture of the condenser itself is too small for the objective. The first situation is remedied by moving the lamp farther away or using a lamp condenser of shorter focal length. The second requires another condenser of higher aperture, unless a top lens can be added to the one in use to decrease its focal length and increase its aperture.

Whether the size of the illuminated field is as large as desired is judged by looking at it through an ocular. The desired size may be that required to cover a negative size or merely the diaphragmed size set by the ocular-objective combination to be used. (A lower power is needed to judge the whole field.) An illuminated disk considerably larger than needed can only add to the flare and loss of contrast. If the illuminated field size is too small, the focal length of the substage condenser is too short. In this case its available aperture will probably be larger than necessary.

The focal length of the substage condenser is ideally selected so that the image of the lamp condenser in the field will have the correct magnifica-tion as the desired diameter. At the same time the aperture of the con-denser must be adequate to fill the desired NA of the objective. This might not be the whole NA as discussed in Chapter 2, but usually the ability to fill it is required. This involves having the image of the light source large enough to cover the opening of the substage iris diaphragm when the latter is open to the desired extent, as observed with the pinhole ocular. Since the NA of the objective and that of the condenser are usually known (or can be obtained from the catalog), this selection of the con-denser can normally be made *a priori*. If the substage iris diaphragm must be closed down to a small hole *just* to fill the objective aperture, a substage condenser of longer focal length would seem to be preferable, unless the field size were adequate and protection from glare required.

The shift to a condenser lens of longer focal length and less aperture can be done in several ways. Many condensers are constructed so that one or more elements at the top can be unscrewed and removed. The new focal lengths and reduced apertures are given in the catalogs of these condensers or on the accompanying direction sheets. For the very best aplanat-achromat condensers, which are necessarily quite expensive, this procedure is not considered very good practice, especially if the separated condition constitutes a large fraction of the use of the condenser. In this case it is preferable to obtain a separate condenser of longer focal length. Unfortunately the market does not support so great a variety of substage condensers for brightfield use as might be desirable. For instance, there is no substage condenser that is well suited to the 8 mm, NA 0.65 apochromatic objective.

Another similar objective can be used as a condenser, but a spectacle lens should then be added at its rear, which will face the lamp, in order to preserve the conditions for its optical corrections. A centerable holder utilizing an objective as a substage condenser is sold by most of the microscope companies, since the double-objective technique is used quite widely to obtain excellent optical conditions, especially with white light. The correcting lens is included in the Zeiss accessories. It should have a negative focal length that will make the rays from the lamp enter the objective at the same angle assumed in the design of the objective for the rays leaving it to form the microscope image. This will be a negative focal length equal to the originally assumed image distance minus the distance of the lamp condenser. This image distance is not the "optical tube length," since it is measured from the shoulder of the objective. For instance, if it is 140 mm and the lamp condenser is 300 mm away, the correcting lens should have a focal length of -160 mm.

Another way of lengthening the focal length of the condenser, in effect, is to place a spectacle lens of negative focal length at the bottom of the condenser. Some microscopes come equipped with spectacle lenses that can be swung to one side. This will, of course, degrade the corrections of a fine substage condenser but it may not be too impractical for the common Abbe condenser. It certainly is useful as an expedient for using a lower-power objective to locate a suitable field before taking the photomicrograph with an objective of higher power.

A third method is merely to separate the components of a condenser. This method, which is used for a commercial condenser of variable focal length, has its chief merit in convenience. Again, it should be perfectly satisfactory for finding fields at lower powers. The microscopist must make a personal judgement as to whether it is adequate for photomicrography with medium- and low-power objectives. The field and apertures will not be uniformly illuminated except as glare light makes them so.

This variable-focus condenser is a simple Abbe type which is used correctly when it is racked together for high-power work.

How well the system can be set up to fill both field and aperture adequately, uniformly, and in a controlled manner really is limited by the **quality of the substage condenser.** This point was discussed in Chapter 3. We saw that the very simple and common two-lens Abbe condenser has such large aberrations that it cannot be expected to furnish good images in both field and aperture. Much of the illumination uniformity apparently obtained is due to light bouncing around to fill both field and aperture, that is, the glare. If the specimen has strong contrast and the finest detail is not important, this may not prove to be so bad as it sounds. For black-and-white work with color filters, an aplanatic condenser is very practical and an enormous improvement. For critical work with white light, especially for color photomicrography, an achromatic condenser that is also aplanatic will be found to be an important link for an excellent system.

Removal of Glare

All light that is not under control and that does not travel from a point in the specimen to its antipoint in the camera image may degrade this image; it does not participate in its formation. This constitutes glare. For its control diaphragms should be present, and accurately so, in both field and aperture planes. Obviously the closer to the front of the system the diaphragm is placed, the better, since it will be reimaged in subsequent conjugate planes. On the horizontal bench with a separately supported light source it is easy to place a field diaphragm next to the lamp condenser which can be sharply imaged in the field with great efficiency. The aperture diaphragm is the substage iris diaphragm, of course, with the compound microscope. Both diaphragms should be closed down while observing with the eyepiece and pinhole ocular.

In addition, the photomicrographer should examine the inside of his microscope, first through the pinhole, and then with the latter withdrawn, for any reflections within the tube, except for the upper part normally covered by the ocular. Such reflections should be removed by covering the surfaces with a matte black. Black Velour paper* is excellent for this purpose if care is taken to remove particles that may fall from it. The pinhole can be used to scan the image plane at the back of the camera for reflecting surfaces in the camera or in its connections to the microscope.

---

*This paper, well known as "coffin paper," can be obtained from Louis DeJonge & Co., Woolworth Building, New York 7, New York.

One of the greatest sources of glare occurs when two parallel surfaces of glass face each other, especially with only air in between. This can easily be demonstrated by first observing a relatively low-contrast specimen or pattern on the open surface of a microscope slide, then covering it with a cover slip fastened at the edges but with only air between it and the slide with its pattern. A pair of such surfaces in common use occurs between the top of the substage condenser and the bottom of the slide with "high dry objectives," that is, 4 mm or 40X objectives. Again, with low-contrast specimens a marked improvement is obtained with a complete immersion system, including an oiled condenser and an oil-immersion 40X objective.

CHOICE OF OBJECTIVES

The principles involved in the objectives and oculars used in a compound microscope were discussed in Chapter 2 and the objectives and oculars that are actually available in Chapter 3. Only a few practical considerations need be treated here.

With a compound microscope a group of objectives and several oculars are available; the stock of each may, of course, be large. Probably most frequently there are three objectives precentered by the manufacturer on a revolving-nosepiece carrier. In this case much care should be taken not to mix the objectives on their mounts; often there is identification of each. Such a battery may be quite sufficient for the purposes likely to be met. On the other hand, if much microscopy and photomicrography is to be done, other objectives and oculars may be advantageous or even required. If certain extremely critical work is sometimes required, it is definitely advantageous to have a selected superior objective that is not used in daily routine. The several manufacturers sometimes have some objectives or oculars that are specialized or especially good. They will undoubtedly require centering into the optical axis. In the author's laboratory it has been found advantageous to fit most of the microscopes of several makes with the Leitz quadruple, centerable, rotating-nosepiece objective holder. Special care must be taken if the tube-length scale is to remain correct.

It is assumed that the first objective put into position for use on the microscope will be one that is of appreciably lower power than that expected to be used for the ultimate examination or photography, unless the specimen represents a familiar routine. Even then the highest power should rarely be used at once. The two objectives normally considered lowest in the power series, that is, the 2X (48 mm) and the 4X (32 mm), are now infrequently employed for visual microscopical examination because the two-objective stereo microscope is available in most laboratories for such work. However, both objectives are still important in photomicrography, especially with 35 mm film, or in cinemicrography. With sheet film and a larger camera, a simple microscope would usually be more appropriate (see Chapter 1).

The best general combination of objectives, for a total of three has become quite standard equipment, is (a) a 10X (16 mm), (b) a 40X (4 mm), and (c) a 90X (2.0 mm) (or a 100X, 1.8 mm). Their quality, principally in terms of whether they are achromat, fluorite, or apochromat, will depend on the available money but normally also on the need. Each is also available in several apertures. At 0.85 NA and above, a dry objective (see *tolerances*, Chapter 2) should have a correction collar.

The first question in selecting an objective for photomicrography is: Is white-light illumination (or a comparatively wide-band color filter) necessary? It is if several colors of a specimen are to be reproduced (but see Chapter 10). If not, the author recommends the use of monochromatic illumination from a mercury arc, as discussed in Chapter 9. This extracts the best performance from any medium- or high-power microscope objective and the improvement may be startling. This is the illumination used for most photomicrography of unstained specimens, including industrial subjects on black-and-white film in the author's laboratory. A better performance is obtained even with an apochromat when monochromatic light is used. However, the performance of an apochromat over an achromat with such illumination is often not sufficiently better to overcome the larger field offered by the achromat with monochromatic light. The Zeiss planachromats plus monochromatic light definitely offer an appreciably larger field.

However, with white-light, including color, photomicrography or with wide-band filters, only stringent economic reasons should deter one from choosing apochromats.

We can say that the user of a microscope objective can avoid the consequences of chromatic faults in his objective by choosing monochromatic light, but even this should occur at or near the wavelength for which the spherical aberration of the objective has been calculated — in the green for achromats and fluorites and in both green and blue for apochromats, as discussed in Chapter 1.

The one optical defect of microscope objectives that the user can correct is spherical aberration. This is done by introducing this aberration with the opposite sign by adjustment of the tube length or a correction collar. The subject is discussed in Chapter 2 and the technique is described near the end of this chapter. The most common causes of this defect in a good objective are (a) use with a cover slip of incorrect thickness, outside the tolerance range, (b) use of an incorrect immersion liquid, (c) use of a wavelength for which it is not corrected, and (d) use of incorrect tube length, which can be substantially corrected by adjustment of the tube length. Moreover, if critical work is to be done, the optimum tube length of the objective should be determined for a cover slip 0.18 mm thick, the correct immersion fluid (including air), and the wavelength (see p. 62). In the author's experience this may not prove to be 160 mm within reasonable tolerance.

The proper immersion liquid for immersion objectives is a moot question. To the author's present knowledge three microscope manufacturers are designing objectives around three different oils. The original immersion oil, and one still recommended, is cedarwood which has been oxidized up to the standard refractive index. Its oxidation, however does not stop there permanently, and cedarwood oil that has been standing in air can vary considerably. Two synthetic oils, Crown and Cargille's (Shillaber's), were devised to imitate cedarwood oil rather than some glass. Both are colorless and much more stable than cedarwood. Moreover, they are more of a lubricant, never becoming the adhesive that causes old cedarwood oil to freeze the mechanical stage if allowed to remain in contact with it. Choice of paraffin oil is poor policy, although the author understands that it is sometimes used on the grounds of economy.

CHOICE OF OCULARS

The different types of ocular available were discussed in Chapter 3, If a microscopist had every kind of ocular on his shelf, he would have a wide choice indeed, differing in type: Huyghenian, Hyperplane, Periplan, compensating, negative, and wide field, each also having a series differing in magnifying power. In this chapter the basis of their practical choice is considered.

One consideration might be the size of the field simultaneously in focus, especially when recorded as a picture. This is limited, as we know, by the extreme curvature of the in-focus surface of the primary image, which is concave toward the objective.

Usually a given image magnification can be obtained by several combinations of objectives and eyepieces. Field sizes can be compared, even at somewhat different magnifications, by expressing the diameter of the *useful* object field. Normally the largest field is obtained with the lowest power objective and highest power ocular for a given final magnification. The gain, of course, is achieved primarily by the lower objective aperture that is used; therefore a limitation on resolution must be accepted. The field size, however, will normally not be equaled when an objective of higher NA is used with the substage diaphragm closed down to a comparable NA, although this artifice should produce some gain in field size. As discussed in Chapter 3, some gain in field size for coarser detail is obtained by use of negative amplifiers of the Homal, Ampliplan type. A real gain is obtained by use of the newer flat field optics, also discussed in Chapter 3.

The important consideration in the choice of oculars, however, should normally be based on their design suitability for the degree of lateral color aberration (variation of magnification with wavelength) of a particular objective. As already discussed, this defect is greatest with apochromatic objectives, and compensating oculars should be used with all apochromatic objectives. They are needed for achromatic objectives of highest power as well, as demonstrated in Figure 3-16. The following table was presented by L. V. Foster (4. 1934) to show the optimum choice of ocular with various objectives to reduce lateral color to a minimum when using white light. Although originally devised for Bausch & Lomb optics, it is generally representative and applicable except for the new plano objectives in which lateral color has been deliberately introduced into the lower power objectives to make it a constant through the power series and therefore to take the same compensation. We should watch for this happening with other manufacturers. The negative Homals

Table 6-1 Eyepiece types for Objectives

Focal Length (mm)	Objective Type	Eyepiece Type	Correction (percent)
48·0	Achromatic	Huyghenian	+0.25
40.0	Achromatic	Huyghenian	+0.25
32.0	Achromatic	Huyghenian	+0.25
16.0	Achromatic	Huyghenian	None
8.0	Achromatic	Hyperplane	+0.15
4.0	Achromatic	Hyperplane	−0.15
1.9*	Achromatic	Compensating	−0.3
4.0	Fluorite	Hyperplane	−0.1
1.8*	Fluorite	Compensating	−0.2
16.0	Apochromatic	Compensating	+0.1
8.0	Apochromatic	Compensating	−0.1
4.0	Apochromatic	Compensating	None
2.0*	Apochromatic	Compensating	−0.5

*Oil-immersion objectives

and Ampliplans are overcorrected to the same extent as compensating oculars.

As discussed in Chapter 3, when conventional objectives and oculars are being used, it is not so important to be careful that the objective and ocular are of the same make *if the mechanical tube length is correct and if the ocular of the correct color compensation is selected.* Some years ago, new compensating oculars, made with a new type of glass that gave appreciably improved image quality over that of the previous type, were introduced. On the whole, the statement of interchangeability of eyepieces among the microscope makers is not applicable when their new and special plano objectives are used. Even then, the corresponding ocular of the competing manufacturer may produce acceptable results.

The images obtained on projection into the camera by oculars of the general Huyghenian type show pincushion distortion. On the other hand, the images formed by the negative-amplifier type show even more marked barrel distortion. These considerations may become important when measurements on photomicrographs are made.

CHOICE OF COVER SLIPS, IMMERSION OIL, TUBE LENGTH

The reason that these three factors are grouped together has already been adequately discussed in Chapter 2. At this stage we are considering the practical application of this knowledge. Now it might be wise to refer back to Figure 2-3 or 2-4 to note the tolerance to cover-slip thickness applicable to the case at hand. Possibly, as with objectives of 0.30 NA or less (10X or less power), the thickness or even presence of a cover slip is not critical at all. For critical work at higher apertures use of cover slips of only No. $1\frac{1}{2}$ class thickness seems to be the best policy. The immersion oil is taken care of when we use that supplied by the manufacturer of the objective, although the author does not accede that it is ever necessary to use an oil that later will become an adhesive, namely, cedarwood. Unfortunately, the tube length is frequently fixed by the manufacturer. Often a tube-length corrector, such as the Watson Tube Length Corrector (see Chapter 3) can be employed. This device is satisfactory except that it does decrease image contrast due to added flare; the lenses of the corrector should be given an antireflection coating.

However, for critical work at high NA's each microscopist should determine the optimum collar adjustment or tube length for each of his high-aperture objectives. This need be done only once for each objective, if the result is recorded, by applying the star test which is described later in this chapter.

FOCUSING THE IMAGE

A. It is very important to have a habitually careful technique for focusing high-power objectives of short working distance. The technique for focusing the visual image when setting up a vertical microscope was discussed on p. 250). It should be reviewed or read if that section was skipped over.

B. Focusing the image in the camera is a subject well worth considering, since many times a photomicrographic system is well set up only to lose the best definition in the photomicrograph because the location of best definition in the image does not coincide with the surface of the photographic film after the objective has supposedly been carefully focused for that purpose.

Normally the image should be focused with a focusing glass magnifier, often called a loupe, on a central clear area of glass bearing some mark, usually a cross. Thus an aerial image is focused by sending it back and forth along the axis until it coincides with the cross, which is viewed simultaneously.

There is a primary question: does the location of the cross on the glass slide of the camera back lie exactly in the plane to be occupied by the photosensitive surface? With a compound microscope, this in general is not a critical matter, since the greater depth of focus now lies in image space but for this very reason the cameras may be carelessly constructed. The author found with depth gauges that the coincidence of ground-glass surface and the plane of the photographic plates in one photomicrographic camera was not sufficiently good for excellent definition in very low-power work.

It is usually recommended that the optimum focus be checked by the method of parallax. This means that only when the cross and the image lie in the same plane will they remain together in movement as the eye is moved back and forth above the magnifier. This effect is easily observed by closing one eye and moving the head while observing two fingers or two pencils pointing at each other in and out of coincidence in a common axis.

With some images, especially weak ones, this method does not seem best. The following is both better and faster in the long run. In the first place it must be realized that no common loupe is chromatically corrected well enough for critical focusing and that, besides this, **the eye plus loupe must be calibrated**. The back of the camera which receives light at very low aperture is *not* a suitable place. The author keeps an illuminated opalglass surface on a bench top near the optical bench; a negative or a transparency viewer will do very well because of the diffusing surface. This ensures that the focusing will be done at high visual aperture and thus more critically. Now the ground glass is removed from the camera and the same or similar color filters placed under the cross on top of the illuminated surface. The loupe is adjusted under these conditions for critical focus on the cross. This must be done independently by each operator who may use this loupe, but it will then be permanently set for each one.

There is a certain depth of acceptable focus in the image, representing depth of field, as illustrated in Figure 6-6 and discussed in Chapter 1. By definition of terms the optical sharpness appears to be equally good within the inner and outer boundaries of this zone, although not so, necessarily, as the distance from the axis increases. The camera plate is fixed, and this image is moved back and forth with respect to the flat plane of the plate, although the converse is illustrated in the figure. It is obvious that the center of the field will be equally sharp for positions *a–a* and *b–b* but that the size of the field will be larger in *b–b*. To ensure the latter condition, the loupe is momentarily moved off the axis to check the direction of focusing that sharpens this off-axis area. The cross is again focused on axis, but the focus is left as far as possible in the direction of the larger field, *without sacrifice of central definition.* When the negative is examined critically, the neophyte photomicrographer can re-train his focusing habits.

Figure 6-6 Depth of focused image.

Obviously, if the center is slightly worse, he has gone too far into the image; if the center is not so good as he saw at the glass, he is too far out.

Depth of Field

As discussed in Chapters 1 and 2 and Appendix 1, the depth of the field that is simultaneously in focus decreases rapidly with increase of the aperture of the system and with increase of magnification. It will therefore decrease whenever an objective or ocular of higher power is substituted for another. It will also decrease with decrease of wavelength. Lack of depth of field naturally is more troublesome in photomicrography than in visual microscopy, in which a slight twiddling of the fine focus becomes a habit. In high-power photomicrography, in which high apertures must be used and short wavelengths are advantageous, the lack of depth of field can be extremely serious. In metallography, in which all is in a strict plane, also normally true of fields of fine particles, the focus must be placed very precisely and the size of the fields will be smaller because of this lack of depth of field combined with field curvature. With biological specimens the specimen thickness is often greater than the depth of field, which may enlarge the field of good focus and make focusing seem less critical. This effect depends on the nature of the detail. If its unit formation is greater than the depth of field, the appearance is poor. In this case we can resort to integrated focusing, whereby a series of exposures at successive focus adjustment is accumulated on one negative, which is then printed. This is really successful only when the wavelength is short enough to be in the ultraviolet and apertures and magnifications are high, but this situation is just the one in which the technique is most needed. This is because the out-of-focus image must go *at once* so completely out of focus that it gives only a general fog. This technique however, is quite practical when used with a wavelength of 365 mμ, which is available to anyone with an apochromatic objective and the ability to use nonstandard tube lengths in the microscope. It is described in Chapter 14, p. 647.

A more advantageous, though more elaborate method of increasing the depth of field has been devised by Dan McLachlan, Jr. (*News in Engineering* **37**, November 1965, an Ohio State University Journal). The full depth is scanned, as in the integrated focus method, but by movement of the specimen along the optical axis toward the objective. However, the specimen is illuminated only by a thin horizontal sheet of light (assuming a vertical optical axis). Thus the portions of the specimen not in focus at the beginning will not be illuminated until the scanning movement raises them into the illuminated plane. Calculations are made so that the thickness of the illuminated plane is no greater than the depth of field at the magnification and aperture employed (McLachlan, 4. 1964). This method is used with compound microscope magnifications up to 1000X, in which case the sheet of light must be very thin. It is accomplished by illuminating from behind a mirror that bears a slit of the desired width cut in the silver reflecting backing. A razor blade is used. The slit is imaged on or through the specimen at the microscope axis by a good lens system such as a microscope. Because McLachlan is interested in minerals, many of his specimens were opaque with a rough surface, and because of the sharp shadowing two horizontal sheets of light, orthogonal to each other, were used.

Obviously this same technique can be adapted to low power work, in which the need for depth of field is so frustrating, and to translucent specimens such as insects. McLachlan reports that "piping" of the light through the interior of a specimen component, to appear elsewhere in an out-of-focus area, may be a problem. This is overcome by coating the specimen.

Some commercial concerns are reported to be interested in marketing a microscope assembly to utilize this method.

INTRODUCTION OF THE CAMERA

In the description of the method for setting up a photomicrographic apparatus given so far in this chapter, except for Case III for the horizontal bench, the specific instructions have been left with the microscope well set up but without describing the placement of the camera. Assumption of the complete system has sometimes had to be made.

1. (former Case III see Index). The horizontal bench almost invariably uses the open-back view camera; and, if the directions earlier in this chapter were followed, the camera should now be integrated into the system and well aligned, since the cross on the camera back was a fiducial mark at one end of the system.

2 (former Case I). Again in the large "universal stands" and in some other commercial stands the camera is an integral part of the apparatus, and the manufacturer's directions for alignment will probably be most expeditious if it is not already an inflexible part of the system. Focusing the image in this type of system is usually done by means of an auxiliary eyepiece and a reticle that is in an equivalent optical plane to the camera back; sometimes direct access to the latter is possible. That is almost always an advantage.

3 (former Case II). With a vertical microscope axis and a view-type camera with an open back, the bellows may be fixed or variable but should be at least 10 in. long for reasons discussed in Chapter 2 (see Index, *image distance*). If the position of the camera can be varied over a microscope already well set up, it is only a matter of ensuring that the axis of the camera coincides with that of the microscope. A pinhole ocular is useful for this purpose but certainly not necessary. With the pinhole ocular in place, it is only necessary to see that the spot of light from the microscope lies in the center of the camera, with any optical setup of the microscope that is complete and well aligned. The same can be done with a low-power eyepiece and an objective with a field smaller than the camera back. Simultaneously we should ensure that the stage and camera back are parallel to each other. This is important. It can be done by using a spirit level on both microscope stage and camera back.

Sometimes a microscope and its illuminant are made up as a unit on a board; in fact, several such units may be in existence, all of which are carried to a vertical camera stand when some photomicrography is to be done. The pinhole-ocular method makes the operation simple if the position of the camera can be easily altered, as if it swung around a post and had some lateral adjustment.

When the position of the camera is inflexible, except for vertical displacement, it becomes necessary to center the microscope to *it* in a preliminary manner before setting it up carefully as discussed in Case III of this chapter. For this preliminary setting the pinhole ocular is especially useful for determining the optical axis of the microscope without accurate centration of the substage.

4. *Cameras Supported by the Microscope; the Eyepiece Camera.* The author is not enthusiastic about supporting the camera on the microscope, especially a fairly large camera that uses $3\frac{1}{4} \times 4\frac{1}{4}$-in. and even 4×5-in. sheet film. The objection is not to the weight of the camera; this thrust is even beneficial and is built in to reduce backlash in focusing. The objection is the ease of transmitting vibration to that part of the junction between stage and microscope by which the amplitude of the vibration is extended by the fact of magnification. Even then support on the microscope is preferable to independent support from a side by a vertical rod of inadequate diameter or without firm bottom support or any other source of instability such as an inadequate horizontal supporting arm. The act of snapping the shutter can be a source of vibration. This is discussed in Chapter 11.

The use of a camera attachment sitting on the microscope and roll-film cameras, especially 35 mm cameras, has become prevalent. These devices are also discussed in Chapter 11.

ACCESSORY TECHNIQUES

Determination of Magnification

As we learned early in Chapter 2, the magnification of a projected image at a bellows distance of 10 in. (or 250 mm), *measured from the eyepoint of the compound microscope*, is the product of the magnification numbers that are engraved on every microscope objective and ocular. The magnification at any other projection distance then becomes this distance divided by the standard distance (i.e., projection distance in inches ÷ 10). This assumes that the focal lengths of the components are as designated and that the tube length between them is correct for the objective. Even with a simple microscope the magnification can be calculated from the focal length of the objective and the projection distance as determined from the focal scale of Chapter 1.

However, in photomicrography with both simple and compound microscopes it is more accurate and usually simpler in the long run to determine the magnification by reference to the projected image of a stage micrometer; when there is a scale along the optical bench or up the vertical camera support, it is not necessary to repeat the determination each time. A stage micrometer is a very small rule or scale engraved or photographed onto a glass slide (see Figure 6-7). It is usually marked in 1.0-, 0.1-, and 0.01-mm units and is sold by all microscope supply houses. Unfortunately, for those who wish to do accurate micrometric work such as measurement of particle size the errors of some of the more inexpensive micrometers can be very great. For accurate work a micrometer certified by the Bureau of Standards is not necessary but acquisition of a micrometer of the same type probably is. However, great accuracy can be obtained by use of a secondary standard. In this case some object, conveniently a micrometer of any quality, is compared *at a given point* with an accurately known distance at some magnification (e.g., against a valid standard that is temporarily available). This specified interval on the secondary standard then acquires validity.

A satisfactory procedure for measuring the projected scale in the image plane employs a transparent glass rule that bears a centimeter scale.

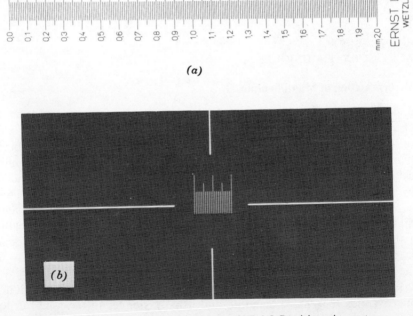

(a)

(b)

Figure 6-7 (*a*) Standard stage micrometer. (*b*) B & L Precision micrometer.

This scale, subdivided into millimeters or even finer, can be purchased. It can also be made by contact printing from one used as a master on a Kodak High Resolution Plate. A negative is obtained first, of course, from which many rules can be printed. A transparent rule ensures no parallax error and is also much easier to use, especially with a dim image, than an opaque (e.g., celluloid) one. The next best technique is to use a pair of draftman's dividers with which the distance from the ground-glass image of a rule is first taken off to compare later on an opaque rule, or conversely, if the magnification is being set to a prescribed value.

If images are being measured on photomicrographs, either negative or positive, either directly or on projection, as is common in the determination of particle size methods, the magnification should be determined similarly from photomicrographs of the micrometer ruling (see discussion in the next section).

It is wise, especially if many measurements are to be made from photomicrographs and if considerable photomicrography is to be done with a given set of objectives and eyepieces, to determine the magnification relation with bellows draw for these optics. This merely assumes a scale parallel with the axis and a pointer on the camera; it need not be coincident with the image plane. This procedure is especially simple with a horizontal bench but it is also practical with a vertical stand. Sometimes a retracting coiled steel tape which comes up and hooks invariably at the same extended length is used with the latter.

Now the magnification is *not* simply proportional to the image position, in spite of statements in the literature, unless the zero of the scale is positioned *accurately* at the exit pupil of a positive ocular. It is much simpler to record the position of a fixed but arbitrarily set pointer along the scale and to graph it on a piece of graph paper of adequate size to give the required accuracy. The relation for a positive ocular is

$$\text{magnification} = ar + k,$$

where r is the scale reading and a and k are constants. In practice the values will probably be read from the graph, thus making the calculation unnecessary. For a negative ocular, such as a Homal, the relation is not even quite linear because of the shift in effective exit pupil with projection distance.

Sometimes it is desirable to have a scale, or possibly only two fiducial lines of known separation, appear in the field of the photomicrograph. In this case an eyepiece reticle is most appropriate, if an ocular with focusing eyelens is available. Actually the reticle can be shoved along the ocular tube on its diaphragm, with a tube of lesser diameter, until it is in sharp focus for any projection distance. Sometimes the availability of the conjugate plane near the lamp condenser is overlooked for this purpose. A reticle on a transparent sheet placed in this plane will be imaged in the field by the condenser and of course again in the image plane of the microscope. If an achromatic condenser is used, or even an aplanat with a restrictive color filter, this method, made up and calibrated with a stage micrometer, is perfectly practical but requires some care that its focus and magnification stay constant from slide to slide.

True versus Apparent Size

When very accurate measurements are to be made of photomicrographic images at high magnifications, certain fundamental limitations and rules must be taken into account.

The true edge of the image of some small object, such as a hexagonal platelet, may be considered an infinitely thin line, but lack of resolution alone will broaden this line to a band of width $\lambda/2\,NA$. which at a wavelength of 436 μ and NA of 1.30 would be 0.17 μ wide. This will be multiplied by the magnification on the photomicrograph being measured. Only half of this widening will occur on the outside of the body. This extra microscopical distance from the "true edge" toward the outside may be called δ.

In measuring magnification, the distance between the images of two of a series of parallel lines on a ruling is measured *from the same side of each ruled line*. The extra distance, δ, with each line, being on the same side, will cancel out and so this phenomenon does not affect the determination of magnification. In measuring the image of the hexagon, or any other particle, however, a rule is laid across it (or an equivalent operation) and it can be seen, at once, that in the measurement of particle size, which must be made across two opposing edges, the extra distance 2δ will be included. This should be of the order of 0.17 μ with blue light and 1.30 NA.

However, a magnified microdensitometer trace across any image edge, or its photographic record, will show this edge to be wedged rather than absolutely sharp, as we would expect. This means that its apparent location will be affected by many factors, including contrast, which have caused opponents of the microscopical method of particle-size measurement to state that the results can be arbitrary.

The author and his co-workers, notably J. J. Duane, investigated the effect (13 Loveland, 1952, 1958) and importance of these factors. The apparent size is very definitely affected by them. A summary of the results is as follows:

1. *Exposure of the negative.* With brightfield illumination, the apparent

size becomes smaller with increase of exposure. With darkfield illumination, the apparent size becomes larger with increase of exposure, but the size of the image is a bit smaller at *low* exposures than in brightfield.

2. *Effect of development.* Changing the contrast of the negative (or gamma, see Chapter 15) by changing the time of development changes the apparent size somewhat but not so much as variation of exposure. The changes in dimension involved are only tenths of microns. For particle-size determination, however, the negatives should all be exposed to obtain a constant density of background at a constant contrast. In the author's laboratory a density of 2.0 was chosen for the background of the negatives.

3. *Effect of objective aperture.* The objective aperture is the most important factor, as might be expected. Both resolution limits and flare factors are affected. Thus both the width of the edge (contrast of the wedge shown by the microdensitometer trace) and the displacement of the edge are affected. It is of interest that no significant or critical effect was observed in the edge displacement curves when the objective aperture was equal to the condenser aperture, although flare effects and visual appearance might be changing rapidly. Flare increases the difficulty and decreases the accuracy of the measurements but does not change the position of the apparent edge.

With darkfield illumination, the width of the edges is affected by the central stop of the darkfield condenser, which is normally larger than the subtended objective aperture. The dependence on the edge width becomes critical as the size of stop is decreased to that just occluding the objective (see Figure 6-8).

An investigation was then made by the author with D. C. Skillman to determine δ, discussed above, in absolute terms. For this purpose the same hexagonal crystals were photographed and measured by both light and electron microscopy, with the crystals lying on a brittle cross-lined grating to standardize the magnification. (The brittle grating would break before it would distort, which could be seen.) Thus the accuracy of determination of magnification with the light microscope could be used with the resolution of the electron microscope.

$$D_\mu = D_p - \Delta,$$

where D_μ = the true size (diameter),
D_p = the apparent size (diameter),
$\Delta = 2\delta$, the discrepancy in size.

The investigation confirmed that the discrepancy was indeed a constant, with a given wavelength and numerical aperture of the objective

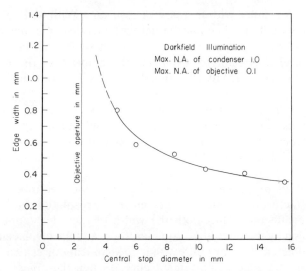

Figure 6-8   Effect of darkfield stop diameter on edge width.

throughout the particle-size range. **If the true size D_μ is plotted against the apparent size on the photomicrograph D_p, the graph merely shows a straight line at 45° just slightly below the true diagonal.** A more instructive graph is the plot of percent error of size determination $100\Delta/D\mu$ plotted against the size on the photomicrograph (see Figure 6-9).

The constant Δ has been determined for the case in which the photomicrographs were made at λ 365 mμ with an objective of 1.25 NA but only in a more preliminary way for the blue wavelength λ 436 and 1.30 NA. The use of λ 365 mμ, however. would be recommended for particle-size determination of small particles (see Chapter 14).

Wavelength	Δ
365 mμ	0.047 μ
436 mμ	0.084 μ

The constant, 0.047 μ multiplied by the magnification, is merely subtracted from the diameter of the measured particle photographed with λ 365 mμ. This correction can be incorporated in the rule. Figure 6-9 will show whether the correction is going to be significant at the sizes to be measured or it may indicate at which size level the measurements should be transferred to the electron microscope. There the error becomes a matter of accurate determination of magnification from field to field and across fields, that is, distortion.

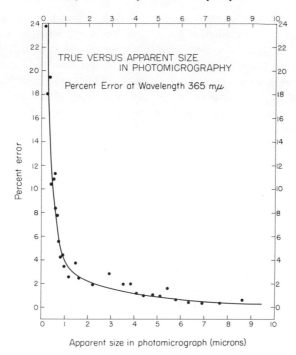

Figure 6-9   True versus apparent size in photomicrography. (Percent error at wavelength 365 mμ.)

It should be noted that enough psychophysical factors are involved that the correction at λ 365mμ is below that expected from considerations of resolving power alone. Moreover, as resolving power decreases with increase in wavelength, the correction required increases more than proportionally. This seems established and is due to judgments of more indefinite edges.

CARE OF THE MICROSCOPE AND MICROSCOPIC OPTICS

All trained microscopists who have had to pay for their instruments out of their own pockets can be counted on to give them good care. This includes a large number, especially in the medical profession. A larger number, however, work with instruments belonging to institutions, including commercial firms, and may even have to share them.

It is important to remember that with poor care the quality of the images obtained from the microscope and its accessories will degrade along with the results obtained from it long before the operator will feel justified in throwing all of it on a scrap heap and starting afresh. Moreover, most people who work with a microscope probably use it most frequently by brightfield illumination with specimens of relatively high contrast and at medium and low powers. This is the condition in which degradation of the surfaces of the optics is least observable. However, when a specimen of low contrast, possibly requiring darkfield or other special illumination, is to be studied, the information obtained will then be limited by the condition of the optics.

Probably a minority of the injuries to objectives and other optics occurs during use. Objectives have been dropped. This is calamity that a good lens cannot tolerate. It is best avoided by careful habits. Much inspection and manipulation of objectives and other lenses can be done a short distance above a surface instead of over the floor. When passing an objective to someone else, a hand should be kept under the article.

The writer has seen fingerprints on lens surfaces, a condition that seems inexcusable. Some bare lenses are used as condensers. They should be held by their edges but will need cleaning more often. Fingerprints, as well as other dirt, will burn in or etch a glass surface in time if the latter becomes very hot. This applies to lamp condensers and the lamp bulbs themselves. The latter should be wiped after insertion in the socket and before turning on the current.

Sometimes some sample material or mounting medium will be found on a dry objective. There seems little excuse for this under normal circumstances if the methods of focusing discussed earlier in this chapter are used and large blobs of mounting medium are not allowed to remain at the edges of the cover slips. The slide should never be moved unless it is known that the objective is *not* in contact with it. With subjects difficult to focus on and some objectives of small working distance, this may require care.

Probably the majority of injuries to lens surfaces, certainly of the scratches, occur while cleaning them. Unfortunately, this condition was compounded by the advent of the thin antireflection coatings on lens surfaces. However, all fine lenses must be treated as if they were made of soft materials. In selecting a material for its optical characteristics, there is usually little opportunity to specify its hardness also. Even certain salts have been used, including rock salt. Common optical glass is softer than bottle and window glass, which furnishes the common concept of glass. Moreover, the blown glass of common experience has a hard surface skin that is absent from ground and polished lenses. Obviously the best procedure involves habits that will reduce cleaning requirements to minimum. The biggest single factor is protection from dust and, in highly populated rooms, from grease that holds it. Dust may contain a certain proportion of abrasive particles, which varies greatly with the location. Here, again, the owner of a microscope will usually be the one who pops a cover over it whenever he is not peering through it, even when he expects to resume use of it soon thereafter, since laboratory life is full of interruptions. The hours a microscope is not in use enormously exceed those of actual use in most cases. Stiff celluloid covers are by far the most convenient.

The vicinity and the instrument stage and supports should be dusted reasonably often with an appropriately moistened cloth or lens tissue. A rubber aspirator bulb with a rubber tube should be handy to blow dust off the mirror and other optics, including the lamp condenser. This will avoid the temptation to blow away the dust with the breath or to use compressed air. Neither method is clean.

Inspecting Lens Surfaces

By far the most effective way of determining the cleanliness of lens surfaces, especially if they are plano (e.g., the bottom of objectives and the top of eyepieces), is to inspect them by specularly reflected light with a magnifier. The following technique is worth learning; from watching others the author realizes that there is a slight trick to it, yet it soon becomes so natural that it is easy to forget this. The objective or ocular lens surface must act as a mirror for a bright surface whose reflection is larger than it: this may be a lighting fixture or a portion of the sky or a cloud through a window. Usually, this bright surface should be above at a fairly high angle. The specular appearance, with the lens surface bright, will flash up when the angle of the surface makes the angle from the light equal to that of the eye. The magnifying lens (up to a 10X loupe) is held close to the eye and the objective or ocular brought up to focus. The surface should be clean and free from scratches. All dirt and scratches will show up accusingly.

Cleaning Lens Surfaces

No matter how shocked the inspection may leave you, do *not* pull out a handkerchief and wipe the lens, if it is a good one. Rubbing with absorbent cotton leaves scratches on the glass, probably from entrained, colorless dirt. All microscopists should have a supply of lens tissue at hand. The author keeps rectangles of it in a box with a hinged lid near the microscope. Its protection from dirt is very important. This tissue used to be called "rice paper." However, if a *clean* microscope slide is rubbed with lens paper and then examined under low-power darkfield illumination, it will show smears from the paper or some filler.

The first operation is to make up the lens paper for cleaning. Fold all the edges in, leaving the center bulging out, untouched. Twist the stem a bit. Now, if the piece is held back of this stem, the latter will act as a spring and no more pressure can be applied to the lens surface than this paper spring allows. One may wish to have several such packets ready if a volatile solvent is to be wiped off before it evaporates or to speed the operation.

If there is dust on the surface, assume that it is highly abrasive and remove it by blowing with the rubber aspirator or wiping carefully with the lens paper before considering any smeared dirt. Then the lens can be wiped with a packet moistened (but not sopping wet) with xylene or purified ligroine. For small front lenses of higher powered objectives the lens tissue may be sharpened to a twisted point, but the principle of the paper spring which limits the force that can be applied should still be used. With some types of smear it is advantageous to clean with distilled water containing some nonalkaline glass-cleaning detergent, followed by adequate rinsing with clean distilled water (again without a sopping paper). "Dry" objectives may be inadequately sealed against the penetration of fluids behind the front lens.

Carrying a Microscope

If the writer had not seen a microscope, otherwise good, which had been permanently damaged because it had been lifted so that the weight was borne by its bearings, he might not have thought of this paragraph. All microscopes formerly had a massive stem in the casting which was the most convenient as well as the safest way of carrying them. Now it may require care to ensure that the microscope is grasped and carried by the support that bears the rest of the instrument.

EVALUATION OF MICROSCOPE OBJECTIVES

One may wish to evaluate the condition and quality of an objective that is being offered for sale, especially if it is secondhand, or one may be interested in the condition and quality of optics inherited from a predecessor in a job and laboratory. In any case, a rather systematic examination is called for. It is wise to jot down the observations as one proceeds.

First, a direct examination of the objective itself, including its mechanical features, is called for while holding it in the hand. This should be done while sitting at a table or desk with the objective only a few inches above its surface.

Has the objective an RMS thread (q.v.)? All professional objectives, except those from some speciality instruments, will have this thread. If the objective is a used one, the condition of its metal case is significant. Older objectives should be checked for tightness of their parts. The lens components of modern objectives are held at the bottom of a series of concentric tubes. If the group is held in by a top screw ring, it is not recommended that it be loosened but rather that it be firmly tight. If there is a correction collar, as in a 40X objective, it should be tested to ensure that it turns smoothly and through the marked range. The interior walls should be black. Now the lens elements should be included in this examination. The surface of the front lens should be examined by specular light with a 10X hand magnifier. This method was described under "Inspecting Lens Surfaces." Any scratches and nicks should show up here. The back lens surface of the objective should also be examined. A magnifier of longer focal length is required to see the top surface of the back lens. Probably the most effective illumination of this rear lens surface showing part of it at a time is oblique light through the front lens. Some special devices are made to show it up by reflected illumination.

Malies (4. 1959, p. 114) recommends making a ring to hold the objective under test in the substage. This ring should have the same diameter as that of the condensers and a flat disk top containing a hole with the RMS objective thread. If the objective is illuminated from below with well-directed, diaphragmed light and examined from above with a low-power objective, each surface can be successively examined and even the condition of the cements often ascertained.

Now the objective should be used as a telescope with the rear of the objective pointing at a specimen, such as a printed page, about 6 in. (a tube length) away and perpendicular to the axis, with at least a 10X aplanatic magnifier as the eyepiece. As the magnifier is moved about, gross defects of the objective lenses will show up.

All of the steps just described will take only a short time, yet most gross defects should be caught. Indeed, such an examination may be sufficient to explain the poor quality observed in images formed by the objective.

The numerical aperture (NA) of an objective is normally engraved on it. Very occasionally, especially on the less expensive objectives, the actual NA is different. Measurement of the NA of an objective is described in Chapter 2, p. 86. One would expect to check the full NA only rarely.

If the objective under consideration is one of high power, the next step should be the star test. This will ensure that the later examination of a test specimen is done under optimum conditions.

The Star Test

The value of the star test for determining the optimum tube length and cover-slip thickness, etc., has already been discovered in Chapter 2. Although the pinhole slide discussed below is the most satisfactory test object, the reverse image of this can be used, that is, a fine, black dust particle, to set for optimum tube length. The same rules apply, but a single black ring will usually be observed instead of the set of bright rings.

A frequently recommended test slide for setting optimum tube length and for other tests is the Abbe test plate, which can be obtained commercially. Although the author prefers the star test, the Abbe test plate can be used for the purpose. It consists of a slide with bars of silver mirror or clear parallel slots in an opaque film. It is covered with a rectangular cover slip of thickness wedged along the length and bearing a calibrated thickness scale. The cover-slip thickness that gives the best edge definition or the tube length for a given cover-slip thickness can then be determined.

Preparation of Pinhole Test Slide. The classical method of making a test slide for this purpose is to silver a microscope slide chemically. This is still a relatively satisfactory and practical method. Formulas for silvering solutions, such as Brashear's, can be found in many chemical handbooks. If the evaporating equipment used for the preparation of specimens for electron microscopy is available, as it may be in a related department of an institution, evaporation of an aluminum mirror on a glass slide becomes almost a routine matter. Several points must be considered in either case. The pinholes should be round and clean, preferably with a range of sizes to correspond with the magnifying power of the objectives to be examined. It is well to have them lie a little above the limit of resolution. The slide should be cleaned fairly well; for example, by rubbing with acid alcohol. The metal coating should be as thin as will give reasonable opacity. This is because a small hole in a thicker coating may behave like a tunnel and cause anomalous behavior. By leaving slides at various distances from the filament in the evaporator, one can be selected as of minimum thickness for opacity.

A superior test slide, bearing truly round holes, can be prepared by dispersing Dow polystyrene beads on a glass slide or cover slip before evaporating an aluminum film over them and subsequently removing them. Beads of about 0.25-μ diameter seem about optimum. In this case it is worthwhile to follow a careful technique that includes an excellent precleaning of the slides.

For most purposes it is preferable to evaporate the aluminum mirror directly on a No. 1½ cover slip selected by micrometer to be exactly 180 μ thick. Thus, when the mirror is inverted and mounted on a microscope slide, it will be known that the effective thickness of cover slip is as specified. With the mirror on the slide there is usually about 20 μ additional optical thickness when moderate pressure is applied.

The cover slips, or slides, should be of a glass that can withstand the drastic acid cleaning. Cover slips, 22 × 40 mm, in the No. 1½ thickness class are most suitable. They can be rubbed clean without breaking if laid on a thick sheet of clean glass. They are best handled during cleaning in glass staining trays that isolate each cover slip or slide. After cleaning with a suitable detergent and *thorough* rinsing in *clean* water, they should be degreased without a separate act of drying by passing, in their trays, through about 12 molar sulfuric acid before dipping in clear fuming sulfuric acid for 3 min or in the standard concentrated chromic-sulfuric acid mixture. They should be passed back into a 12M sulfuric acid bath before being dipped in a water bath and *thoroughly rinsed with clean water*. Drain off all excess water.

The dispersion of polystyrene beads should be diluted 50,000 times from the concentration obtained from Dow. An entire tray of slides or cover slips is dipped into the dilute dispersion of beads to coat them. To dry, the entire tray is put on a blotter in a dessicator and evacuated.

Alcoa High Purity Aluminum, 99.9998% Al, is needed to minimize danger of bloom. Ionic bombardment of the glass slides is used for final cleaning after the bell jar has been partly evacuated. It is best to have the slides above the source, which can be a conical tungsten wire basket containing the aluminum. The slides should be quite far away, at least 10 times the diameter of the aluminum source, to avoid undercutting the spheres. An optical density of at least 3.0 should be obtained; at much higher density the beads become too difficult to remove.

The simplest, most effective way to remove the beads is in a bath of neutral xylene in an ultrasonic cleaner. They can also be carefully brushed off.

A coated cover slip is mounted face down on a standard glass slide.

The star test is a dynamic test; constant observation is required as the focus is changed back and forth through "best focus." Practice and personal experience are required, but each observer can acquire them himself. The test is discussed from the standpoint of determining the nature and extent of the spherical aberration, which is its most important application; other types of observation are discussed later.

Technique of the Star Test. A strong light source is needed; the test is easier with monochromatic light (e.g., from an intense mercury arc with the wavelength of 546 mμ isolated by a Wratten 77A filter). With an incandescent source, a green filter should be used; the narrower the cut, which still leaves sufficiently bright illumination, the better. A high-power ocular, even as high as 25X, should be used. It is important to examine the smallest pinhole from which there is sufficient light to examine the diffraction rings. Examination of a hole that is too large seems much easier but the result may be in error; usually a somewhat shorter than optimum distance is found with an excessively large pinhole.

It is sometimes glibly stated that when spherical aberration is absent the appearance of the diffraction pattern is the same on each side of focus. This is rarely true. Usually the determination of the optimum conditions for low-aperture dry objectives, except for their insensitivity to rather large tube-length changes, is simple. The criteria for optimum tube length become more puzzling as corrected objectives of high NA, especially the oil-immersion objectives, are examined. The pattern at the various tube lengths will actually vary appreciably with the objective. It is wise first to find the focus and tube length at which the simple spot looks quite good and relatively free from rings before starting the more critical test for the presence of rings and equality of going to haze on each side of focus. The expansion of rings on each side of focus is the best criterion of the latter. With a very small spot, the central bright dot will come and go (replaced by a dark spot) as the focus is moved in one direction. This is not true, coming up with the focus, for the larger pinholes. With some high-aperture objectives it is possible to find a tube length at which better focal symmetry seems to exist with very strong rings at all focus levels. The first step of finding the best general in-focus level will avoid this trap.

A beginner can best introduce himself to this test by the seemingly negative approach of recognizing the presence of bad spherical aberration, both positive and negative (see Chapter 1) and then gradually removing it by change of the tube length of the microscope or the thickness of the cover slip. A Watson Tube Length Corrector (Chapter 3) is useful for those microscopes without tube-length adjustment. It can be calibrated in terms of tube length and cover-slip thickness (4. Spinell and Loveland, 1960). To start with, an objective, such as a 4mm dry, might be used without a cover slip, assuming that it is corrected for use with one. The extreme asymmetry of focus of this **overcorrected system** cannot be missed, in this case **with strong sharp rings on the inner side of focus** (i.e., as one focuses down with an objective or up if the stage is the moving part), **but only a hazy appearance is seen on the other side of the best focus.** The reason is clear from Figure 6-10, although this represents a case of strong undercorrection, with the sharp rings being outside of focus (downward with a vertically focusing objective). The side of focus on which the sharp rings lie can be reversed by introducing the opposite state of correction, by increasing the tube length, or conversely according to the need.* Both conditions should be examined.

---

*The tube-length correction for a 4 mm objective (biological type) and no cover slip is too long to be practical.

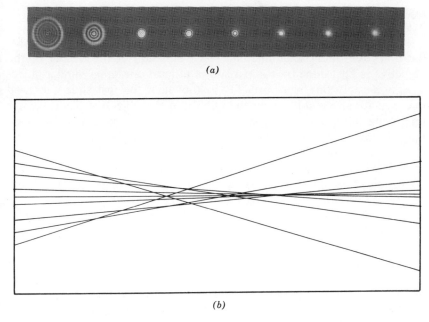

(a)

(b)

Figure 6-10 Star image structure: Extreme undercorrection.

These useful rules may be repeated and summarized (see Chapter 2). **A tube length shorter than optimum introduces undercorrection, as does too thin a cover slip; too long a tube length or too much thickness of cover slip introduces overcorrection. Therefore to compensate for too thin a cover slip, lengthen the tube and conversely.** Also, **lengthening the tube tends to make the marked rings appear as one focuses upward**. On the correction collar of a 4 mm dry objective this is equivalent to turning the collar to lower numbers, since these are values on the scale at which to set the collar for known cover-slip thickness, hence are compensating, not direct, values for adjustments. In practice, unless the specimen is mounted directly on the cover slip, *at least* 0.02 mm should be added to the cover slip's assumed thickness.

To study the star test further, using a pinhole test slide, mount a No. $1\frac{1}{2}$ cover slip on the slide with immersion oil, if it does not already bear such a slip, and make the tube length long enough to produce over-correction. Then the tube length should be shortened (or the correction ring turned) while focusing back and forth through best focus. It will be noted that the intensity and contrast of the diffraction rings are reduced as symmetry on each side of focus is approached. By this time the observer will be using the small pinholes for the best criterion. The pinhole under examination should be very close to the center of the field. The best "in-focus" image of the pinhole should be noted.

The observer will now have determined the optimum tube length for the objective under examination. The evaluation of the quality of the objective is best done by comparison of at least one other similar objective, preferably one whose quality is known. If the test slide is a good one, much can be told about the quality of the objective from the star test, especially when confirmed by a test with a specimen discussed later. Pinholes that are isolated, clean, round, and in a thin but opaque coating are needed. The illumination should be well centered back of the test slide, since undeviated oblique rays from the larger pinholes are confusing. The appearance when out of the best focus and even with incorrect cover slip or tube length may be revealing.

If one or more of the lenses of the objective are not concentric with the optical axis of the rest, this will be evident by the lack of concentricity of the rings.

A rarer defect is to find that the rings are misshapen rather than smoothly round; they may even be elliptical, with the direction of the major axis rotating as the image goes through focus. Most likely such a condition is due to strain on the objective lenses in their mounts. Careful loosening of the retainer ring may help but this is really a case for the expert or factory repair shop.

R & J Beck Ltd. of London sell a testing device that consists of pairs of disks to isolate several zones of the objective aperture when they are mounted back of the objective.

As the pinhole is moved from the center of the field, its image will become unsymmetrical and finally show a tail, usually of ring segments. With the fine uniform holes made from micro beads scattered over the slides this appearance and its pattern over the field can be seen without moving the slide. This is normal for all microscope objectives, but the extent of the field with specific objectives and oculars can be inter-compared.

The effect of color can also be studied. Although with white light the relative position of the colored rings can be seen (with sufficient light), it probably is more instructive to examine the pinhole images through blue, green, and red filters successively and determine the optimum tube length with each. With achromats it may be found that the tube cannot be sufficiently shortened for the optimum length with blue light. The differences observed are caused by longitudinal chromatic aberration. This is related to the lateral chromatic aberration which is more directly observable in images from objectives and more usually discussed. The longitudinal error is greater than the transverse by a factor equal to the magnification.

Use of Test Specimens

The final and best test of a microscope objective is made, of course, when a suitable specimen is examined with it by an experienced microscopist. The amount of experience required is considerably diminished if the test can be made a comparison of two or more objectives. Moreover, this examination is performed better just after the star test has been made with the objectives, using for the star test the same oculars to be used, at least in part, in this test. Again, it is best first to examine the specimen with a narrow band of green light (or possibly blue with an apochromat) before using white light.

The nature of the most suitable specimen can be a matter of debate. There are microscopists who would choose diatoms. If a microscopist is especially familiar with a certain type of specimen, it should certainly be included. However, the author is of the opinion that biological stained specimens, such as sections, do not usually make the best test objects. The power of the objective will determine the fineness of detail that should be present, but critical tests are most likely to be made with at least moderately high-power objectives.

A colorless specimen seems most suitable; certainly it is simpler for study of contrast and color defects. The difference between two good objectives most often turns out to be one of image contrast; the image of one will be crisper. Therefore the most suitable test specimen should have fairly low contrast in itself, in which such contrast differences are most noticeable and good illumination is required. It should be flat to study field size.

Metallography abounds with suitable specimens; cast irons need only be mentioned. The metallographer will know and probably have at hand the most suitable specimen for himself.

The user of transparent specimens has a slightly greater problem. The specimen must be one with which the quality of the image is recognizable at once. This rules out most cross sections of appreciable thickness except for medium or low powers. Some specimens of pond life can be good for someone very familiar with them. The author votes for colorless and tabular crystals of a precipitate with a large range of crystal sizes, but he may be accused of prejudice on the ground of familiarity. The simplest of these test objects to prepare is made by coating the crystals of a suitable photographic emulsion on a microscope slide. Unfortunately for this purpose, the films and plates bearing large hexagonal and thin platelets are not so common as they once were in commercial photographic materials. The most suitable are made from films or plates of great photographic speed. A piece of the material should first be soaked in water for at least 10 minutes and the emulsion scraped off into a small beaker or other suitable dish. Very little more water should be added before carefully melting this gel mass with stirring. It can then be smeared onto the surface of a glass slide with the side of the stirring rod. If the gel does not melt before it becomes very hot, a *very* little nitric acid should be added. Suitable precipitates can be made on the microscope slides by the methods described by Chamot and Mason (4. 1958, pp. 395–418). A glance at the photomicrographs in their book will aid in the selection of a compound. Lead iodide may be mentioned as a possibility, although it is yellow.

The first test made with such a specimen will probably be of **resolution** and as a comparison between objectives. For this purpose the two should be precentered to make it easy to transfer to examination of the same field by the alternate objective or objectives. A grid can be drawn on the gelatin-smear slides with India ink and is quite helpful for relocating fields. Examination of two very fine particles which can just be resolved is often recommended for comparison between objectives and is a good test. The writer likes to use the case in which the straight edges of two tabular crystals come together in a very narrow V. The extent to which one can see into the apex of this V is a good measure of resolution. This should be done with several color filters as well as white light. A micrometer disk in the eyepiece will aid the memory while changing objectives.

Chromatic aberration will be present in all achromatic objectives and to extent in apochromats, but in the latter it should be negligible; as should be the case for low-power achromats. It will limit the definition in black and white photomicrography unless a narrow portion of the spectrum is used, as with a contrast filter. For good color photomicrographs excellent color correction is obviously important. The principle, important in optics, of relaxing the specification of the nonessential should be kept in mind. The author obtained an objective rejected by another microscopist after an examination confined to white light. It certainly exhibits bad chromatic aberration not mentioned in the maker's advertisement. With monochromatic light, however, its behavior is superb in resolution, contrast, and flatness of field; and that is exactly how it is used, since in the author's laboratory most colorless objects, including particle dispersions, are photographed with monochromatic illumination.

Some idea of the state of chromatic correction of the objective will already have been obtained by the differential focus of the pinhole test with the three color filters. This was direct observation of longitudinal chromatic aberration. Now the lateral chromatic aberration and the very common defect of differential chromatic magnification will be seen directly as the result of the combination of objective and eyepiece, discussed in Chapter 2; therefore the combination significant for the objective and the intended work must be tested together. Since this error increases with distance from the center of the field, near the edges of the field we should look for differences in color in a spectrum in what should be thin, colorless edges in the specimen. The blue outlines may be of low contrast to the eye; it must be remembered that they may be of higher contrast to the photographic film. Actually, a stage micrometer makes a good additional test object for this test if it has sharp thin lines. Then we should project the image of such a micrometer and focus successively through a Wratten Filter No. 47 (blue, C), a Wratten Filter No. 58 (green, B), and finally a Wratten Filter No. 25 (red, A), measuring the magnification each time. A further test with a Wratten Filter No. 29 (far red, F), or Wratten Filter No. 92 (far red) may be added. This would be a very severe test for an achromat.

After an examination by well-centered illumination the effect of oblique illumination should be tried, leaving the focus of the objective untouched until it has been noted whether adjustment is necessary. It should not be necessary if the correct cover-slip thickness and tube length are in use. For this reason this test is advantageously done with an Abbe test plate, since shifting it along the stage will determine the effect of cover-slip thickness and find the optimum. The oblique illumination should be produced by obscuring part of the aperture (or displacing the iris diaphragm) quite accurately in the focal plane of the substage condenser and at right angles to the length of the lines. This is provided for in many high-grade microscopes. Determination of the true aperture plane of the condenser has already been discussed. A number of defects will become more noticeable with oblique illumination. With an Abbe test plate the edges of the silver stripes may become unsymmetrically hazy or colored. With an achromat there should be a thin edge line of color, yellow-green on one side, purple on the other. If the band is wide and has a range of colors, the achromat is poor. The apochromat should show negligible color at the edges.

By this time enough observation will have been made through the several objectives under comparison to lead to general conclusions, including the contrast of the visual images.

Chapter 7

Illumination

Since illumination is such a fundamental factor in microscopy, it is not surprising that it must be considered in the discussion of the different phases of photomicrography. Therefore it is also not surprising that it has already been considered in Chapters 1 and 4, in which illumination by reflected light is discussed, or in Chapter 5, in which are presented the methods of illumination appropriate for low-power photomicrography by transmitted light. In Chapter 5 and its appendix we learned that **the size of the light source should be appropriate for the case and that, in general, the higher the magnification, the smaller can, and should, be the size of the light source.** We shall see in this chapter that this principle can be extended, but a source may be too small in spite of an old publicized belief that the best light source for microscopy is a point of light.

ILLUMINATION TYPES

As we have seen, and discussed in Chapter 1, the type of illumination may be subdivided as (a) **reflected** by the object field or (b) **transmitted** by the field. Just as important a dichotomy of illumination types is that of (c) **darkfield** and (d) **brightfield.** This is a transverse division to the first, since it applies to both reflected and transmitted illumination; these types of illumination apply at all magnifications, although both apparatus and technique will vary greatly with them. In darkfield illumination direct rays from the source do not enter the observing eye or objective unless it is first deviated to it by an element of the specimen field by reflection, refraction, or diffraction. An empty field (hence, usually, the background) is dark. On the other hand, in brightfield illumination the light source, which should be large with respect to the field of view, sends its rays directly into the objective or it may be reflected *specularly* into it by the specimen acting as a mirror, as it does in metallography. Direct observation of our environment is usually by darkfield. On the other hand, because of the high illumination requirements of the higher magnifications of the compound microscope, and the prevalence of translucent specimens for such examination, brightfield has become so common and familiar that the discussion of the apparatus and technique for darkfield can be relegated to a separate section. This chapter deals primarily with brightfield illumination.

In brightfield illumination some element of the illumination system, functioning as a source, must be imaged in the specimen field for satisfactory results. In practice, good illumination for brightfield microscopy can be divided into two types, which are consistent with the fact that the illumination, optics, compound microscope, and camera form a relay optical system, as discussed in Chapter 2.

1. **The light source may be imaged in the specimen field,** and the lamp condenser, if there is one, is conjugate to the objective aperture. It is illustrated in Figure 7-1.

2. **The light source is imaged in the rear focal plane of the substage condenser** and again into the objective aperture; the substage condenser is focused to image the lamp condenser in the specimen field, hence also

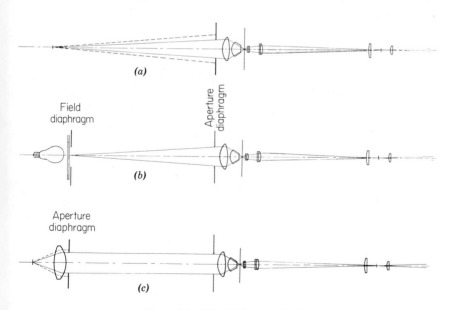

Figure 7-1 Abbe-Nelson method.

into the camera field (see Figure 7-2). **The light source may also be imaged into the substage condenser** (see Appendix 3C).

The first method is often called the **Abbe-Nelson method** and was the first one used. It was customary at one time to image the edge of a lamp flame into the field. When the early forms of electric light bulbs became available, they seemed unsuitable, but A. Köhler of Zeiss showed that very satisfactory illumination could be obtained if Method 2 were adopted. In spite of an original outcry against it as incompatible with the best microscopical image definition, it is now not only recognized as furnishing images of fundamentally good quality but **Köhler illumination** has been touted as being far superior in theory as well as practice, which is going to another extreme. **The Köhler method is the prevalent system of illumination;** control of the size of both aperture and field planes is quite convenient with suitably placed iris diaphragms. Its disadvantage is that with a light source of appreciable size at high magnification the diameter of the lamp condenser utilized, hence its relative aperture, may become quite small so that the system becomes inefficient photometrically. This point is discussed later in this chapter and may be deduced from Appendix 3C. Method 1 also has its troubles on this score. Although a lamp condenser is used at higher relative aperture and with the source filling it, if the source is large the light will spray out from the condenser if the latter collimates it or almost so (see Figure 1-11). Method 2 is discussed on p. 300 et seq.

Forms of Abbe-Nelson Illumination (Method 1)

Primary Sources. **The substage condenser alone, focused to image the light source in the field**, makes an excellent illumination apparatus *if the light source is one of uniform brightness over its area* (Figure 7-1a).

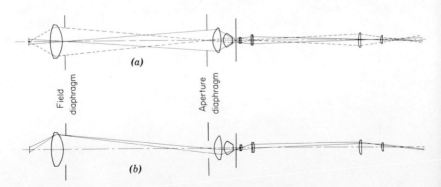

Field diaphragm

Aperture diaphragm

(a)

(b)

Figure 7-2 Köhler method.

The substage diaphragm functions simply and efficiently as the required aperture diaphragm.

As emphasized before and illustrated in Figure 7-1*a*, the solid angle (i.e., the aperture) of the light cone from the condenser, subtended by the specimen, is equal to that passed on to the objective, except for the wider angle that represents light scattered by the specimen. The substage diaphragm should be closed down at least to a diameter that would limit the cone to that accepted by the objective, and preferably somewhat more, to allow the cone of forward-scattered light to fill the objective completely. If the substage condenser is a corrected lens, aplanatic or even achromatic, as it profitably should be, the light source should be at the distance corresponding to the corrections of the specific condenser. (This distance value furnished by the manufacturer should be carefully kept and utilized.)

If the condenser is corrected for infinity, as Zeiss states their aplanatic condenser, NA 1.4, to be, the correction lens at the bottom of the condenser should be a positive spectacle lens whose focal length is equal to the distance of the light source; that is, it is a collimator. Fortunately, as with a corrected lens for a camera, there is a range of distances that can be considered efficiently available.

The problem in this simplest system is to match the size of light source with the desired size of illuminated field. If the source is too small, the penalty is an inadequately illuminated field with only a central bright spot of inadequate size; if it is too large, the penalty is an illuminated area that is too large, with much flare and low contrast. This point is discussed later in the chapter. Such lamps as the ribbon filament and the tungsten arc and some mercury arc types possess **reasonable areas of uniform brightness. When this area is just correct for the desired illuminated field,** reduced into it by the substage condenser, **the illumination is excellent indeed,** as pointed out by Needham (4. 1958), because of the self-diaphragming function of the size. The image is crisp and with little flare. Unfortunately, such light sources are the correct sizes for only a few magnifications, even with change of focal length of the condenser; for example, by unscrewing a top lens or changing it altogether.

Secondary Sources. Use of a secondary source, imaged in the field by the substage condenser, is the solution to the more general application of this method. This **merely consists in illuminating a diffusing surface from behind with a variable aperture directly in front of it** (toward the microscope), thus limiting its size to the desired diameter. However, certain rules must be obeyed to obtain the best system. A word should be said about diffusers.

Diffusers. A completely diffusing surface, such as a sheet of opal glass, will diffuse a beam of light rays through a 360° solid angle so that as much goes forward as backward. Naturally, if such a sheet is interposed into a beam of light entering some optical aperture, the effect is great photometric *inefficiency*, since the loss in intensity will be equal to the original solid angle of the beam, which may be very small (zero for a collimated beam), divided by 360°. The effect with opal glass is one of diffraction. The effect of a matte surface on a transparent sheet (e.g., a sheet of glass made matte by sandblasting) is somewhat different. The light is scattered by refraction, as by small prisms that are larger than the wavelengths. As with lenses, if put in the plane of an object which is sending rays in all directions anyway, such a surface has no effect; we are all aware that a matte surface can be laid on a printed page without appreciably affecting the legibility of the printed matter. Lifting an efficient matte surface very slightly, however, will completely blur the print. Therefore several sheets of matte film or glass held parallel and slightly apart form an efficient diffuser with somewhat less loss than the opal glass.

It has been found that a sheet of White Translucent Plexiglas (W-2447), $\frac{1}{16}$-in. thick, can be used instead of flashed opal glass with appreciable gain in transmitted illumination, both in quality and quantity, yet complete diffusion is obtained.

A second characteristic is important. All ground-glass-type surfaces not only chiefly scatter forward, which is good, but **they allow a very considerable portion of the light to pass through without appreciable deviation** so that it is imaged as before. This is usually bad if the diffusing surface is expected to act as a light source or if photometric measurements are to be made when the diffuser is in the image plane. When a piece of matte film, such as Kodapak Matte Sheet, or ground glass is placed in front of a light source or in a beam of light, only part of the illumination on the other side of it acts as if it originated at the diffusing surface. Examination through a microscope of any matte surface, including ground glass, will reveal that a surprising area of the surface is still lying in the plane of the whole sheet. Experiments in the author's laboratory showed that **the most efficient diffusing surface was made by putting on glass a thin coating of a crushed powder of high refractive index and high concentration in a vehicle of low refractive index.** This was definitely preferable to a dispersion of optically comparable spherical beads of equivalent size. The next best was **sandblasted glass,** compared with that obtained by grinding with an abrasive.

A third important characteristic of diffusers is the wavelength effect, that is, **the increasing yellowness of their appearance as the degree of diffusion is increased.** This effect is caused by the fact that the shorter the wavelength, that is, toward the blue end of the spectrum, the more the light is scattered. This can be expressed by the statement that **the shorter the wavelength, the broader the polar curves of equal brightness in front of a diffusing screen** with a light source behind it. This has nothing to do with any absorption of the material of the screen that would be superposed on this effect. The bluish series of Kodak Light Balancing Filters can be used to compensate this yellowness, although the spectral neutralization is not exact. **The Kodak Light Balancing Filter No. 82A will balance out the yellowing effect of a sheet of flashed opal glass** reasonably well. The White Translucent Plexiglas (see Index) yellows the light less so that **a Light Balancing Filter No. 82 is sufficient to balance out its defect**.

Use of a Matte Diffusing Sheet to Make a Secondary Light Source

The result of the second consideration discussed above is that when a diffusing sheet of this type is placed in an illumination beam of a microscope the latter will be illuminated, in effect, by two sources — one at the location of the primary source and the other at the location of the secondary source. This may not be good. **The diffuser should definitely be either in a field plane or an aperture plane,** not in the vignetting space in between.

One way to solve the problem is to **image the source from behind onto the matte sheet with a lens.** The two sources are now in conjugate planes. However, the method usually furnishes light at a level too low, for photomicrography, at least, with lenses that are practically available. The source is four focal lengths away from the sheet for a 1 : 1 image. The lamp can be brought much closer to the diffusing sheet than it could to any lens with much gain in brightness. Although there would be no diffusion if the lamp were in the plane of the sheet, as mentioned above, the distance of the radius of the lamp bulb gives sufficient diffusion for most lamps. At this distance the error of the two source locations seems inconsequential; with the separation very much greater it may not be so. The seriousness of flare depends on the type of specimen.

This type of lamp (primary source plus diffusing screen plus diaphragm) is not usually suitable for use with an exceedingly small primary source, the very type that is most efficient for high magnification with Method 2; the area of uniform illumination at the screen is too small. On the other hand, it is most suitable for the larger lamp bulbs, especially the miniature type, and the original source need no longer be uniform over its area; a lamp with parallel filaments can be used efficiently, although the filaments *can* be too widely separated. When a large lamp bulb must be used, there is a method (discussed later) for reducing its size with condensers, but the author prefers this method. When a No. 1 photoflood lamp is to be used for making photomicrographs with a Type A color film, it has been found preferable to illuminate a diaphragmed matte sheet, rather than to use the bulb directly, also with a diaphragm, even though it has a diffusing envelope. The illuminated area is more uniform (see Figure 7-3).

The photoflood lamp is held in a sheet-metal housing that allows the tip of the bulb to be near its front edge. A perforated tin can would do, but a number of commercial lamps hold household-type bulbs—including lamps for reading in bed. A water cell of about $\frac{1}{2}$-in. (1 cm) depth is made (see Chapter 9); its glass plate away from the bulb is a fine matte on the outside. A small iris diaphragm is held by a brass plate that could be slid over the frame of the

Figure 7-3a   Adaptation of photoflood lamps for photomicrography.

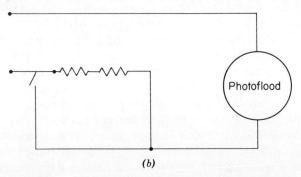

Figure 7-3b   Resistance circuit for visual and photographic use of photoflood lamps.

water cell to center the iris aperture being held down by spring studs with locks. Another piece of very fine glass or a piece of matte film is held directly behind the iris, making the separation of the two diffusing surfaces about $\frac{1}{4}$ in. Two 600-W radiant-heater units are in series with the photoflood lamp, except for the actual photographic exposure when a toggle switch (double-pole, double-throw) is used to provide a short circuit around them. When this lamp was used with a microscope setup with a 10X objective and 15X eyepiece and with a 35 mm camera bearing a 45 mm lens above the microscope, there were slightly more than 2.0 ft-cs in the image plane of the camera. There was no lamp-condenser lens and the iris diaphragm was imaged in the field by the substage condenser. This would require an exposure of $\frac{1}{20}$ sec with Kodachrome II Film, Type A. When an $f/1$, 60 mm, aspheric condenser was interposed at a focal distance away and the substage condenser refocused to again image the lamp diaphragm in the field (Figure 7-1c), the illuminance in the camera image plane was exactly the same. The lamp diaphragm opening now was less than 3 mm in diameter, which is inefficient with this type of lamp. Hence the condenser was of little help in this case. These values of illuminance in this plane compare with an illuminance of 32.5 ft-c in the same plane obtained from a lamp containing a ribbon-filament lamp and an $f/1$ condenser. Allowing for the filter factor to convert to the correct illumination quality, an exposure of only $\frac{1}{200}$ sec would be required.

In the general case it is usually best to set an iris diaphragm in front of the primary lamp, each individually centerable. If the iris diaphragm is held in a vertical metal plate, the matte diffusing sheet can be held on the back of the plate with clips; it can rest on a ledge or other stop. Thus the diffuser can be lifted out temporarily so that the primary lamp can be revealed for adjustments or replacement. A small lamp iris diaphragm sold by Bausch & Lomb* will screw into the hole in a metal plate and is most conveniently held. A larger iris diaphragm, sold by the American Optical Co.,† can be used for larger areas of illuminated diffusing screen and is useful for low-power work with larger fields. For homemade equipment an efficient variable diaphragm can be made from two sheets of metal sliding over each other, each bearing a horizontal V-notch at one edge to form a square diamond aperture of variable size.

Use of Secondary Source (Diffuser) with Flash Lamps

Probably the most useful application of a diaphragmed diffuser (such as matte Plexiglas) as a secondary source is to increase both the efficiency and convenience of the use of flash lamps for color photomicrography. This, in turn, makes probably the most convenient illuminant appropriate for color films. The total method is the simplest and easiest for color photomicrography for many microscopists not set up for routine photomicrography, who may sporadically wish a color record of a field that is being examined. This is discussed on p. 605.

*Their catalog No. 31-58-28.
†Their catalog No. 354.

Characteristics of Köhler Illumination (Method 2)

Illumination Method 2 utilizes the principle that in concentric space (an infinite series of spherical shells) surrounding any *relatively* small light source, or in all successive planes directly in front of a large flat light source, these equidistant areas are uniformly illuminated. Hence an aperture of any reasonable size in front of most light sources will be illuminated quite uniformly. (It is easy to show that depth in the light source is the factor most likely to cause nonuniform illumination of this aperture; e.g., in a double layer of coiled filaments the rear coils are shadowed to produce stripes of darker or lighter illumination.) **Aside from depth-of-source effects, the nature of the pattern of the source will not affect the uniformity of illumination in the aperture plane in front of the source to an appreciable extent.** As observed in Chapter 1, the aperture can be filled with a lens without affecting the illumination distribution of the plane, except as lens thickness produces some out-of-plane effects. Although the method had been used before, it was August Köhler, head of the Optical Department of Zeiss, who systematically recommended imaging this plane of the lamp-condenser aperture in the field with the substage condenser and supported his argument with a mathematical development of the case. This method also includes the procedure of imaging the light source in the lower focal plane of the substage condenser so that a cone of collimated beams will pass through the object plane and the light source will again be imaged in the rear focal plane of the objective.

It can easily be seen there by merely removing the eyepiece and looking down the tube. **A pinhole eyepiece is a great help for this important act.** Here the pattern of the light source, although it is no longer directly seen by the eye when using the microscope to examine its field, is definitely important and is discussed later. The Köhler method has two highly desirable advantages which have caused it to become the more used method: (a) the light source itself need not be essentially a uniformly illuminated area; (b) it is practical and relatively simple to provide adequate diaphragming for *both* field and aperture planes. Since the hot source itself is no longer imaged in the field, a diaphragm against the relatively cool lamp condenser becomes the field diaphragm. The light-source image at the substage iris diaphragm is the efficient aperture diaphragm for both condenser and objective.

Unfortunately most microscope lamps that project a beam toward a microscope do not have their lamp iris diaphragms at or very near the lens. The Wild Universal Microscope Lamp, however does have its iris diaphragm near the condenser lens, a real advantage. When the lamp and its lens are separately supported, as is simple and advantageous with a horizontal illumination system, the field diaphragm can readily be placed almost against the lens. It was found that the flare light in the field was reduced about the same amount, whether the iris was placed just behind the lens (between it and the lamp) or just in front of it. However, **when two iris diaphragms were used, one on each side of the lens, the flare light was reduced with cumulative effect.** This is taken to mean that the source of the flare that is removed is different in the two cases.

Quality of the Lamp Condenser

Even today the quality of the lamp condenser is too often neglected as unimportant. Actually, relatively good optical corrections for the lamp condenser are quite advantageous, and a completely uncorrected, single lens is a definite detriment to final quality. The specific defects that can be introduced by a poor quality lamp condenser are the following

1. Nonuniformity of the illumination of the field.
2. Degradation of image quality by inadequate resolution.
3. Degradation of the color quality of the illumination in the image plane.
4. Introduction of flare.

If both an entirely uncorrected "bull's-eye"condenser and a corrected lens of about the same focal length are available in a form in which they can be used with a lamp, it then takes only a few minutes to discover the troubles encountered with the simple lens. The four troubles listed are discussed in order.

1. *Increased Difficulty in Obtaining Uniformly Illuminated Field.*
Correction of spherical aberration is the most important improvement because of the apertures often involved, and fortunately it is the one most frequently made. **The simplest and cheapest, hence most popular, method is that of dividing up the lens into components, each a simple spectacle-lens type** whose total power in diopters (= l/focal length, see Chapter 1), as modified by the separation factor d is equal to the desired condenser lens. Since spherical aberration and some other optical defects are increased with the curvature of the lens, that is, the angle at which the light rays depart from being perpendicular to the surface, this artifice can really improve the quality of the image of the light source. Its principal disadvantage is the number of air-glass surfaces introduced, with consequent increase in flare from the interreflections of the light. **If the lenses have an antireflection coating on them, the gain is definitely worthwhile** — in fact, may be striking. When assembling a lamp condenser from simple lenses, remember the principle of the best quality with minimum devi-

ation of the rays on entering or leaving each surface; **arrange the lenses, usually with the strongest in the center, so that there is no one surface with rays flatter to the surface than can be avoided.** Usually, also, **the surface toward the light source should be plano or a meniscus lens with the concave side toward a small source.** Since spectacle lenses of uniform diameter may be obtained to specification from any optical shop (source for most opticians) and coated by several firms,* such a condenser becomes a practical improvement.

The alternate method of correction for spherical aberration is to make an aspheric lens. If this is well designed and produced, a very satisfactory lens is the result (except for possible chromatic troubles discussed later). The Bausch & Lomb 60 mm *f*/1 aspheric condenser is such a lens. The author has a group of them which has proved to be a great general utility. Unfortunately Bausch & Lomb no longer sells a microscope illuminator that uses these lenses but at this writing they are still available.

2 *Inadequate Resolution.* With a simple "bull's-eye" condenser, it will be found difficult, if not impossible, to place a well-defined image of the light source in the rear focal plane of the objective *to fill it well*, with no light image outside the plane of the substage iris diaphragm where it is well controlled. This is seen by examination down the microscope tube through the pinhole ocular. The central area can be filled with a blob of light, or one can have an annular ring with a darker central portion. This, of course, will affect the NA-resolution relationship and the quality of the image.

3. *Effect on Color Quality.* The characteristic of **uncorrected condensers** that seems to be least comprehended is the effect of such collectors on the spectral quality of the final image plane. However, we need only to consider the matter, which is important, since almost all lamp condensers for photomicrography have zero chromatic correction. The lamp condenser is visually focused to produce an image of the light source at the rear focal plane of the substage condenser. **Actually a series of images of the source, stretching along the optical axis, is produced.** These images range from ultraviolet to the infrared, with the ultraviolet image being closest to the lamp (see Figure 7-4). The substage condenser is a short-focus, very high-aperture lens. It can accurately focus **only one** of these images in the plane of the objective aperture, which is quite a small hole, filled with a lens. The other images are somewhat out of focus, hence larger. The eye is most sensitive to yellow-green, and in attempting to focus the white-light image it is not surprising that **the yellow-green**

*Such as Evaporated Films, Ithaca, New York.

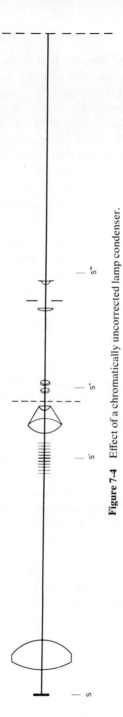

Figure 7-4 Effect of a chromatically uncorrected lamp condenser.

image is focused best and gets through best. It is a most revealing experience to lead a portion of the beam of light, by good quality mirrors, from the lamp to bypass the entire microscope system and end at a groundglass set up in the same plane as that of the camera. It may seem difficult to understand why the yellow-green cast of the image plane had not been evident before. This defect is, of course, worse with lamp condensers of long focal length. Yet the author found that the yellow-green cast disappeared when an achromatic lamp condenser, as well as a substage condenser, was used; moreover, the illumination was more uniform. This important effect for color photomicrography is discussed in Chapter 19, Volume 2.

4. *Introduction of Flare.* The flare introduced by the use of uncorrected lamp condensers is a more obvious factor and a real and important one. The light that is not imaged in the planes of the controlling diaphragms is vignetted and poorly controlled. Flare is discussed in more detail later in this chapter.

Focal Length of Lamp Condenser.

As in every case of a lens and its object, the focal length of the lamp condenser should be great relative to the used length of the light source (with a ribbon filament, this is its width, i.e., about 2 mm). **As the light source becomes very small, the relative size (i.e., the aperture) of the lamp condenser required becomes larger.** With Köhler illumination this produces better photometric efficiency and shorter camera exposures for the same microscope image magnification, but **it also requires condenser lenses of better quality.** However, as just discussed, the better lenses are advantageous anyway in achieving excellent illumination quality. It will be noticed that extremely small commercial light sources, such as the zirconium-arc lamps of lower than 100 W, are furnished with excellent but often expensive lamp condensers, and even they may not be chromatically corrected.

The important statements of the preceding section have been rather baldly made. The background and supporting evidence are better discussed with consideration of the size of the light source in a succeeding section and in Appendix 3C.

Lamp Distance, Actual or Optical

If a substage condenser of good quality is being used, it will specify the proper distance of the lamp because substage condensers, like all other lenses, must be corrected for a specific object-image distance. In this case the lamp condenser is the object for the substage condenser and it images it in the object field of the microscope itself. The lamp therefore should be at this distance. Table 7-1 shows the distances assumed for some of the substage condensers.

Sometimes there are reasons for the actual distance to the lamp to be different from the assumed value. In the later discussion of efficiency and flare it will be seen that from that standpoint the lamp distances should be adjusted to make the magnified image of the filament *just* fill the aperture of the substage iris diaphragm after the latter has been set to fill the chosen numerical aperture set by looking at the back of the objective, unless this opening is known a priori. With experience it will be so, but it will mean that a somewhat variable distance of the lamp will depend on (a) the size of the light source, (b) the focal length of the lamp condenser, and (c) that of the substage condenser.

What if a well-corrected substage condenser is being used and the lamp distance is *appreciably* different from that assumed by the designer of the substage condenser? Since "object distance" for a lens is really a matter of the angle of the incident rays from the object, the latter **can be brought to the correct distance optically with an auxiliary lens against it** by using the simple lens formula (Equation 3 of Chapter 1). The German manufacturers tend to assume a 30-cm distance, the American, 10 in.

Example. Assume a lamp distance of 1 m is being used with an achromatic substage condenser made for a 30-cm distance. The actual lamp is 1 m away (object distance for auxiliary lens), whereas it should seem to the original condenser to be 30 cm away (image distance). The image is virtual, of course.

$$\frac{1}{f} = \frac{1}{1000} - \frac{1}{30} = \frac{30-100}{30000} = -\frac{7}{3000} = -\frac{1}{429}. \qquad (1\text{-}3)$$

A spectacle lens of $-1000/429$ or minus 2.33 diopters is needed. Zeiss furnished a -3.0 diopter spectacle lens for their achromatic condenser.

Zoom Optics for the Illumination System

It is obvious that to use the Köhler illumination system with great efficiency the whole illumination system should be refocused with each appreciable change of image magnification, since this involves a change in the size of the field and in the required aperture of the illumination cone. This requires change in the lamp distance and, with great variation in magnification, change in the lamp condenser for focal length or the lamp filament itself to maintain optimum size in illuminated field and

Table 7-1 Substage Condenser (Brightfield) Specification

Manufacturer	Type or Name	Designation or Catalog Number	Assembled		Top Off		Both Upper Elements Off		
			NA	E.F. mm	NA	E.F. mm	NA	E.F. mm	Distance
A-O	Abbe	C 233	1.25	10.49	0.40	33.0			10 in.
	Wide angle	C 234	1.40	7.78	0.61	17.9			10 in.
	Achromat	C 235	1.30	6.13	0.57	14.1			10 in.
	Achromat	C 236	1.40	9.78	0.61	22.4			10 in.
Bausch & Lomb	Abbe(2-element)	31-58-74	1.30	10.2	0.40	32.3			
	Abbe(3-element)	31-58-72	1.40	7.70	0.73	19.0			
	Achromat	31-58-83 -85	1.40	8.85	0.70	19.4	0.45	37.4	
							0.30	41.6	440 mm from stage
Leitz	Two-lens swing out	80 81 70	0.77	18.5	0.38	53.6			∞
	Two-lens swing out	95	1.20	10.4	0.53	29.8			∞
	Two-diaphragm swing out	76 77	0.95 With immersion cap 1.30	10.4	0.22-0.30	25.7			∞
	Berek			6.0	0.22-0.30	25.7			∞
	Aplanat-Achromat	ilpen ilpa 72	1.36	9.35	0.63	20.5	0.34	41.9	∞
	Three-lens	96 (ilona)	1.40	7.24	0.75	17.6	0.43	36.5	∞

Bu-12 and Bu-13 with phase aperture.

Maker	Condenser	Number							Distance
Watson	Universal No. 1 achromat	7C 180	1.25	10.0	0.54	20	0.87[a]		
	Holoscopic	7C 184	1.00	5.5	0.4	12	0.33[b]		
Wild	Wide field	6020	1.30	48.1	0.60				
	Swing Out	0.40	0.20	0.2	48.1				
	Two-lens aplanat	6013	13.35	12.1	0.65	22.8			34 cm
	Four-lens aplanat-achromat	6030	1.30	13.8	0.70	23.6			34 cm
	Special top lens	6030	1.30	14.6	0.95	16.1			34 cm
	Universal	6071	0.89 Darkfield	10.6					
	Universal	6071	1.26–1.43	14.6					
	Phase	6070	0.9	28.9					
	Long working phase	6073	0.52						
Zeiss Prewar	Achromat	11-42-30	1.0	11.5	0.45	26			30 cm
	Aplanatic	11-41-40	1.4	10.5	0.4	37			∞
	Achromat	11-42-24	1.4	9.5					30 cm
	Achromat-aplanatic	IV/Z7							∞[c]
Zeiss[d] (Oberkochen)	Achromat-aplanatic	V/Z	0.9	10.4	0.63	15	0.32	30.7	∞
	Achromat-aplanat	IV/Z7 and V/Z with lens II	1.4	6.77	0.63	15	0.32	30.7	25 cm[c]

[a] Middle lens out only.
[b] Both lenses off.
[c] An auxiliary lens, 25-cm focal length, is furnished (Lens II).
[d] Other condensers (not listed) are not corrected, hence are less critical for distance.

aperture. Few microscopists do this. Therefore, it was attractive to construct an illuminator that incorporated a lamp condenser of variable focal length or to change its distance optically by a combination of negative and positive lenses that utilized variable distance between them to change power (zoom principle). Thus the lamp condenser could be kept in focus in the field with variable magnifications for its image just to cover it; simultaneously, the filament must be focused in the substage aperture with the magnification adjusted to have its image just fill the properly adjusted condenser aperture. Two pieces of equipment were marketed for this purpose—the Zeiss Pancratic Condenser and the Panfocal Illuminator of Bausch & Lomb. Actually both items were complete illuminating systems, including the lamp. There is so much to be accomplished automatically that neither system was perfect. Both depended on some diffusion by the lamp bulb and the lamp condenser lens.

The rather famous Zeiss Pancratic Condenser is unfortunately no longer available as a separate item, but it is incorporated in some present microscope models for built-in illumination [the Zeiss (Jena) Lumipan LP and the Nf models] and the principle is frequently mentioned by name and in the literature. A small precentered Osram lamp is used. The conveniently small diameter and length of the total tube is a limitation. Centration is not always perfect with such a small lamp condenser of such short focal length, with no centration adjustment provided. The desired aperture value is set by turning a knurled ring to the appropriate scale reading. This refocuses the illumination by altering the effective focal length of the zoom lens system, so that **the filament image, somewhat diffused, is in good focus in the aperture diaphragm and at the proper magnification to just fill the prescribed aperture.** This diaphragm is independently adjustable to set the admitting aperture suitably. There is no field diaphragm at the lamp condenser lens, since the small size of the lens and the variable magnification allow control of the field illumination. However, an iris diaphragm here would be most advantageous, especially since **the condenser is not focused accurately in the field** except by separate focusing of the whole unit; for example, when it is used in a carrier operating with a rack and pinion. The field is uniformly illuminated only because of the diffuser.

The Panfocal condenser is very convenient to use because it is all contained in a unit held vertically under the stage. It also has a simple adjustment that allows the use of darkfield, which can be most advantageous. The opposite design decision was made, however. There is a **centerable diaphragm against the lamp condenser which is accurately focusable in the field** to give uniform illumination. In this case **the lamp filaments are not well focused in the aperture planes at all magnifications;**

this defect becomes worse at high magnification with the accompanying high apertures. With the Pancratic Condenser the lamp diffusion was required to illuminate the field uniformly. With the Panfocal, **diffusion is required to fill the aperture; even then it is much brighter in the center at high apertures.** The greater photometric efficiency of the first method, accurate focus in the aperture plane, is strikingly illustrated here. The image-plane illuminance of the two illuminators cannot be accurately compared because a "daylight color filter" is built into the Panfocal condenser with no clear aperture. However, at an image-plane magnification of 400X at 0.85 NA, the illuminance of the image is about 170 times brighter with the Pancratic condenser (with both lamps at the same color temperature) but would probably be only about 35X brighter without the daylight filter. Both devices were tested against a ribbon-filament lamp that was set up on an optical bench with the same color temperature and with *both* aperture and field of ideal magnification and focus; the image at the same magnification was almost four times brighter than with the Pancratic condenser, yet all planes had well-controlled illumination. The image illumination obtained with the Panfocal condenser is probably inadequate for much photomicrography.

TYPES OF INCANDESCENT LAMP

Anyone doing a considerable variety of photomicrography should be familiar with the types of incandescent lamp that are commercially available, particularly if the work includes much done at low magnifications at which the specification for the optimum size of the light source changes rapidly. The principal sources of such lamps in the United States are the General Electric Co. and Sylvania Electric Products, and from Europe Osram and Phillips. Japanese bulbs are competing, and the Japanese are selling some of the compacted rectangular grid lamps that are so effective.

The most important types for photomicrography are projection-lamp and miniature bulbs. Most miniature lamps operate at relatively low voltage. Many projection lamps operate at standard voltage (115 V), but those lamps seeking to furnish a small concentrated light source of high intensity use low voltage to prevent arcing within the filament. **Miniature lamps** are designated by number; for example, GE No. 1630 is the 23-cp, 6.5-V, 2.75-design-ampere lamp of 100 hr rated life used by Bausch & Lomb in their illuminators for the Dynazoom microscope. All American **projection lamps** are designated by an ASA code of three letters; for example, lamp CDS is a 100-W, 115-V lamp in a T-8 bulb with a single-contact pre-focused bayonet base of candelabra size; it is used by the American Optical Co. in their No. 735 lamp.

The types of filament that are available and of possible interest to photomicrographers are shown in Figure 7-5a, taken by courtesy of the General Electric Co. from their catalog. A letter in front of the number tells whether the wire is straight (S), coiled in a spiral (C), or the latter also coiled in a spiral (coil-coil, CC). A coiled wire not only allows greater length of resistance but cuts thermal conduction losses, since the spiral acts like a wire of the same diameter in this situation.

For some years now a relatively new type of incandescent filament structure has been on the market, which was designed especially for microscopy to fill a round aperture efficiently and of a size to fill both aperture and field over a range of magnifications; yet at present this structure has no standardized code designation. It consists of a compact grid of filaments that forms a rectangle facing the end of the bulb through which it is normally imaged. The first of these lamps came from England as a Siemen's No. 99A "48-watt Solid Source," shown in Figure 7-8, No. 8, and listed in Table 7-2. It is now sold by the A. E. I. Lamp Lighting Co., Ltd. It consists of a set of straight coiled filaments, each similar to the C-6 construction but close enough actually to touch each other to form a rectangle, $3 \times 2\frac{1}{2}$ mm on a side. Apparently this stimulated a whole set of competing lamps with a rectangular filament grid facing the end of the bulb that is a definite advance in photomicrography. However, in all of these other lamps the compact rectangular grid is formed by a single, usually rather thick, round filament that is wound into a flattened cylinder (see in Figure 7-5 a 6V 5A). They do not achieve the actual contact in the filaments of the "Solid Source," although the image of the incandescent filament may show little space between adjacent rings of the coil. The

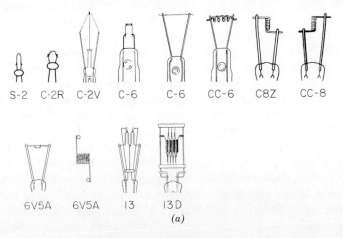

Figure 7-5a   Filament types.

space in the center of the coil is only about twice the thickness of the filament wire. They do vary in the compactness of the coil, a 15-W version being a bit more open. The first of these lamps seem to have come from Japan and their lamps are quite prevalent. Now they are made by most of the lamp companies but are unmarked, being publicized only by a catalog number of a microscope manufacturer or his dealer. These lamps include Phillips (No. 13347, 6-V, 15-W), G.E. (No. 1649, 6½ V, 2.75 A S-8), and several from Leitz and Zeiss of different wattages up to 12 V, 100 W. They are also discussed again in relation to various illuminators in this chapter (see Index, *lamps*).

Bulb shapes are illustrated in Figure 7-5b. Tubular bulbs are designated according to their diameter in eighths of an inch; for example, bulb T-8 has a diameter of 1 in.

About the only sizes of base that are important in this field are the *candelabra* and the *medium*. The common household bulb uses a medium base, which may be either screw or bayonet. The latter can also have an arrangement for the correct orientation and focus of the filament. The candelabra base may have a single central contact (SC) or two contacts (DC).

The Shape and Structure of Light Sources

The structure, including the shape and pattern, of a light source, is very important in photomicrography, even with Köhler illumination. The image of the source is impressed directly on the round aperture of the substage condenser, and only that portion of the image with that area and shape will enter it. It can be seen, with Köhler illumination, by looking down the tube; it is convenient to use a pinhole eyepiece. The pattern is the most obvious feature if the image of the source does not fill the aperture.

According to many texts, the pattern should be one of uniform illumination, as shown in Figure 7-6a, whereupon the substage iris is closed down to 0.9 aperture. This is somewhat of a carry-over from the days of the oil lamp, when so much was studied and written about this subject. Now, only a fraction of the used light sources can give this pattern; one of them

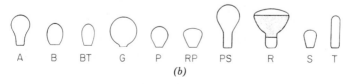

(b)

Figure 7-5b   Bulb types.

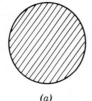 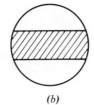 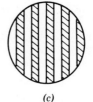 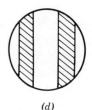

(a)   *(b)*   *(c)*   *(d)*

Figure 7-6   Pattern of filaments in aperture.

is the ribbon filament, whose pattern in the objective aperture *may* look like that of Figure 7-6*b*. Obviously the lamp should be removed to a further distance until the refocused filament image is wide enough to fill the substage opening. A lamp condenser of shorter focal length may be needed if it is impossible or undesirable to move the lamp. The limiting dimension of the ribbon filament is its width, which is 2 mm. The arrangement represented by Figure 7-6*b* might give inadequate resolution in one direction.

Sometimes the pattern may consist of parallel filaments, as represented by Figure 7-6*c* or *d*. Sometimes, again, the pattern may not adequately fill the aperture in one direction; in Figure 7-6*d* the two filaments are most likely a pair from a grid of four. Here, again, the lamp should be removed and refocused to recover a pattern, including some of the four filaments and more nearly spanning the aperture. This pattern gets far away from the often-advocated uniform filling of the aperture, as in *a* of the figure. This filling now seems unnecessary in the light of the virtues of anoptral and phase contrast (see Chapter 2), in which *some* of the aperture should be left clear of the image of the illuminant to be filled by light scattered from the specimen. However, the two contrast methods cited utilize the fact that most of the diffracted light is scattered forward in a narrow angle and cannot be counted on to fill the objective aperture when the patterns are like Fig. 7-6*b* and *d*, *except for especially strongly scattering specimens*. In industrial microscopy in particular the specimen may contain elements that act as lenses on the stage. These will form an image in the image plane of the aperture which contains an image of the light source. If the latter consists of parallel filament strands, the appearance of the specimen may be that of having detail that is really due to incomplete imaging of separate filaments.

One should expect that the troubles just described could be obviated by utilizing the structure of a biplane filament, illustrated in Figure 7-8*a*, item 4. It is used especially for picture projectors, but its weakness is emphasized whenever an attempt is made to use a biplane-filament lamp in a photometric type of instrument that involves a large field or aperture. If the field is bright, the eye is deceived into believing that it is uniform. If it is scanned with a photometric probe of small aperture, it will be found to be ribbed with a nonuniform brightness. Moreover, this ribbing of illuminance will be found in the space in front of the biplane lamp before it enters a lens system. It is caused by the variable shadowing of the rear filaments by those in front of them as the probe passes horizontally in front of the lamp. This nonuniformity usually shows up photographically only in a relatively bare field when the photographic contrast (gamma) is high but then may cause lost time in running down the trouble. Therefore it is safer to use the monoplane-filament lamp unless it is known that this fundamental nonuniformity will not bother the case in hand. The axial distance of the two planes of the biplane-filament lamp is most likely to bother at the higher apertures and powers in which it is least likely to be used.

The structure safest from many troubles is that of a uniformly incandescent solid plane of limited area. Such a source also is not bothered by flare from semifocused reflections from the back of the glass lamp tube, which often definitely do introduce flare. It is imaged in one plane definitely. Some arc lamps belong to this type and are discussed later.

The shape of the source chiefly determines the portion of the light-source image that cannot enter the round aperture of the substage condenser. For one thing, this effects the wattage efficiency of the brightness of the image, hence the required wattage for a given use. More importantly, it contributes to the flare in the field because the light in the excluded image has passed through all of the air-glass surfaces up to that point. The following experience illustrates this situation:

Ribbon-filament bulbs can be obtained either with a horizontal or vertical filament, as illustrated in Figure 7-7. The horizontal filament is folded back on itself so that its ends and the cooler binding posts are concealed from the front. It was found that no variation of current or

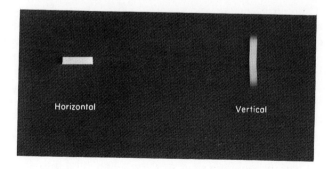

Figure 7-7 Horizontal and vertical ribbon filaments.

photometric filter could be used to make the light quality obtained in the image plane with the vertical-filament lamp (with its mixture of color temperatures) the same as that from the horizontal-filament lamp. Obviously, excluding the images of the cooler ends of the vertical filament *at the substage diaphragm* did not exclude all of the light originating from them.

The very long narrow light sources should have a large flare factor, according to this criterion. It is true that such sources as the ribbon filament have a larger factor than more symmetrical sources of the same width, but it is also true that very long sources such as the H3 mercury arcs (Figure 7-8*b*, items 22 and 23) do not have a flare factor proportional to their length. Apparently after a certain distance little of the extraneous image gets into the field.

The Size of the Light Source

The optimum size of the light source itself is a great variable with large variations in the magnification of the final image and the size of the required field. Use of a source that is appropriate for these factors is most desirable, under pain of **excessive flare, if the source is too large,** and **either a field size or a filled aperture that is inadequate if the source is too small.** This is obvious when the source is directly imaged in the field. With Köhler illumination we have seen in the discussions, and it is obvious from the equations of Appendix 3, that the source plus its condenser form an optical unit (source size × lamp-condenser aperture) in which either the size of the lamp or the size of the condenser can be varied to obtain the desired control. Practically, of course, there is more of a limitation to the extreme size of the lamp condenser, but its effective size, *within* its available range, is varied with the lamp diaphragm and used for this purpose. In fact, as must be continually emphasized, **a lamp-condenser diaphragm should be in use and properly closed down.** However, with a given lamp condenser, located and focused to *just* fill the aperture, **the size of the field will be proportional to the size of the source.** Low-power work has the largest illuminated fields and therefore demands the largest size source.

For an individual who will be engaged in photomicrography covering an unusually wide range of magnifications **it is definitely advantageous to have the illumination system consist of a light source and lamp condenser independently supported on a horizontal optical bed.** This implies either a horizontal bench or a vertical microscope and camera axis with a horizontal illumination beam (Case A of Chapter 6). This excellent but now less-used method **encourages substitution of either light source or lamp**

condenser whenever a different item of either of these components would be more appropriate. The two, source and lens, can even be separated to use Case 2 of the methods of photomacrography described in Chapter 5.

At one extreme a large light source for large fields can be obtained by the use of a diffusing screen illuminated from behind by a bare lamp at an adequate distance from the screen to ensure uniform illumination over the desired source area. **A sheet diaphragm must be placed against the front surface of this secondary source to make it satisfactory.** A ground-glass or matte-film surface may let through too much undeviated light to be satisfactory. A selected flashed opal glass or opal methacrylate sheet (e.g., White Translucent Plexiglas) is almost always satisfactory, except perhaps for low brightness, unless there is a very bright illuminating bulb behind it. This method was discussed early in this chapter under *Secondary Sources*. Fortunately at very low powers less brightness is needed.

The next satisfactory common light source, in order of decreasing size, is the monoplane- or biplane-filament projection lamp with a group of parallel filaments, which are usually coiled, that forms a square or rectangular grill. These lamps are listed as items 1 to 4 inclusive, in Table 7-2 and are illustrated by the same item caption in Figure 7-8. The problems of this type of light source, because of its open-grill pattern, have already been discussed and illustrated as Figure 7-6. The biplane lamp has also been discussed.

The solution to the problems of both types of lamp would seem to have been met by a relatively new lamp type. The low-voltage "Solid Source" lamps, sold by A.E.I. Lamp and Lighting Co., Ltd., of England, are illustrated in Fig. 7-8, items 5, 6, 7, and 8. These decrease in size until the 48-W "miniature" bulb made for microscopy is reached. This lamp and the others also having a rectangular grid were discussed (see Index, *lamps*). The filament grid of the A.E.I. 48-W lamp may be slightly large for conveniently efficient use at the highest magnification, although excellent for low and medium power. Another extreme of this type of grid is the Japanese lamp (distributed by Swift Instruments) rated at 6 V, 5 A, which has a flat grid 1.6 × 2.0 mm. It is most easily used at the high magnifications.

As smaller light sources are utilized, as is desirable with increase of magnification, the actual brightness requirements increase. The alternatives are more compact filament lamps with low voltage to prevent arcing across and higher current, or various arc lamps.

Before considering these light sources specifically it seems appropriate to discuss the criteria for the optimum source size, which will include the background for some of the statements of preceding sections. The more detailed development and equations are given in Appendix 3C.

Criteria for Lamp Characteristics and Arrangement

One of the characteristics of Köhler illumination is that the whole of the lamp condenser may not be used because its image in the field would be larger than the visible useful field. The image size is usually limited by curvature of field; it frequently is no larger in diameter than 4 in. (100 mm). The size of the object field is this image field divided by the image magnification; the size of the object field has been denoted by Q_3 in the equations for the compound microscope. To increase the utilized aperture of the objective, if that is inadequately filled, one would merely withdraw the lamp to make the image of the light source, for example, a tungsten filament, larger on the substage diaphragm; but this withdrawal of the lamp, with its condenser, makes the object distance of the condenser (i.e., lamp condenser to substage condenser) longer and therefore decreases the size of the image of the lamp condenser, which is the illuminated field. This can be expressed by the general equation for all corrected optical systems:

$$s \cdot A_L = Q_3 \cdot A_0 \qquad (7\text{-}1)$$
$$\underset{\text{lamp}}{} \qquad \underset{\text{microscope}}{}$$

where

$s =$ size of light source,

$Q_3 =$ size of object field, as limited only by the illuminated area of the lamp condenser

A_L or $(NA)_L =$ aperture of the lamp condenser, expressed as NA.

A_0 or $(NA)_0 =$ aperture of the objective, expressed as NA.

This expression shows very well the necessity for increasing the lamp size in either filament or condenser to obtain a large illuminated field without sacrificing objective aperture which up to a limit determines image definition. Since the acceptable final picture size is usually severely limited in diameter, only in relatively low-power work are the larger field sizes, hence larger source sizes, needed. Use of this equation is discussed in Appendix 3C, in which a form more convenient for use is given. In this appendix this same equation is designated as No. 4.

Not only is the inexorable relation between the limitations to the maximum size of the field of the microscope and the aperture of the objective shown by this relation, especially as discussed in Appendix 3C, but the need for decreasing source or lamp aperture at higher magnifications is illustrated (see examples, Appendix 3C). The maximum useful field may be set by curvature of field and may be 4 in. (100 mm) in the image plane; it may be the size of a camera frame, as in 35 mm-film cameras. At any rate, since this is usually constant or nearly so, irrespective of magnification, **the object field actually used decreases in size as magnification increases.** With high magnification the diameter of the used field will probably be only 0.1-0.2 mm in diameter whereas the diameter of the area of field that is illuminated, as the reduced image of the lamp condenser, will normally be much larger than that unless the lamp condenser is diaphragmed down or the light source is very small indeed.

Another relation that can be noted is the efficiency with which the light source is used. Since Köhler illumination can be very inefficient when the lamp condenser is diaphragmed down to a small *f*-number or used at a small *f*-number per unit area of field, **the use of a source of smaller area but of the same unit brightness** will not decrease image brightness but **is used at higher efficiency at the higher *f*-number.** A condition to this statement is discussed below.

Effect and Control of Flare

Flare light in the image plane, recorded by a photographic film or plate along with the image, has been cited so frequently as a penalty of an optically inefficient photomicrographic system that it is time to consider it directly. Light rays that pass through a point in the object field but do not pass through the corresponding image antipoint may be caused by (a) the lack of optical perfection of the objective or ocular or (b) by flare. In the first case the rays tend to come near the image antipoint and directly affect definition; in the second case the flare light may be scattered over the whole field. However, this is not a reliable distinction and **flare** certainly can at least indirectly but **definitely degrade definition. Its most-marked and well-known effect is to reduce image contrast.** It would usually occur with even a perfect objective or ocular.

Flare light spread over the entire field can actually be desirable; more frequently its removal does not give a markedly superior-looking image. A case in which it is helpful occurs when a specimen has excessive contrast, too great to be reproduced well in a photographic print, and no delicate detail. A second frequent case is much the same, a contrasty specimen in which the fine detail is perhaps relatively unimportant. In the first case the image, and at least its photomicrograph, will appear to be poorer with the flare removed. However, flare in general produces the equivalent of fog over a photographic film or plate. This will not only lower the contrast of the image but bury fine image detail at both brightness extremes of the image, though for different reasons. An important secondary effect of flare is that **the microscopist may well tend to recover image contrast by reducing the illumination aperture with the substage diaphragm and may close the diaphragm below the value set by resolution**

requirements of the magnification. There is, of course, little leeway at high magnifications.

One photographic component of the flare effect is often overlooked. It is well known in scientific photography that a general controlled flash exposure of a photographic plate or film made before exposure to the image, during exposure, or after the exposure produces an effective increase of speed in most photographic materials. This is, of course, limited by the general fog produced and is at the expense of the distortion of the reproduction curve which lowers contrast and veils shadow detail but in spite of which it has sometimes been useful. It can be seen why a photomicrographer who has carefully removed all flare, after possibly using no field diaphragm at all, suddenly finds that the photographic exposure required is somewhat greater.

It will be assumed, however, that flare is not desirable but rather that "crisp images" are wanted.

In microscopy and photomicrography by transmitted light **all light not going from the source, without reflection, directly to be properly imaged in the aperture and field planes** (and their conjugates) is very likely to end as **flare light**, hence the importance of proper focusing of all elements of the system and the use of both field and aperture diaphragms.

The gross flare should certainly be eliminated, but this may require some deliberate precautions. Dempster (3.1944) lists the important sources. All air-glass surfaces will cause reflections (about 4% each time); the reflected ray may be rereflected into the system as flare. This likelihood increases with the angle of the rays to the axis, especially of converging light. The beneficial effect on flare elimination of a double diaphragm both in front of and behind the lamp condenser has already been reported (see Index, *field diaphragm*), although it is usually impractical.

Light scattered in the image beam is especially bad. This is quickly demonstrated by the terrible degradation that occurs when a dry cover slip is placed over a specimen. Tube flare and flare from lens mountings are usually inexcusable, yet not uncommon. **Remove the eyepiece and look down the microscope tube with strong** brightfield **illumination**. Except possibly for the extreme top of the tube, covered by the ocular, **there should be no observable gleaming rings of light** on the tube or around the lenses. The limitation of the rays should be field and aperture diaphragms earlier in the system. In some cases, for example, with strongly scattering specimens or darkfield, it may be worthwhile to line the offending portion of the tube with the extremely black matte paper sold as Black Velour.*

*See Index, Chapter 6.

The horrible flare with a dry cover slip is a reminder that all parallel glass surfaces introduce flare into the system. Therefore, if a specimen has unusually low contrast, yet also has fine detail, it is wise to immerse all surfaces completely in water or better still in optically homogeneous oil, from slide to objective and from condenser to slide if a flat-topped condenser is used. This should be done, whatever the magnification and whether or not it is called for to fill the required aperture. Immersion objectives or "immersion caps" for dry objectives will be called for. Reducing the angle of the cone of illumination to the minimum that allows adequate illumination is the greatest help, of course. It is extremely slanting yet convergent rays that introduce the most flare.

Finally there is the flare from the object field outside the area directly used in the image plane. In considering image formation, phase illumination, etc., we learned that object points generally scatter the illuminating light forward in a widening cone; this light scattered by the useful field must be collected with care. However, all of the object plane covered by specimen scatters light similarly, but that from outside the useful field is flare and undesirable. Obviously the greatest trouble will come from the illuminated areas just outside the accepted area. Hence a well-placed, well-focused field diaphragm must be closed to the very edge of the accepted field for maximum protection from flare. It is this important source of flare that is most frequently neglected.

A comparison of Figures 1-29*a*, 1-29*b*, and Figure 5-4 provides a good illustration of the effect of flare. In the first case, in which the specimen was mounted on a piece of opal glass, with the illuminated field considerably larger than the used field, the flare was so great, with consequent diminishing of contrast, that the objective aperture was reduced to regain the contrast. In Figure 1-29*b* elimination of the outer ring of illuminated field beyond the used area removed this source of flare and allowed the objective aperture to be opened. Further gain in flare elimination, more obvious in the visual image and the photograph, was obtained by a more controlled specular optical system.

PHOTOMETRIC COMPARISON OF LIGHT SOURCES

Although the brightness of most of the light sources used in photomicrography is given in the handbooks, this quantity is measured by looking at the bare lamps and does not even rate the lamps in the same order as by the illuminance that is obtained when they are used in an optical system. This illuminance is affected by their different sizes and shapes. For this reason, and to tie together the factors that have just been discussed, that is, size, shape, and flare tendencies, a systematic comparison of most of the light sources used for photomicrography was made by determination of the actual illumination that was delivered to the ground glass of a photomicrographic camera.

The conditions of the test were as follows: the optical system was set up for 400X magnification using a 4 mm apochromatic objective. The lamp condenser was a 60 mm $f/1$ lens. The source and condenser were located so that the image of the source *just* filled the aperture of the substage condenser. The substage iris was closed down to 0.92 of the objective aperture. To determine the flare factor the field diaphragm against the lamp condenser, which was first omitted, as is done too often in practice, was then closed down until seen in the visual field.

All but the flashlamps were measured with a barrier-layer photoelectric photometer. All flashlamps and some others were measured with the General Radio Integrating Photometer, which operates by making integral counts, each of which is a unit of exposure; that is, it measures light intensity × time, thus allowing comparison of the continuously burning sources with the flashlamps. The spectral responses of the barrier-layer cell and the integrating photometer were not the same, but all ultraviolet radiation was filtered out of all measurements. Mercury arcs can be compared with incandescent sources by comparing their relative brightness in the image plane with green light, using a Kodak Wratten 58 filter for the incandescent sources and isolating 546-mμ wavelength with a Kodak Wratten 77A filter.

The results have been gathered in Table 7-2. All of the light sources have been given an item number in the table, which is also their indentification in Figures 7-8a, b, and c. The latter show photographs of the light sources made at the same magnification, except for the flashlamps of Figure 7-8c, which were made at half of the magnification. The table and accompanying photographs will allow a systematic discussion of these types of light source for photomicrography. Some data on other incandescent lamps now used for photomicrography are given in Tables 8-1, 2, and 3.

Projection Lamps

This test, which utilizes a compound microscope set up for 400X magnification, is unfair to the large projection lamps, Nos. 1–4, which have their field of special usefulness for low power, especially with the simple microscope discussed as Case 2, Chapter 5. However, their inclusion does illustrate their characteristics and provides a nice example of the use of a much-too-large light source which unfortunately is not infrequently done. The open grill of standard projection lamps, Nos. 1 and 2, can be seen and has been discussed. A series of photographs of a biplane-filament light (Item 4) at a series of angles from the perpendicular axis would have to be made to illustrate the nonuniformity problem encountered with them, as discussed earlier in the chapter. Lamps 5–8

Table 7-2 Comparison of Light Sources for Photomicrography

1	2	3	4	5	6	7	8	9	10	11	12	13	14	15
					Illuminance in Foot-Candles			Integrated Illuminance (I·t)		Flare Factor with No Field Diaphragm			Phot.	Rated
Item	Light Type	Source Kind	N.D. (millimeters)	Amperes or Volts	Color Temp	White Light	Green Light	Total Flash	1/100 sec.	Illumi-nance	% at f/1	α_L	Effic-iency X1000	Life (hours)
1		500 W	13	120 V										
2	Incandescent Tungsten	Projection 100 W (CDS) (PH/100T8SC)	5	120 V ac	3025°	1.00	0.13			(1.95)	95	8.7	10	50
3		300 W (CLS)	7.5	120 V	3170°	5.1								25
4		Projection 500 W (CZX) biplane	8.5	115 V 4.35 Am ac		5.08				(9.85)	94	8.7	10	25
5		Solid Source 400 W	8	24 V 16.65 Am ac		4.25	0.49			(8.55)	102	8.7	11	
6		Solid source 250 W	5	24 V 10.4 Am ac		3.90	0.46			(7.90)	102	8.7	16	
7		Solid source 150 W	4.25	12 V 12.50 Am ac		4.10	0.47			(7.30)	78	6.1	27	
8		Solid source 48 W	2.25	6 V 8.0 Am ac		2.80	0.37			(3.75)	32	3.8	58	
9		Zeiss 38.01.77	1.6	6 V 2.52 Am	2840°	1.61	0.23			1.89				
10		Zeiss 38.01.77	1.6	8 V 2.95 Am	3160°	4.01	0.62							
11		Ribbon filament (horizontal)	2	6.8 V 17.70 Am ac	3000°	2.70	0.36		3	(3.40)	20.6	3.0	22	

Table 7-2 (*continued*)

1 Item	2 Light Type	3 Source Kind	4 N.D. (millimeters)	5 Amperes or Volts	6 Color Temp	7 White Light	8 Green Light	9 Total Flash	10 1/100 sec.	11 Illuminance at f/1	12 %	13 α_L	14 Phot. Effic-iency	15 Rated Life X1000 (hours)
						Illuminance in Foot-Candles		Integrated Illuminance (I·t)			Flare Factor with No Field Diaphragm			
12		Coil filament (cylinder)(CPR)	2	5.2 V 16.5 Am ac	3000°	4.04	0.55	6	6	(4.70)	13.6	3.0	47	50
13		Coil filament (cylinder)(CPR)	2	6.1 V 18.0 Am ac	3200°	6.28	0.86	10	10	(7.30)	16	3.0	57	50
14	Incandescent Arcs	Point-O-Lite	2.5	120 V 1.5 Am dc	2800°	3.05	0.34	5	5	(4.10)	34	4.1	17	
15		Tungs-arc	3.5	115 V ac		6.40	0.89		11	(9.85)	54	5.5		
16		Zirconium arc 100 W	1.5	115 V Rectified dc 6.2 Am	3200° (app.)	10.2	1.40		17	(12.4)	21	2.3	102	
17		Zirconium Arc 40 W	0.75	115 V Rectified dc 1.60 Am		7.40	0.89		11	(7.40)	Very low	2.3	185	
18		Zirconium arc 25 W	0.5											
19		Carbon arc	3	4.5 Am dc	3645°	21.0	3.0		35	(28.5)	36	5.5	39	
20		Carbon arc	4	10 Am dc	3820°	21.0	3.1		35	(42.4)	102	8.7	18	
21		Carbon arc, high intensity		45.0 Am dc	5800°				(178)	(85)			(53)	

322

No.	Category	Source		Volts/Watts								
23	Mercury Arcs	CH4	3	85 W ac 115 V		0.87	0.24	7	(2.29)	12	3.0	24
24		Compact	3	100 W ac 220 V		25.0	5.7		(1.00)	15	2.9	11
25		H6		250 W ac 12 Am ac		38.0	6.2	100	(31.5)	26	3.0	100
26	Xenon continuous	XBO150		115 V	6000°	96.0	12.9	165	(44.0)	16	2.9	27.5
27	Xenon continuous	Pulsarc (cont.)	3			1.08	0.10	3	1.41	(115)	20	
28	Flashpulse	Pulsarc at 40 msec	4			600		150				
29	Flashpulse	Pulsarc at 135 msec	4			1900		140				
30	Flash	SM	34		3300°	26 (1/200)		52				
31	Flash	No. 5	34		3800°	130 (1/50)		65				
32	Flash	M-2			3800°	123 (1/100)		123				
33	Control	Photoenlarger Lamp No. 111	34	115 V		0.04						
34	Flash	Xenon F114	4	900 V 120 W-sec		170 (1/1200)		1400	(250 count)	47	5.5	
35	Flash	Xenon 78M9T	16									
36	Flash	Ascor		300 W-sec		340 (1/600)		2100				

Note: Magnification: 400X. Condenser N.A. filled: 0.95. Diameter of object field: 0.32 mm (5-inch image field).
1. Measured with 29.5 mm $f/2.3$ lamp condenser. Diameter of object field 0.2mm.
W = watts
V = volts

The following comments refer to column numbers, given in parentheses. Hence the formulas show how the items of a column were calculated.

Column   Comment

4   The narrowest dimension in millimeters, that is, its width.

5   The figures given are the significant value or that which specifies the lamp. With a few commercial power supplies the voltage across the lamp and the amperes may not be known. The voltage given for the commercial arcs is for the input of the power supply; the voltage across the lamp is much lower.

6   The color temperature is that of the lamp tested. There may be some variation of these values with the incandescent lamps.

7   The Kodak Wratten Filter No. 2B was used to eliminate the ultraviolet.

8   Filter: Kodak Wratten 77A for items 22, 23, and 24; Kodak Wratten 58 for all other sources.

9   This is exposure E, in intensity $\times$ time. It is a measured value on a *relative* scale and is not simply related to Column 7. Measured with an integrating photometer of different spectral sensitivity. The figure in parenthesis is the reported time of the flash in seconds' duration.

10   These exposures ($E = I \cdot t$) were measured with a calibrated shutter for determining the time. The scale, that of the integrating photometer, is relative. *The values in this column for the flashlamps were calculated proportionally from the actual time of flash and are for comparison of relative intensities only.*

11   This is in foot-candles, with no diaphragm at the lamp.

12   Col. 12 = [(Col. 11 − Col. 7)/Col. 7] × 100. This value represents the proportion of the illumination that is flare because the illuminated field is larger than that used.

13   α_L = focal length/diameter of lamp condenser for a 0.32 mm field. The diameter is that cleared by the field diaphragm.

14   Phot. efficiency = illuminance/wattage = (Col. 7 ÷ wattage) × 1000.

represent the **"Solid Source" lamps** manufactured in England. The placement of their filaments in the aperture should not be so critical as Items 1–2 and the nonuniformity of the biplane filament, No. 4, caused by shadowing of the rear filaments, should be absent. These lamps have proved quite useful for relatively low-power work. Note No. 8 which has been made especially for microscopy. It is a 6-V, 48-W lamp utilized by Baker (Cooke, Troughton, and Simms) in their '*High Intensity Lamp for Microscopy.*' Note from Columns 7, 11, and 12 Table 7-2 that in the larger sources (Nos. 1–6) the amount of flare light obtained at this magnification, when using no field diaphragm, must be about equal to that of the light forming the image. However, the lamp condenser is being used only at $f/8.7$ compared with $f/3.8$ with the smallest Solid Source (Item 8).

Microscopical, 6-V, Filament Lamps

The last three projection lamps, the 48-W Solid Source, the ribbon, and the single-cylinder, coil-filament, 6-volt lamps are quite properly com-

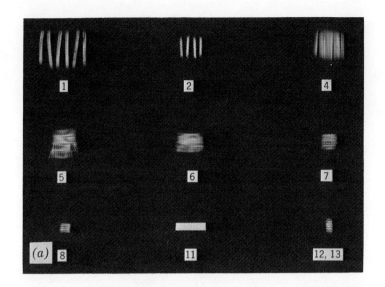

Figure 7-8a Incandescent lamps.

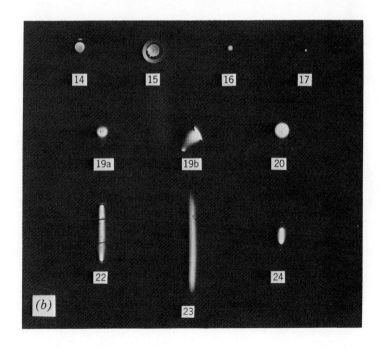

Figure 7-8b Arcs.

325

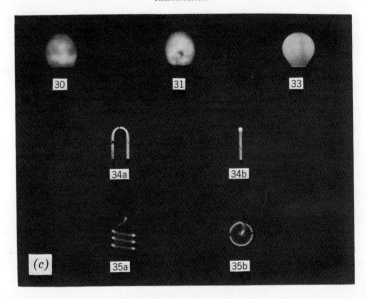

Figure 7-8c   Flash lamps.

pared as competitors. It was necessary to compare the cylindrical coil-filament lamp with the ribbon filament under two sets of conditions, one at equal color temperature (Item 12) and one at the rated current.

The horizontal ribbon-filament lamp is usually operated with the transformer sold with it as an ac 108-W (6-V, 18-A) source. It is very popular because it is unique in the uniformity of its illuminated area for its size, coupled with fairly good stability. It can be successfully used with Abbe-Nelson illumination by imaging it directly in the film plane if the photographic gamma is not high. It is neither mechanically rugged nor very reproducible for color temperature for a given current. In fact, a ribbon filament will sag if the lamp is inclined so that its plane is far from vertical. The filament has such low thermal capacity that the brightness fluctuates with the 60-cycle alternating current. This is not visually obvious directly, but trouble was encountered when taking brightfield photomicrographs at $\frac{1}{200}$ sec exposure; it could give trouble at $\frac{1}{100}$ sec. It does not have a very efficient shape for an optical system, hence has less than half the luminous efficiency of the other two lamps. Probably there is more flare with it *after* diaphragming down the lamp condenser, but there is no measure of that in this test.

Note the excellent performance of the rugged, squat, cylindrical-filament lamp, CPR, also rated as 6-V, 108-W; it has relatively low flare with high brightness in the image plane. It can be burned directly at 3200°K, although its life there may not be extended, but the ribbon filament is utterly impractical at that color temperature. It has considerably more life and reproducibility at 3000°K than the ribbon-filament lamp. It can be considered a good source for medium- and high-power photomicrography.

Tungsten Arcs

The English Pointolite lamp (Item 14) is a relatively old source but still a good one (see footnote, p. 393). It is sold, with a special rheostat circuit box, for use on 110–120-V dc house supply. The light source is a tungsten sphere that is incandescent because of the arc between it and the cathode, which is a wire above it. It is relatively uniform and of a size that is somewhat of a compromise for a large range of magnifications. The author has been told that it is still very popular in England. Its principal disadvantage is the fact that the ball of tungsten moves over and into position after the arc is struck and does not always come back into the identical position each time. Although it has no structure compared with the coil filament, just discussed, it is otherwise less bright and has less efficiency. This seems to be due partly to the fact that it is burned at less current so that it has a long life at lower color temperature (2800°). If the circuit were stepped up to give equal color temperature, it might be comparable. It is a general rule that the efficiency, as given in Column 14 of Table 7-2, increases when the color temperature is raised with a given light source. The life of the source goes down more than correspondingly.

The Tungs-Arc (Item 15) seemed to be a rather clever solution to the position variations of the preceding source. The observer is looking at the end of an incandescent tungsten rod which is stable in position. This rod is one electrode in an ac arc that is burned at a higher color temperature and is therefore a more intense source. It has two serious disadvantages, however, and is no longer being pushed for photomicrography, although it can be obtained. First, and more serious, is its very short life as a uniformly bright source over its area, since sublimation of tungsten from the filament soon degrades its uniformity with accelerating rapidity. Second, the annular filament acting as the other electrode of an ac arc is also bright and of large diameter so there is considerable flare.

Zirconium Concentrated Arcs

The zirconium concentrated arc lamp, which was a development made during World War II, has been justly acclaimed as an outstanding contribution to the roster of photomicrographic light sources. The 100-W arc (Item 16) is much used; it has high photographic efficiency and brightness and a diameter that can be used over a range of magnifications. It also has good color quality, which is discussed in the next chapter. Most stable filament lamps cannot, of course, burn at the color temperature or brightness of the arcs, and one turns to them when the inefficiency of illumination method demands an unusually bright source, as in darkfield or even phase illumination.

The 40-W zirconium arc (Item 17) is used in Appendix 2C as an example to illustrate the advantage of a very small source at very high magnification, with consequent reduction of flare. The reader therefore is familiar with the fact that at the higher magnification of 1000X the illumination brightness will be as great as that with the larger 100-W lamp if the latter is properly diaphragmed down to the used field. The 40-W lamp, however, could not be used with the same lamp condenser (e.g., the $f/1$, 60 mm, aspheric lamp condenser) employed with the other sources, so that the flare factor could not be determined on a comparable basis. The larger magnification of the source required an achromatic condenser.

One used to hear that the perfect photomicrographic source would be a "spot source." However, there are now modern concentrated sources, for example, a 25-W zirconium-arc lamp whose incandescent area is less than 0.5 mm in diameter, that prove the fallaciousness of this assumption; both aperture and field must be filled. To magnify the image of the 25-W zirconium lamp sufficiently to fill a substage condenser for 1000X image magnification, and without using excessive bench distance, would have taken an $f/1.8$ lens of 15 to 20 mm focal length and one of good quality. This is, of course, not impossible nor even impractical when it is justified.

The zirconium lamps have several outstanding advantages. The effect of their small size has been discussed. Then there is the characteristic, common to a few others, that all of the light is thrown forward so that there is less flare from the lamp bulb. **As source size gets smaller, the increase in contrast efficiency shows up undesirable features such as striae** that exist in the optical system, particularly the lamp bulb. However, the zirconium-arc lamps may be obtained with a special window of flat glass without striations. The higher magnification required of these sources makes use of a higher aperture of the lamp condenser and makes one of good quality especially advantageous. **The greatest disadvantage of this light source**, with which such excellent uniformity of brightness can be obtained over the object field, **is the nonuniformity with time,** that is, its flicker. Although both zirconium arcs gave comparatively high efficiency and low flare factor, a new, and it would seem unnecessary, source of flare was encountered in the inner reflections between the bright and flat anode in front of the arc and back lens of the lamp condenser. Without this the flare should decrease to at least that of the coil filament.

Carbon Arcs

When pushed for image brightness, the photomicrographer has tended to resort to the carbon arcs. Justification for this is found in columns 7, 8, and 11 of Table 7-2 for Items 19, 20, and 21. Automatically feeding carbon arcs are now made for each of these current ranges, in which the electric motor feeding the carbons is controlled by the arc itself. When using the carbon arc, the author has found it worthwhile to prebake the carbons at 140°C for one hour to desiccate them and then to keep them dry in a desiccator, since this eliminates the annoyance of "spitting." In fact, this is such an improvement in convenience and quality that it provides almost a new type of source. The carbon arc has unusual quality in terms of excellent reproducibility of color quality and can be chosen on this basis for color photomicrography. It is about the only continuous blackbody source with which a specified light quality can be reproduced easily in any laboratory by specifications wholly concurrent. This point is discussed in Chapter 8. **The high-intensity carbon arc (Item 21) gave the greatest image brightness of any of the sources tested, except the new high-pressure xenon lamps and the expendable flash lamps,** and was used for motion photomicrography of organisms at high magnification by dark-field. For this purpose only the brightness with no diaphragm in the field condenser is shown (Column 11).

Mercury Arcs

Mercury-arc lamps are especially useful in photomicrography, either because of their high ultraviolet component or **as a source of isolated monochromatic wavelengths** with superior definition, or both. Because of some of the relatively new, extremely small intense sources that have become available as mercury arcs, they may be chosen for this feature. The application to ultraviolet photomicrography is discussed in Chapter 14.

It does not appear to be generally realized how much **improvement in definition is realized with monochromatic radiation, even with apochromatic objectives,** and how easily it is obtained. Especially marked improvement is shown with planachromatic objectives. Most of the black-and-white photomicrography in our laboratory is now done with mercury-arc lamps, including all photomicrographs for particle-size studies. For this purpose the important feature of these arcs is the availability of this monochromatic radiation, with which, by the mere choice of filters, monochromatic green, blue, or ultraviolet images can be obtained.

Although the discontinuous spectral distribution emitted by mercury arcs (see Figure 9-26) makes direct comparison of the "white light" of the mercury arc with that of the sources previously discussed rather meaningless, it is possible to compare the brightnesses in the image plane of all the lamps that use only green light. This is done in Column 8 of Table 7-2. The performance of the mercury arc becomes more impressive when it is remembered that the green light of the incandescent sources and the other arcs is the relatively broad spectral band isolated by the Kodak Wratten B filter, whereas that of all the mercury lamps is monochromatic λ 456 mμ. The very intense high-pressure mercury arc, H6 has such a high component of continuous radiation that the Kodak Wratten B filter was used with it. In order to obtain a narrow spectral band of even somewhat comparable quality with a continuous source it would have been necessary to use the Kodak Wratten 74 filter. But this filter would have diminished the image illuminance for panchromatic films by more than 8X and for orthochromatic films by 6X. On the other hand, when the Kodak Wratten B filter is used with a tungsten source to get just the right degree of photographic contrast in a black-and-white photomicrograph (as with a preparation strained with hematoxylin and eosin), the monochromatic green would probably give too much contrast.

Notice that the BTH Compact Mercury Arc (Item 24) gives a more intense image than either of the two common carbon arcs and it is both steadier and easier to use. With both the compact arc and the H6 (Item 25) the areas of the arc stream focused into the optical system lie between two much brighter spots at the end. The long narrow form of the CH3 and the CH4 are far from the ideal shape, but see the previous comment on them (Index, *flare from long sources*).

Xenon Arcs

The high-pressure xenon arc has become an important source for photomicrography. Osram, of Germany, manufactures these lamps in a series of wattages, including 5000 W, but the unit most used for photomicrography is their XBO 150, which operates at about 150 W on a 110-V dc supply. Being European, their ac unit operates on a 220-V supply, but an auxiliary transformer built into the power pack may be obtained. To burn steadily the supply voltage must be well regulated. A voltage regulator greatly improved the stability of the arc in this laboratory.

The lamp has great intensity, since it is so concentrated. Like the mercury arc, there is no area of truly uniform intensity and the decision how large the image of the arc should be, just to include it, well focused in the substage aperture, is somewhat a matter of judgement. The illuminance value in the image plane in Table 7-2 for the xenon arc cannot be so accurately compared as can most of the rest, since the measurements were made on an arc in a Wild microscope lamp with its own condenser (26 mm, $f/1.1$), but the discrepancy is probably not great. It can be seen that this arc has the greatest illuminance *to the photocell and photographic film* of any of the continuous sources, and because of its small size has a very small flare factor when well diaphragmed. The excellent diaphragm of the Wild lamp lends itself to this also. Since it is brighter than the carbon arc, yet easier to operate, it should be most useful for metallography and other applications in which the carbon arc has been used.

Part of the apparent special brightness of this light source is due to the fact that its quality is such (color temperature of 6000°K, see p. 389 that a larger fraction of its total energy affects the photocell and the photographic film than is the case with the sources of lower color temperature.

One word of warning! Anyone who receives such a lamp or has not personally operated it for some time should read the manufacturer's directions before assembling or operating it. The lamp itself should not be picked up with the fingers or at least should then be washed carefully with alcohol. There is some danger of explosion. The first time a lamp is used it should be with a relatively stable voltage and left to burn at least one-half hour. The arc tends to burn into a preferred position, which will help its stability.

Flashlamps

The expendable combustion-type flashbulbs are best used in photomicrography, not to stop motion, but as a **convenient way to obtain optimum illumination for monopack color films**, and are discussed to that end in this chapter. Here it is assumed that they will be used behind the diaphragmed matte diffusing screens. It must not be forgotten that **they can also be inserted directly behind condensers of appropriate focal length at the magnifications at which their size is also appropriate**. On the whole, however, they cannot be imaged satisfactorily directly in the field; their illuminated area will be found to be nonuniform. On the other hand, electronic flashbulbs can be used to stop motion, although again it must be remembered that apparent motion goes up linearly with magnification and the appropriate time of exposure required becomes similarly shortened. Therefore, with common commercial electronic discharge lamps, it is difficult to obtain adequate exposure by darkfield above magnifications of 500X, as would be useful in liquid colloid samples, but now this can be done by phase microscopy. Some loss of speed and especially of contrast occurs with extremely short exposures with many films but the contrast can usually be regained by extending the development time by 50 percent.

Column 9 of Table 7-2 shows a relative exposure, furnished by the flashlamps, as counts of the integrating photometer which represent intensity multiplied by time. The duration of the exposure in seconds is also given. The counts are purely relative. However, the count of 1.0 for the highlight allowed a good negative to be obtained with Kodak Super-XX Pan Film. Other illuminances can be compared with those of the flashbulbs through Column 10. The counts were obtained directly by exposure through a shutter at $\frac{1}{100}$ -sec exposure time; in the case of the flash the illuminance was calculated for this time to allow inter-comparisons.

Item 34 is an FT-114 flashtube. It is a U-tube which is most efficiently focused into microscope system by turning it on edge as shown in 34b.

PRACTICAL CONSIDERATIONS

A general discussion of light quality is reserved for a separate chapter. Maintenance of quality in an incandescent lamp may require special care, and incandescent lamps are most common. At any rate, some check of quality constancy is wise; it is almost essential for good color photomicrography. Therefore **a voltmeter or an ammeter should be included in the circuit of the light source.** Of the two the ammeter is preferable; it is the more sensitive indicator of light variation, since the light output varies as the square of the current and the fourth power of the voltage in an incandescent source. The size of the scale and the sensitivity of the available meter, however, may influence the choice.

Power Regulation

A fluctuating supply voltage is unfortunately a frequent problem. From this viewpoint the sources may be classified as follows:

1. Direct current   (a) standard voltage 110–120 V,
                             possibly 220 V
2. Alternating current (b) low voltage, 24 V or lower

Regulation of standard-voltage direct current is the greatest problem, if it is necessary. The author has a specially built regulator, operating from 220 V, dc, that puts out a 110–120-V supply, regulated constant for either voltage or current and can handle 50 A. It actually regulates only a fraction of the load, which is then added to the whole. For small loads and low voltages a bank of storage batteries is ideal. They can be on continuous charge. For a greater load an ac regulator with rectifier is a commercially available combination, although not inexpensive.

Regulation of alternating current is simpler. The simplest and least expensive method is the use of a saturable autotransformer such as the Sola constant voltage transformers. Because of their principle of operation, that they operate on the flattened top of the current curve, they must be chosen for the load and used at full load for good regulation. Sometimes they adversely affect the "power factor" when used with other transformers in power supplies of commercial lamps. In practice this means that they may be useless for regulating the system. Such transformers, however, are the most commonly used and, if chosen well, may be expected to perform fairly well.

Electronic regulators can give considerably more precise regulation. There has been a marked improvement in them in recent time. Unfortunately the commoner less expensive models do not handle large currents. The type, such as the Nobatron, that can handle as much as 100 W tends to be both rather large and expensive.

Connections for Low Voltage

From his observation of the use of 6-V sources, the author feels that a warning is required in many cases. A 6-V potential is so low that little resistance can be tolerated between the power supply and the lamp, including the wires and connecting links. Even an additional few hundredths of an ohm in the lamp circuit can give trouble. The writer has seen individuals in trouble with inadequate light quality for color photomicrography with a ribbon-filament lamp, although they were using the commercial lamp and transformer system with its wire and connections as received; yet appreciably less than 18 A was passing through the lamp. Common household wire for 115-V supply is inadequate for the 6-V secondary side of the transformer to allow reasonable leads from a transformer to a lamp on a bench. The wire connections to the lamp may corrode and increase the resistance unless kept very tight.

It is sometimes difficult to persuade a practicing electrician to consider that low voltages may require heavier than normal copper wire; that is, that negligible voltage drop in the required length of wire may be the test of adequacy rather than safety from overheating. To be liberal assume a 10-ft wire lead from transformer to lamp:

Gauge No.	No. 16	No. 10	No. 6
Ohms (10 ft, 68°F)	0.040	0.01	0.004

The addition of 0.040 ohm is definitely too much.

Since the cord supplied with 6-V transformers is definitely inadequate, there are two remedies that can be applied.

1. Projector cord may be used. This may be asbestos-covered, size No. 6. To complete the reliability of this system the ends should be soldered to the lamp-socket terminals and heavy lugs provided for any variable resistance or ammeter that is included in the circuit. The variable resistance is to lower the current through the lamp for periods of stand-by and visual use only.

2. The voltage at the power supply may be raised with a variable resistance or transformer plus ammeter to ensure that adequate but not excessive current is carried which would burn out the lamp too quickly. This method really "blasts through" wiring and connection resistance.

Since the 108-W, 115/6-V transformers are relatively inexpensive, two of them may be used with the primary leads plugged into the house circuit in parallel and the two secondaries wired in series.

A variable autotransformer, such as the Variac or Powerstat autotransformer, may be placed in the primary lead to the 6-V transformer. This may require the variable transformer to be set to deliver somewhat more than 120 V from the house supply.

The following arrangement is often used to control a lamp for photometry or quantitative photography (sensitometry). This excellent arrangement allows delicate and stable control but does require two of the more expensive variable autotransformers, one of which acts as a vernier for the other to give smooth voltage adjustment. This second variable transformer uses the whole rotation angle of the Variac or the Powerstat, etc., to add or subtract the voltage increment represented by the secondary voltage of the filament transformer. The latter can be chosen to be adequate for the current carried by the lamp circuit (see Figure 7-9 in which no attempt has been made to suggest the actual circuit of the variable autotransformer). The circuit can be varied somewhat according to the characteristics of the lamp. As shown in Figure 7-9 the lamp should be one operating at or near 100–120 V, since otherwise it could easily be burned out by pushing up the Variac too far. A step-down power transformer can usually be added before the lamp when the latter is a low voltage type, but note the comment at the beginning of this section.

Life of Incandescent Lamps with Current and Voltage Variations

The photographic efficiency of an incandescent lamp goes up very fast with an increase in the current through it or voltage across it. Unfortunately the life of the lamp is shortened at an even more rapid rate

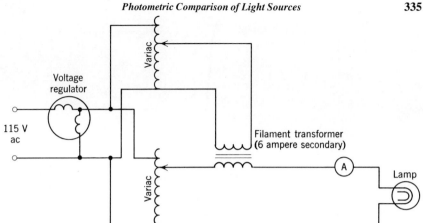

Figure 7-9 Circuit for vernier control of voltage.

with increase of current and voltage. This is illustrated by Figure. 7-10, which shows the effect of voltage variation in terms of percent variation. This can be applied to most incandescent tungsten lamps if their rated voltage is known. The rated life of a household lamp is usually 1000 hr, that of projection lamps, much used in photomicrography, 25-50 hr, whereas miniature lamps are most dependent on the particular lamp. Some rated life values are quoted in Table 7-3. A few are also given in Table 7-2. It must be remembered that this is a statistical concept, not followed exactly by any one lamp but closely enough for the relation to be useful.

Usually a lamp of known quality for color photography is precious for the inherent calibration involved in it. There is normally little reason for the lamp to be kept at its photographically optimum current value except during photography; a lamp thus used will last a long time. Figure 9-9 shows the life of the lamp as affected by current and voltage variation and as related to color quality (color temperature).

Means of modulating the current through the lamp have just been discussed. It should be so convenient that the lamp will promptly be turned down – except during photography.

A somewhat more extensive review of light sources, especially of incandescent lamps, is given by F. E. Carlson and C. N. Clark[2. Kingslake (Ed.), 1965, Vol. 1, Chapter 2].

Heat Filters

In photomicrography by transillumination a large diverging cone of radiation emerging from the light source is redirected by the optical

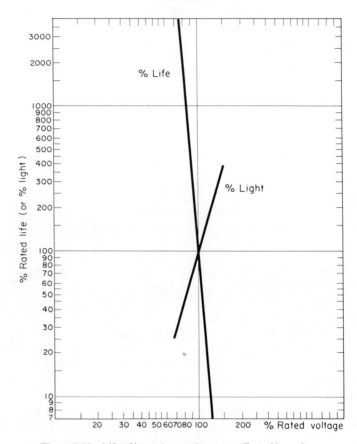

Figure 7-10 Life of incandescent lamps as affected by voltage.

system to a convergent beam, transmitted by the optical system itself and the specimen. Moreover, the intensity level is normally brought up so that photographic exposures in the final image plane are conveniently short. The concentration at the plane of the specimen, therefore, is tremendous, that is, the square of the image magnification more than in the image plane; for instance, at 1000X magnification the radiation is one million times more intense in the object plane, assuming no losses. It is still very strong going through the substage condenser and the objective. The objective, at least, is a valuable lens with cemented components. In the older ones the cement can be presumed to be balsam; now other resins are also used.

The fact that the heat component of the radiation from an incandescent tungsten lamp is many times that of the visible component is obvious

Table 7-3 The Rated Life of Some Incandescent Lamps

Lamp	Rated Life (hr)	Rated Volts
100G16½/29SC	200	115
CDS	50	115
CMV	25	115
#1634	200	20
#1630	100	6.5
#1497	100	6.5
#1493	100	6.5
#1460	100	65
#1130	200	6
#870	50	8
#419	50	8
#88	300	6
#55	500	6
CPR	50	6
#14	15	2.47
Horizontal ribbon filament-108 W	150	6

from such graphs as that of Figure 7-11. Only that portion of the spectrum below the limit of about λ700 mμ (or 0.7 μ) is used by the eye or most photographic films; the rest, namely, the infrared, is most effective, however, in carrying heat and raising the temperature of receiving material. The quantity of the energy from the source is given by the area under the curves. Water is the most effective single material for filtering out the heat radiation without affecting the visible and photographic spectrum. This is shown by its spectral transmission curves in Figure 7-12. It will be noticed, however, that reasonably thick layers of water do not absorb much of the heating infrared below about 1.15 μ, leaving an important window above the end of the visible spectrum and about λ1.2 μ. This gap can be plugged by using a water solution of a copper salt (such as copper sulfate), more effectively with less absorption in the visible spectrum than with any other material. The concentration of the copper sulfate will depend on how much far red with consequent cyan coloration one can accept. For black-and-white photography or for color photography, in which a cyan color filter is needed anyway, a minimum of a 2.5% solution of crystal copper sulfate* should be used. Copper sulfate is an ideal cyan and in very strong solution can be made a complete minus red.

*For chemists this is 0.1 molar $CuSO_4$ or 2.5% $CuSo_4 \cdot 5H_2O$

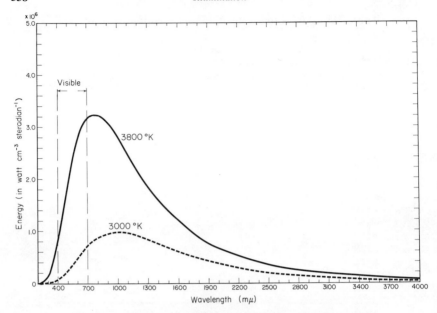

Figure 7-11   Spectral distribution of radiation from tungsten lamps.

Although the absorption-concentration relation is expressed in percent-salt solution in a 1-cm layer, a thicker cell such as the common 3-cm cell or any other thickness can be used to obtain the same absorption by dividing the concentration per centimeter by the layer thickness that is to be used. It can be put into the common water cell with a gain in the absorption by the water. A 3-cm (1⅛-in.) layer of water effectively absorbs all significant infrared from lamps beyond 1.5 μ (Figure 7-12).

Although there is no ideal heat absorber with a sharp cut at λ 700 μ that absorbs all longer wavelengths, and a solution of a copper salt is quite good, there are times when a solid filter is required. A glass filter is available and is now used in many slide and movie projectors. One of these is the 2043 glass filter made by the Pittsburg Plate Glass Company. Like a copper sulfate solution, its absorption is proportional to its thickness, which must be specified. Its absorption is shown in Figure 7-13. The Corning Glass Filter 1-69 is quite similar.

Because a heat absorber, such as the glass, comes to equilibrium and reradiates heat as fast as it absorbs it (at a longer wavelength at ordinary temperatures), it should not be placed against the substage condenser. **Its absorption ceases beyond λ 900–1000 mμ at which that of water is starting, so that the latter is still really needed** for many light sources. It is good practice to put the heat-absorbing glass directly into the cooling water.

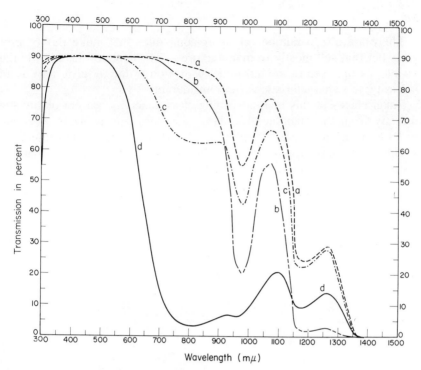

Figure 7-12 Spectral absorption of liquid heat filters:
(*a*) water, 1 cm; (*b*) water, 3 cm; (*c*) $CuSO_4 \cdot 5H_2O$–0·25% water solution, 1 cm; (*d*) $CuSO_4 \cdot 5H_2O$–2·5% water solution, 1 cm.

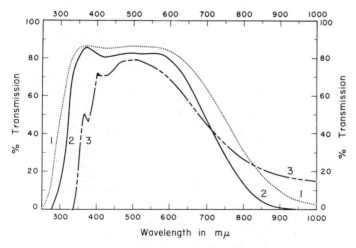

Figure 7-13 Spectral absorption of Pittsburgh Plate Glass heat filters: (1) 2043, 2-mm (1) 2043, 2-mm thick; (2) 2043, 6-mm thick; (3) Solex, 1/4-in. thick.

There are also a number of other companies that make this type of glass but that sell mostly to manufacturers. The depth of the color in the visible, as opposed to the infrared, absorption is due to ferric iron as an impurity; this characteristic varies considerably.

Small pieces of this heat-absorbing glass generally are not obtainable directly from the manufacturer. However, they are available as round disks by ordering spare or replacement parts of 35 mm slide projectors; for instance, at this writing the following are available from the Apparatus Parts Service of the Eastman Kodak Company.

Projector	Part No.	Size (Diameter)
Carousel	835596	$2\frac{1}{2}$ in.
Supermatic	835490	$2\frac{1}{16}$ in.
500	827161	$2\frac{3}{4}$ in.

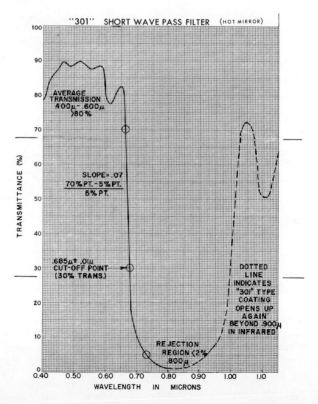

Figure 7-14   Spectral absorption of Kodak Infrared Cut-Off Filter.

"Solex" heat-absorbing glass, made by the Pittsburgh Plate Glass Co., is sold by the square foot by dealers for that company. Its spectral absorption curve is included in Figure 7-13.

An entirely new type of dry-heat filter was introduced with the advent of **interference filters**. This type of optical filter is *discussed in Chapter 9*. The edge of its transmission band is even sharper than that of any absorption filter with all of the radiation being reflected that is not transmitted. These filters can be made to transmit the visible, reflecting the infrared or conversely, to make either a "cold mirror" or "hot mirror." The transmission of Kodak Infrared Cut-Off Filter No. 301 is shown in Figure 7-14, as taken from a data sheet of the Eastman Kodak Company. By comparison with the preceding figures, it can be seen that **the cutoff at the red edge of the visible spectrum is the sharpest of any filter**. On the other hand by λ 1000 mμ in the infrared these filters again transmit freely. Hence for effective protection from heat radiation this type of filter must be protected by a water cell of adequate thickness. Such infrared filters have been made on heat-absorbing glass; then the absorption edge lies further in the visible, but is sharper than that of the glass alone.

Chapter 8

Lamps

GENERAL CONSIDERATIONS

Unfortunately, in the English language, particularly as applied to this field, the word *lanp* may mean either the light source itself, for example, an incandescent electric bulb, or the illumination assembly both optical and mechanical, for example a "research illuminator for the microscope," as furnished by several companies. The first category, the light-source lamps, we have already considered but certainly not exhaustively. A brief preliminary discussion of the types of illumination units is worthwhile.

1. The lamp may be a bulb in a case with a diffusing window. Very common are the small boxes meant to be placed below the microscope with no mirror. These are normally neither suitable for photomicrography nor for visual work either, really. There is no control of field glare at all. However, a box type with an opal diffusing glass, or double-ground glass, **behind a diaphragm of the correct size** or a variable iris diaphragm **imaged directly in the field by the substage condenser** is both practical and often quite suitable. Between the field diaphragm against the diffuser and that of the substage there is control of both aperture and field. This arrangement is discussed under "Secondary Light Source" in Chapters 7 and 13 and really constitutes the only efficient method of introducing a diffuser into the system.

2. Somewhat better than the simple box plus diffuser is a box containing a bulb plus a lens on top of it to go directly under the substage condenser. This is very common. Usually there is no control of flare with adequate field diaphragm, so it is not so suitable for the best microscopy or photomicrography. On the other hand, it can shade into a complex optical system, although it rarely does because space is so limited.

342

3. Very common nowadays on the more expensive microscopes for research and photomicrography is the **built-in illuminating system**. Often great claims are made for it; it certainly is convenient. Frequently it is hard to appraise the virtues of such a system without extensive personal trial because almost no design details are given out about these systems, but they are normally quite satisfactory for most visual work and can be very good. Whether flare is under control and the field is really uniform will show when the specimen has very low contrast and high photographic contrast is used. The designers are quite correct; their systems are certainly superior to those the average microscopist sets up. The designer has a problem; to be so very convenient he must set up a relatively small system in quite a short space and that imposes limitations, including the selection of the lamp bulb. Still it seems to the author to be somewhat an admission of failure to add diffusion to a lens surface, as is done by both Leitz and Bausch & Lomb and probably others, even though, given the conditions, he might have to do the same thing. This assumes that really corrected lenses are too expensive for the purpose. In a large system, such as the Zeiss Photomicroscope, there is more length and a good "standard" design can be incorporated. A little more can be said on this type in the next section. The small bulb used may be inadequate for many types of photomicrography, including color work with daylight-type color films because of the high filter factor. Provision is usually made for inserting a mirror below the substage and using an outside light source. In several cases (Leitz and Zeiss) xenon arcs can be added to the built-in system.

4. The light source and condenser are separately supported, necessarily on a horizontal optical bed. The microscope axis may be horizontal or vertical. The author considers this to be the best arrangement whenever it is appropriate and practical. It is not usually the most convenient with a microscope having a vertical axis and so is not used often. Therefore it is discussed in detail last.

Frequently, these days, a so-called "photomicroscope" is purchased with the light source and illuminating system incorporated in it. Although all of the principles of this chapter apply to this arrangement, there is usually less freedom of choice and arrangement and most decisions have already been made. The maker's direction sheet is not only the most valuable guide for usage but should be thoroughly read and preserved. On the other hand, such outfits usually make provision for using an external light source through an auxiliary mirror or prism. This arrangement is most useful when the built-in illumination is inadequate for a given job, either in intensity or quality.

In selecting a separate lamp, including its condensing lens, both optical and mechanical considerations are involved. The selection of the appropriate light source, in regard to intensity, quality, size, and structure, has already been discussed, although there is more to be added about quality.

Definitely, a good lamp should allow centration of the source to its condenser with a relatively convenient mechanism, yet one that is stable. Unless there are locks, easily manipulated centering screws, which should be large but with a fine thread, allow the lamp to go out-of-center too easily also. The lamp base should be stable. It may be made for use on an optical bed or table top and should be quite heavy to give stability to the hot lamp. It should possess good ventilation, yet no direct rays of light should emerge. Tungsten sources and arcs demand that the housing be a good radiator. Modern aluminum castings with fins or deep corrugations are an important modern improvement in this respect. On the other hand, some of the lamps do not allow easy access to the bulb. Another frequently lost feature is direct viewing of the lamp and condenser from behind. That is the simplest and most efficient way to center the lamp assembly, as explained in Chapter 6.

A mechanical feature with a very important optical function is the **iris diaphragm of the lamp**; the field diaphragm has been discussed. It **should be against the condensing lens** but rarely is. It must either be well centered to the condensing lens or **centerable** to it. This is also a neglected feature, which is, unfortunately, sometimes needed.

A purely optical feature is the quality of the lamp condenser. That is also discussed elsewhere. To summarize, a completely uncorrected lamp condenser, that is, a single-component spherical lens, will cause poor quality in the microscope. The most important correction is that of spherical aberration. The condenser can be an aspheric lens or a group of spherical lenses without sharp curvature. For color photomicrography an achromatically corrected lamp condenser may improve the color reproduction with fewer correcting filters.

COMMERCIAL LAMPS, EXAMPLES

1. One of the first lamps to consider for brightfield work is the **Ortho-Illuminator** (see Figure 8-1), now made by the American Optical Co., formerly by Silge and Kuhne, because it corresponds to the type of secondary source already discussed. In this case, however, a condensing lens is used; the diffusing screens lie on a selector disk as round apertures about $\frac{7}{8}$-in. in front of the lamp filament. They control the intensity of the light by the amount of diffusion they introduce. There are four $\frac{3}{4}$-in.

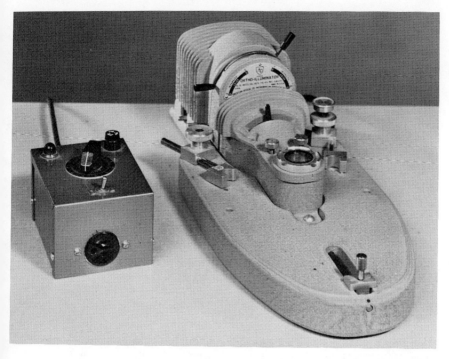

Figure 8-1 A-O Ortho Illuminator.

apertures; one is a clear hole, the second contains a ground-glass disk, the third, two ground disks, and the fourth, a disk of flashed-opal glass. Therefore some change of light quality occurs as the different disks are changed. The lamp is said to burn at 3200°K directly on the house supply; it is the BMY miniature lamp, 100 W, standard voltage (115–120 V). With a flip of a switch it will burn with 25 ohms in series at a reduced brightness that will greatly extend its life. If a Kodak Wratten Filter No. 78C were permanently added to the aperture containing the flashed-opal glass disk, the light would have closely the same quality as that from the clear hole.

With a clear aperture in the selector disk, the filament image, which has an almost square pattern of four lines, hardly fills more than half the aperture of 1.3 NA. When one of the diffusing apertures on the selector disk is interposed in front of the lamp, the whole objective aperture becomes filled. The flare does not seem excessive. The illuminator is very convenient for visual use and allows much photography to be done.

2. Nearly all microscope companies now furnish some models of microscopes with **built-in illuminators** in a variety of unpublicized de-

signs, some of which are inadequate for photomicrography under adverse conditions. Probably the most helpful comment therefore concerns the systems that are most efficient for photography and the light bulb that is used.

The Bausch & Lomb Dynazoom and Dynoptic microscopes can be furnished with either of two illuminators, 31-33-69 or 31-33-20, of which the second is optically the more efficient. Both use the same bulb, the GE 1634, a 20-V 1.0-A, miniature S-8 bulb with a CC-6 filament. It is rated as 24 cp. The author has been able to make photomicrographs at 100X with the -69 illuminator and film of ASA speed 50 with an exposure of $\frac{1}{2}$ sec. The following values of the color quality and brightness *of the bare lamp* are published by Bausch & Lomb:

Table 8-1   Characteristics of No. 1634 Lamp

Transformer switch set	Tap 1	Tap 2	Tap 3	Tap 4	Tap 5
Lamp volts	12	14	16.5	20	25
Color temperature in degrees Kelvin	2400	2540	2700	2900	3150
Luminance candles/cm²	130	190	400	750	1475
Relative luminance	0.17	0.25	0.53	1.00	1.97
Life in hours	60,000	20,000	2500	200	10
Recommended filters for Type A film			82C+ 82B	82A+ 82B	82+ 82A
Recommended filters for Type B film			82C+ 82	82B	None

The best and most efficient built-in illuminator for the American Optical Co.'s MicroStar Microscope, Series 10, is their No. 1036. Its optical system consists of a triplet lens of three simple convex components of decreasing curvature, the strongest component being, of course, near the lamp. Its rear surface (toward the filament) is concave; its front surface is matte. This effectively lowers the color temperature of the lamp by about 100°K because of the differential scattering discussed near the beginning of Chapter 7. There is a first-surface mirror and an inconel density of 1.0 (10 × brightness factor), which is merely flipped out of the way with a lever. Apparently many people are using the density in the beam almost all of the time and then turning on the lamp at its highest voltage, which is an overvoltage provided for temporary use during photography. Even the direction sheets that come with the microscope imply that the density will be in the beam for most photographic exposures. This is unfortunate and has brought complaints of unreasonably short lamp life. The General Electric No. 1460 lamp used has a rated voltage of 6.5 V for a rated life of 100 hr at 2.75 A and then furnishes 23 cp. It has a 6-C filament and is very similar to miniature lamps Nos. 1493 and 1947, described in Table 9-2. The No. 1460 lamp, when overvolted to 7.5 V, furnished sufficient illumination for ordinary brightfield at 225X magnification to allow an exposure of $\frac{1}{5}$ sec through a Kodak Wratten Filter No. 58 (green) with Kodak Panatomic-X Film (135). Therefore it should also be adequate for photomicrography with color roll films at the fractional second exposures, which they should have, when the illumination is simple brightfield and white light.

Table 8-2 Miniature Lamp No. 1460 in A-O Illuminator No. 1036

	Transformer Setting		Color Temperature K°	Relative Illuminance in Plane	
	5.5	(5.6)	2800		1.00
Bare Lamp	6.5	(6.65)	2990	1.0	1.75
	7.5	(7.55)	3140	1.47	2.6
	6.5		2890	1.0	
Total Illuminator	7.5		3000	1.46	

These color temperatures were determined at a line voltage of 120 V, which is probably somewhat higher than normal without a regulator. A check with a voltmeter (given in parentheses, Column 3) revealed that the transformer settings probably indicated the voltage of its output for 115-V input.

The relative illuminance was read with a selenium photovoltaic cell directly in front of the A-O illuminator but should also represent the ratio that is observed in the image plane of a photomicrographic camera. It should be noted that the increase of brightness with voltage corresponds to the ratios of the cubes of the voltages rather than the fourth power. This factor also expresses the relative illuminance that will be obtained at the other settings.

The "built-in illuminator" of Unitron's large photomicroscope is actually held in an outside trunion and clamp but is probably integral with the whole optical system. It is tubular in form, has ventilating fins, and boasts a four-element condenser system that contains no diffusion. It utilizes a rectangular grid wire filament lamp (Japanese), Catalog EL-1B, which Unitron Instrument Co. states was designed for this illuminator. The grid is about $2\frac{1}{2} \times 4$ mm, which is longer than those cited on p. 310. An ammeter provided with the instrument has a red mark at 4 A to show the highest amperage recommended for continuous burning; the lamp is at 11 V when drawing this current. Unitron literature quotes the following values:

Table 8-3   EL-1B Lamp

Color Temperature — Amperes				
Amperes	3.4	3.6	3.8	4.0	4.2
Color Temperature (°K)	2653	2747	2838	2910	2983

Some of the "visual" Unitron microscopes, such as their BM-1C and BRM-1C, use frosted bulbs.

Leitz employs several built-in lamps with its several large microscopes. The Ortholux has an illuminator, coded EYMZE, that fits as a unit into the rear of the base, although one lens is incorporated near the internal mirror in front. It utilizes a miniature lamp, LINID, whose color temperature characteristics are listed in Table 9-2. This is the Osram 8110 lamp, 6 V, 5 A (30 W), which has a fairly compact rectangular cylindrical coil filament. A rather strong aspheric condenser lens is placed just in front of the lamp, whose far surface (from the lamp) is frosted. This is the surface that should be imaged in or near the object field in the microscope so that rays deviated by the frosting can be acting as the source. Unfortunately, however, for the size of this "source" the diaphragm does not lie against it.

The Labolux microscope also carries a built-in illuminator, which is somewhat simpler and uses a 6-V, 2.5-A (15-W) lamp with the code LINOP. Under this code, however, either of two lamps has been obtained, both miniature concentrated wire filament — one, the Osram No. 8017, the other, the Phillips 13347. The latter seems somewhat preferable for photography. The Phillips lamp has the flat square grid described on p. 310; the Osram lamp is a compact cylinder with a horizontal axis, similar to a C-6 filament. The color temperature characteristics, given in Table 9-2, are for the Phillips lamp.

The most impressive lamp produced by Leitz is the Universal Model No. 250 when used in connection with the EYMZE, although this does make an expensive combination. The No. 250 lamp has a rectangular lamphouse with insulated centering and other control knobs. It can be fitted with the xenon XBO 150 lamp or a mercury HBO 200. A tungsten illuminator, notably the EYMZE, can be incorporated from the side with a swing mirror to take its beam in or out of the microscope system. Thus, with the xenon bulb as alternative to tungsten, one would be well equipped for many types of work, including the use of color film for daylight quality. See Figure 8-2.

Carl Zeiss (Oberkochen) has a low-voltage lamp (Cat. No. 38.01.77) rated at 6-V, 15-W, which is supplied in the simple substage lamp, with little control, as previously discussed, in a separate lamp on a clamp stand and in a built-in illuminator with a train of lenses. The lamp condenser has been split into three components to minimize maximum inclination of the rays to the surface, as discussed in Chapter 1, with a strong meniscus near the very small filament for the same reason. The color temperature obtained with this bulb is listed in Table 9-2. The filament coil (cylinder) of this lamp is about $1\frac{1}{2} \times 2$ mm, thus forming approximately a square projection; the lamp bulb is used end on. The problem with this system, as with all others of this type, is its inflexibility for covering all magnifications and field and aperture sizes efficiently. It will be most efficient at the relatively high powers. Note by comparing

Figure 8-2   Leitz Universal 250 Light Source.

items 9 and 10 of Table 7-3 and also in Table 9-2 how efficient this lamp becomes when overvolted from 6 to 8 V. It furnishes appreciably more illumination than the ribbon filament and at a sufficiently high temperature to be efficient with color films made for artificial light. In Figures 7-10 and 9-9, however, it can be seen that the expected life of the lamp has been reduced from about 100 hr to less than $1\frac{1}{2}$ hr if burned at 8 V continuously. If a good supply of the lamps is available and provision is made to maintain 8 V *only* during the photographic exposure period, this lamp could be used for most brightfield photomicrography *if* properly imaged for aperture and field.

A stronger light source for the built-in system of the Zeiss (Oberkochen) *Photomicroscope* and *Standard Universal* stands is supplied as a 12-V, 60-W lamp (Cat. No. 38.02.16). The author has not made a direct photometric comparison between these two bulbs. Zeiss state that it "has twice the luminous density of the 15-W lamp"; it would seem as if it might be overvolted and burn at 3200°K a little longer than the 15-W bulb. The somewhat larger filament may have some advantage at lower magnifications.

3. Next we must consider what have been called **spot lamps**, often with a tubular lamp housing equipped with cooling fins and usually with a low-voltage lamp which has a small filament and of a type that can be focused on a field with but slight diffusion. The whole lamp may rest on a clamp stand or other support which allows good manipulation of position and direction. The A-O Universal Lamp shown in Figure 4-9 has long been a most useful example of this type. Although advocated as a lamp for brightfield microscopy at all magnifications, such lamps seem to be a poor choice over the larger, more specialized outside lamps for microscopy. These lamps may not be smoothly focusable during use, nor be easily centerable, nor have sufficiently strong illumination.

Lately it has become less easy to generalize. Lamps have appeared of this type, that is, tubular, in relatively small diameter and on supporting clamps to give versatile use, but with strong filament lamps, spherically corrected and centerable condensers, and iris diaphragms. The Japanese have tended toward them (e.g., the Nikon Universal Lamp) and therefore they have appeared in the apparatus of those American companies whose products are made in Japan. However, the Wild (Switzerland) Low Voltage Lamp No.7000, which uses a 6-V, 30-W bulb, is also of this type. The author has not worked with these lamps, although he has used them briefly in apparatus displays. It would seem to him that he would wish to compare them critically with the bulkier outside lamps before making a purchase principally for transillumination with a compound microscope.

4. Some laboratory supply houses classify the bulkier outside lamps for microscope illumination as "**professional microscope lamps**". This is a good term if it causes one to look for all of the characteristics of adequate and efficient microscopy and photomicrography. Then, with a permanent set-up, the greater bulk is immaterial.

The American Optical Co. has chosen to utilize lamps that burn at standard voltage without a transformer. Their No. 735 lamp is famous. It is shown in Figure 8-3, set up to utilize the *double-condenser method* of incident illumination described in Chapter 4. Such an illuminator is very useful in many ways, but for most critical photomicrography the author would perfer a low-voltage lamp with a flat or compact coiled filament and a corrected condenser with an iris diaphragm close to it. Smooth focusing and easy but stable centration of the lamp to lens are also necessary for a "professional" illuminator.

The author mourns the passing of the Bausch & Lomb "research lamp," even though it definitely was not the perfect microscope illuminator. The

Figure 8-3   A-O No. 735 Illuminator equipped for double condenser method.

back of its rectangular box housing could be lifted to allow centration of the lamp by the reflected images from the lenses, a good and easy method discussed in Chapter 6. A number of lamps are used in it — ribbon filament, the CPR coil filament, and mercury arcs such as the CH-3. This lamp was completely centerable to the lens. However, the diaphragm was not sufficiently close to the condenser and was not centerable.

The present B & L professional lamp, PR-27, still uses the ribbon filament, frequently the most satisfactory microscope lamp. It is an illuminator designed to be most convenient and requiring the least trouble for efficient use. The strong aspheric condenser immediately in front of the lamp is fixed in position. It is *not* frosted. The lamp socket sits on a plate which has some forward and back movement along the optical axis, possibly $\pm\frac{1}{2}$ in. Considerable care is taken in assembling the lamp to center and orient the lamp socket correctly; prefocused lamps, as supplied by G.E., are always supposed to sit in their sockets so that the filament will occupy the same spot in the optical space. The forward and back motion of the base plate for the socket affords some longitudinal leeway. A weaker and larger lens at the front of the lamp has forward and backward travel by a smooth rack and pinion. In its extreme forward position it provides an adequate field and aperture for lower power, specifically as required by a 10X objective and 7.5X eyepiece. With the front element at the extreme rear of its travel (a few inches), the conditions are such that will fill both field and aperture for any high magnification with any available high-power, high-aperture objective.

If we should wish to check whether the lamp is completely aligned, we would find that it is impossible, with simple focusing, to throw an image of the filament at any distance, such as on a wall. However, by racking forward the base plate that holds the lamp to its extreme position this can be done, but would have to be returned to its former position for normal use of the lamp. Once it was found that the bulb was out of center, beyond the tolerance that the author would normally accept. In that case nothing could be done about it except hunt for a better bulb.

With Bausch & Lomb equipment and with appreciable photomicrography to be done, the author prefers the lamp furnished with the B&L L camera, shown in Figure 3-8. Admittedly it is more expensive but still less than half the cost of some of the foreign illuminators just discussed. It can be used with most of the lamps that utilize normal sockets, the ribbon filament, the CPR 108-V coil filament lamp, and the concentrated mercury arcs of the CH series. Its iris diaphragm is still not close to the lamp condenser.

The Unitron Instrument Co. makes substantial claims for its LKR Research Illuminator, which, as already mentioned, first looks like the spotlight type standing on a swivel clamp on a flat base for table support. It has rack-and-pinion focusing of a single-element aspheric condenser, which is a change from its first design. It is now calculated to be used at a distance of 170 cm from the substage condenser. A clamp, extending to the front, locks the swivel clamp after easy adjustment. The iris diaphragm is near the lens, which is good. It utilizes the El-RKL tungsten lamp, which has a rectangular compact grid, and is rated for maximum continuous burning at 8-V, 5-A. The lamp is not centerable but uses lamps that have a precentered base. A voltmeter is provided with the transformer power supply. (see Figure 8-4).

The Multipurpose Microscope Illuminator of Carl Zeiss (Oberkochen) is a professional outside lamp equipped to take a variety of sockets, hence light sources. Some are discussed later under "Special Lamps." It is a spherical, finned lamp with a condenser of relatively short focal length, that compares, except for appearance, with the much used but now discontinued B&L Research Illuminator. It is prepared to take the 12-V, 100-W incandescent tungsten bulb discussed a page or so back.

Zeiss (Jena) has announced a microscope illuminator that uses a 12-V, 100-W lamp with a flattened compact coil, though rectangular rather

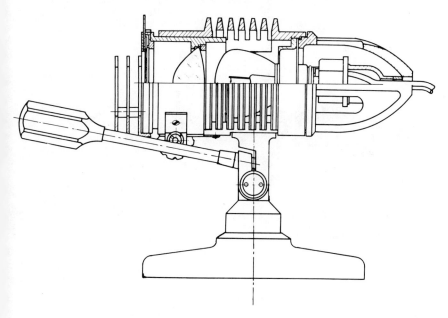

Figure 8-4   The LKR Illuminator of Unitron.

than square. The lamp housing is rectangular. In both cases the lamp house and its fittings are carried by a vertical support rod clamped to the side of the housing and having a flat disk-type base suitable for a flat table but not for an optical bench. The spherical lamp of Zeiss (Oberkochen) becomes a "built-in" source for the Ultraphot II Photomicroscope by attaching directly into the frame by its face. A special central support is provided for the Zeiss (Jena) lamp to fit the optical bed of triangular cross section. The author prefers such a support. The back of this lamp is removable so that the lamp can be centered by back reflections as described in Chapter 6. It might be necessary for this purpose to cut slots in the back cover to clear the centering screws.

The Wild Universal Lamp No. 7100, also built to accept the sockets of a variety of light sources, looks somewhat like the Zeiss Multipurpose Lamp. **Its field iris diaphragm is inset to be close to the condenser lens**, as the author believes it should be (see Figure 8-5). In a new model, available in 1966, the concave mirror is rigidly mounted and can be aligned with four screws. A socket is provided for the ribbon filament and some other tungsten bulbs as well as the xenon arc. It does make a rather expensive lamp as a tungsten illuminator but a very nice one. The two-component

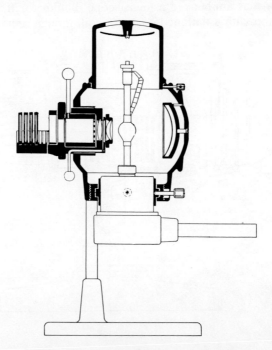

Figure 8-5   The Wild Universal Lamp 7100.

lamp condenser is made of quartz for work with fluorescence. When using this lamp with the xenon arc, the author found the focal length of the condenser to be inconveniently short for some purposes. The arc is only 19 mm from the nearest lens. The Widl-Heerbrugg Instrument Co. specifies a distance of 31 cm (12 in.) between this lamp and the substage condensers of their microscopes (see Table 7-1).

Reichert also has a professional type illuminator with the iris diaphragm near the lamp condenser. It uses a low-voltage 30-W bulb. The lamp is supported by a side clamp to a vertical rod with a large flat base.

The Model XL-6 "Research Micro Illuminator," produced by Swift Instruments, is a tubular lamp supported at the side by a swivel clamp from a ring stand with a base for a table top. It has an aspheric condenser with a focused beam and uses the 6-V, 5-A Japanese lamp with the compact square grid described at the beginning of this section. It is claimed that the lamp will reach a color temperature of 3200°K directly. It should not be burned long at that quality (see Figure 9-9). Sales catalogs are often over enthusiastic about the color temperature of a lamp burned on a standard transformer setting.

5. **Special Lamps.** There is a question now whether lamps utilizing xenon arcs should still be called "special lamps." They have become extremely useful for photomicrography and will be more so as daylight type color film becomes about the only one readily available.

The Leitz "Universal Light Source Model, No. 250" has already been mentioned and illustrated in Figure 8-2. This model can be considered as a generally available "professional lamp" with an extension optical system that can be shoved under the stage of any microscope. Like the Ortholux illuminator, it shines a cone of light vertically up into the substage condenser with an internal mirror.

This lamp can utilize the XBO 150 xenon lamp or the HBO 200 mercury arc, which have been considered elsewhere as light sources. The fact that the auxiliary tungsten lamp. EYMZE, attached from the far side in Figure 8-2, can be quickly alternated with the xenon or mercury arc makes this an effective and versatile source but an expensive one (more than $1000). This Leitz lamp is featured for fluorescence photomicrography, chiefly with the mercury arc HBO 200, because of the intense ultraviolet. However, the xenon arc XBO 150 is excellent or even superior for much of this work, as discussed in Chapter 20 (Vol. 2) and is most appropriate for the exposure of color films made for daylight quality, both in intensity level and quality (with a Kodak Wratten filter No. 1A).

The Reichert "Fluorix" illuminator has most of the characteristics of the Leitz No. 250 lamp with its auxiliary tungsten alternative source, but it is not so obviously applicable to other equipment and features only the HBO 200 bulb for fluorescence.

Obviously the Zeiss and Wild lamps can also use the XBO 150 and HBO 200 arcs. They are professional outside lamps without the incorporated swing-out mirror and accessory tungsten lamp.

The **No. 505 BP Pulsarc** (see Figure 8-6) of Nems-Clark, which is marketed by the American Optical Co., is definitely a special lamp that can become very useful when color photomicrography with daylight film is contemplated along with much visual observation. The lamp itself is a concentrated small source, characteristic of an arc rather than the long tube-filling source of a standard flash tube. The arc can be operated continuously and does so at a rather low level (see Item 27 of Table 7-2). When a photograph is to be taken, the tube is "flashed" by increasing the voltage. Four duration periods for the flash may be chosen with a central knob, from 40 to 135 milliseconds. Thus, unlike the high-powered xenon flash tube, the duration is much too long to be useful to stop motion. Moreover, since the source is burning continuously, with the "flash" just a momentary peak, a synchronized shutter set at "M" (20-msec delay) must be used to protect the film. This is easily done with an Ilex or Kodak shutter.

The literature for the Pulsarc stated that an increase of illumination of 36X from the continuous level would occur during a flash. With our lamp an increase of 70X was consistently observed. The exposure values obtained (illuminance × time [duration]) are given in Table 8-2 as Items

Figure 8-6 The Pulsarc.

27a and 27b. It will be seen that relatively heavy exposures were obtained always remembering that this is due to the duration of the exposure. In photomicrography these flashes could be used as units to count up to the exposure needed. Column 10 of Table 8-2 can be used to ascertain how this compares with other sources.

The electronic flash lamps especially designed for photomicrography are considered in Chapter 13. Both Leitz and Zeiss market such lamps. Leitz has a new electronic flash apparatus that is appreciably smaller than the Mikroblitz 300 which it replaces. It is usable with no accessories on their Labolux and Ortholux photomicroscopes into which it fits. The flash tube is a straight cylinder with a cylindrical mirror backing. An accessory low voltage tungsten lamp, with its own condenser, is incorporated as a control lamp for visual observation and exposure determination. The brightness from the flash is modulated by a series of neutral filters to control exposure.

The Zeiss (Oberkochen) "Microflash equipment," is also located in a tubular lamp house with a 15-W incandescent lamp as a side arm with a split-beam mirror. The whole can be used as an attachable source or as a separate lamp. It would seem that this would give better controlled illumination with less flare than the Leitz illuminator. The author has actually used neither but has had experience with devices similar to both.

6. The discussion of **zirconium arcs** in Chapter 7 is well illustrated by the characteristics of the commercial lamps that utilize it. A much-advertised lamp is the "Zircon Arc" sold by Fish-Schurman Corp., which utilizes the 40-W zirconium arc whose characteristics are given in Table 7-2 and the corresponding image of the illuminated face of the arc in Figure 7-8b-17. This source is so small that considerable magnification is required to make its image fill the aperture of the substage condenser. To do this well requires an excellent lamp condenser, although the small "field" requirements of this source simplify the optical specification. However, when one has paid for this lamp with its excellent condenser, one really has a most excellent system that gives superior results *if* the substage condenser of the microscope is also excellent. **It is almost an optimum light source for high-magnification photomicrography**. The equations in Appendix 3C show that it is unnecessarily small if a larger field is desired, as in medium- and low-power work.

The 100-W zirconium arc seems to be the best choice as a general microscope illuminant. The bulb for this arc, like those for other wattages, can be obtained for illuminating through the end of the bulb and with a special flat optical glass window at the end (Type C100). This is the type most suitable for photomicrography and may be procured from Sylvania Electric Products Co. and its dealers or from Geo. W. Gates & Co.,

Franklin Square, Long Island, N.Y. The latter company also sells a power supply for it; in fact, the author notes that their power supplies are used by the commercial lamps. The arcs operate on direct current but the power supply includes an adequate rectifier from the usual ac-line supply. A standard four-pin radio socket is used for the base. It is quite practical and, with machine-shop facilities, it is relatively simple to set up this lamp for photomicrography. One lamp made this way is described on p. 362.

7. The **carbon arc lamps** should not be neglected. Just as other efficient and more convenient competitors (xenon and zirconium arcs) have come along, the carbon arcs have become much more convenient. Most of the microscope companies featuring metallographic equipment also sell an electronically controlled carbon arc. Desication of the carbons prevents "spitting." It is still the most intense source that will fill aperture and field. See p. 329, Table 7-2, and Figure 9-11 for the spectral characteristics.

8. A treatise of special lamps for photomicrography must mention two that are more potential rather than in much active use, although both have been explored in this field experimentally.

a. **Lasers** have been considered, at least pseudo-scientifically, for photomicrography even before they were practical for any illumination. Development has been fast, but simultaneously it has become recognized that the extreme **coherence** of the laser output is a positive disadvantage in the formation of highly magnified images, greatly exaggerating all of the deleterious diffraction effects discussed in Chapter 1. On the other hand, the fact that the laser beam can be extremely monochromatic, yet very intense, can be a great advantage if the available wavelengths are desirable for a purpose.

They should be usable, if the requirements and conditions of a situation warrant it, by the device of rapidly spinning an opal glass disk in the lower focal plane of the substage condenser (with adequate diaphragming). This method is necessary to make the light incoherent; simple diffusion is not adequate and the motion must be fast, in micro dimensions, with respect to the wave travel of the light. Ground glass passes too much undeviated light, and for once the differential wavelength scattering of pot opal glass is not operative.

There are two principal types of laser, the seeded solid rods and gas lasers. With the present stage of development, except for far infrared micrography, gas lasers would seem most applicable. They will not constitute a convenient, small, and simple source. In fact, *lasers are dangerous* and should not be empirically tried in this field except by individuals who have acquired sufficient background in their characteristics. Cooperation with such an individual and laboratory might prove interesting.

b. The relatively **new quartz-iodine** and quartz-bromine **incandescent tungsten lamps** are now available for use as an outside light source. The question merely seems to be that of their advantage in photomicrography. Probably the most applicable field is that of cinemicrography.

The upper limit of the life, brightness, and color temperature of a standard incandescent tungsten lamp is set by the sublimation of the tungsten which then condenses on the relatively cool walls of the glass bulb to alter all characteristics. As stated in Chapter 19, this dark film is not even a neutral filter but is greenish by transmission, which causes the light to become less blackbody in quality. It has been discovered that if some iodine or bromine vapor is included in the lamp, tungsten iodide or bromide is formed which eventually decomposes but is seeded to deposit tungsten back onto the filament if the walls are sufficiently hot. In this case, it is desired to have the lamp small, with the walls close to the hot filament. The walls may reach 700°F. Quartz is the only otherwise suitable material that can stand this heat, and therefore small quartz tubes and bulbs are used.

The absorption of the iodine or bromine vapor effects the spectral quality of the emitted light and of course, alters it somewhat from that of a blackbody. The deviation is not very great, however.

The author plotted the spectral distribution of a quartz-iodine lamp on reciprocal wavelength paper, plotting the expression $(\log J_\lambda + 5 \log \lambda)$ against $1/\lambda$. This gives a perfectly straight line in the visible region when the distribution is that of a blackbody; the slope of the line is proportional to $1/T$. Here, J_λ is the relative output of the lamp, λ = wavelength, and T = color temperature. The line was perfectly straight except for a small region near $\lambda\,600\,$mμ, where there was a deficiency of radiation amounting to 0.1 log unit. Because the eye is so sensitive in that region, the effect is definitely noticeable visually. It probably is more so than in color photography.

At this stage, at any rate, there is appreciable variation in the iodine content of individual lamps. This variation has already decreased, since the lamps were new, and will probably decrease still further. For exacting work, such as colorimetry, lamps are selected for minimum iodine content.

If a Kodak Color Compensating Filter were available whose maximum transmission was at exactly $\lambda\,600\,$mμ, a CC05 filter would be almost enough to counterbalance the spectral defect of the lamp tested. A CC10 filter is more than enough. No such filter is available. The existing magenta or red filters are somewhat less efficient, so a slightly deeper filter is required with some spectral alteration of the resulting quality.

This spectral deficiency, however, occurs about where the green and red sensitivities of the color films overlap and a neodymium filter may be added to block it out. In other words, this "defect" may be a positive advantage in improving the rendition of such biological stains as eosin and fuchsin if it is noticeable at all. In practice, it can be neglected.

It might be noted that if the color temperature of such a lamp is being determined visually or if the two-filter method is being used photoelectrically and the second filter transmits at λ 600 mμ, a somewhat incorrect value will be obtained unless this defect in the light from the lamp is corrected.

Some quartz-bromine lamps are being made and may become more important in the future. Their absorption is in the short blue and ultraviolet and so they are visually whiter. They may look yellower photographically.

For black-and-white photomicrography there should be no special problems. No filter is needed and color temperature need not be quoted.

The lamps are now available, but because they are so hot it seems wise to use them as an outside source. In spite of popular conception the radiation should not have excess infrared except as the quartz reradiates, and the ultraviolet passed by the quartz, more than the usual glass lamp, will be absorbed by the glass of the standard optical system. It is true that a "cold mirror" that reflects the visual and transmits the infrared should be useful. Such a mirror should be useful with any strong incandescent lamp, the bulk of whose radiation is in the infrared.

General Electric separates its *Quartzline* incandescent lamps into those of low voltage, handled by the Miniature Lamp Department, and those using standard line voltage, just as it does other incandescent lamps. Among the low-voltage type, the GE No. 1962 would seem to be most directly applicable to photomicrography. Its filament, which has the C-6 form shown in Fig. 7-5a, is a cylinder 0.100 in. long, 0.100 in. diameter (2.54 mm × 2.54 mm). Therefore it will be projected onto an aperture as a square whose size is determined by the lamp magnification. Its other characteristics are listed in Table 8-4.

The G.E. No. 1962 is designed for 8.5 V but can have a much longer life by maintaining this voltage only during photography. Its base is

Table 8-4

Voltage	Candlepower (mean spherical)	Watts	Average Life Hours	Candlepower Maintenance at 70% Life	Light Center Length
6.0	25	37	2500		0.285 in.
7.0	45	46	500		3.3 mm
8.5	80	62	50	85%	

special and should get no hotter than 350°F. It is a wire terminal with a special collar.

9. Carl Zeiss (Oberkochen) has added a lamp for color photomicrography too recently for it to have been tested by the author. This is the CSI 250W mercury arc which contains "metal halides" as an additive, probably principally caesium iodide. The Zeiss agency states that it is a lamp developed by Phillips at the request of Zeiss to provide a really intense light source for use with color film for artificial light that is more efficiently adaptable to Köhler illumination (filling *both* field and aperture efficiently), since the light-emitting area is both larger and more homogeneous than in the other concentrated arcs, such as the XBO 150 or the HBO 200. The luminous area is 5×2.5 mm². The socket position is somewhat adjustable so that it can be aligned accurately into the optical axis, whereas the bayonet base and groove allow only the correct orientation.

The luminance, 15,000 candles/cm , is about the same as that of the XBO 150 and also the 5-A carbon arc. Because it operates with alternating current and exhibits the usual doubled frequency flicker rate, it is not normally suitable for cinemicrography. Its quality is discussed near the end of Chapter 9.

Lamps Appropriate for the Horizontal Bench.

10. The author has already stated that for considerable photomicrography (it may not be "routine") he prefers the horizontal bench when the specimen, usually a dry microscope slide, allows it. If a vertical microscope, hence camera, is to be used, the author would then mount **the illumination system on a horizontal slide** or optical bench to give the system designated as Case II A near the beginning of Chapter 6 and illustrated in Figure 6-1.

If this arrangement is chosen, it is the author's opinion that for most purposes **a better lamp arrangement can be constructed than can be purchased**, if a good shop is available, for relatively little effort and at no more cost than the commercial lamp, probably much less.

In the first place he would **support the lamp separately.** With this arrangement the lamp can be centered first into the axis with no optics intervening, as discussed in Chapter 6. With a good bench and clamp system, it becomes very simple to switch precentered lamps or condensers to suit the case or to try out for satisfaction. The supports for the elements should be sturdy $\frac{1}{2}$- or $\frac{5}{8}$-in. rods.

One of the simplest yet most satisfactory methods is illustrated in Figure 6-1. The vertical rod to support the lamp ends in a horizontal table or shelf. The appropriate lamp socket is screwed to it so that the lamp filament will be on the vertical axis. Some way of laterally centering the lamp must be provided. This may be at the base of the clamp on the bench or built onto the shelf, for instance, a double-slotted shelftop with screw clamps. The front of the shelf should be cut closer to the lamp than the rear to allow the condenser to approach the lamp. **The lamp housing now merely sits** on the table, held in place by cleats; it may have a light lock well. It is often a mere cylinder or box with appropriate apertures. The point is that **the housing is no longer important in the centration of the lamp**; it is now merely a light lock. **The lamp condenser is held on a separate clamp** and **should be centerable and focusable**, as discussed in Chapter 3.

An associate, Bernard Spinell, has designed some more sophisticated and very satisfactory centerable supports for lamps. One of them is shown as Figure 8-7 to illustrate the use of a zirconium arc. The "body" of the lamp is a simple tube with adequately thick walls to take the two centering screws and vertical centering pins at each end. These pins and screws support a metal disk at each end, each a simple plate. The rear disk supports the radio socket for the zirconium lamp; the front disk, which must contain a hole, making it a ring, supports the lamp condenser. In the lamp in Figure 8-7 the condenser lenses are carried in a simple tube which has a sliding friction fit into a sleeve that is screwed to the front centerable disk. Thus the lens can merely be pulled out when we wish to align the light source alone (as discussed in Chapter 6) or if we wish to use the lamp without lens for Case II low-power illumination (Chapter 5). The diameter of the outer tubular jacket can be varied according to the diameter of the lamp condenser lens to be fitted inside. The springs for each centering screw set should be stiff enough to provide stability. This sleeve and friction-fitted tubular condenser are no longer available from Bausch & Lomb. However, the rack-and-pinion condenser mounting and the condenser itself from the lamp for the L camera (see Figure 3-8) can be obtained separately and mounted on a plate for the front of this lamp. It provides rack-and-pinion focusing. Moreover, instead of a straight $\frac{5}{8}$-in. rod support, the rod containing an inclination joint can be obtained to allow the lamp to be tilted for illuminating a substage mirror. In both cases these parts should be ordered as components of the lamp for the Bausch & Lomb 42-47-64 camera. The rod can be supported either by a bench clamp or a flat stand for a table top.

A complete lamp utilizing the C100 concentrated zirconium arc may be purchased. The Romicron zirconium arc illuminator (Paul Rosenthal, 300 Northern Boulevard, Great Neck, N.Y.) contains this lamp. It has a ribbed aluminum lamp house with a vertical rod support at one side.

Figure 8-7   Lamp Mounting for zirconium arc lamps by Spinell.

Figure 8-8 Lamp mounting for compact mercury arc by Spinell.

The condenser is an achromatic cemented doublet of 40 mm back focus and better than $f/1$ aperture. The lamp can be centered to the condenser. No iris diaphragm is provided, but one could be fitted to it by the user and would be as close to the condenser as those of most lamps for photomicrography. Mr. Rosenthal states that if the arc is imaged each time just to occupy the *used* field of the microscope, a diaphragm will not be needed. That is true, normally. The author still contends that to utilize the excellent characteristics of this lamp a diaphragm close to the lamp condenser is needed.

The outer tubular jacket of the illuminator, as used for the zirconium lamp, shortens to make a ring for the illuminator illustrated in Figure 8-8, which was made to hold the Compact Mercury Arc discussed earlier in this chapter. One set of centering screws at the rear holds the lamp; the other set in front holds the disks, which in turn carry the condenser lens mount.

Chapter 9

Quality of Illuminants

SPECIFICATION OF QUALITY

By quality of light we usually mean its color, as discussed in Chapter 1 and determined by wavelength. Only with **monochromatic** light, that is, single wavelengths, is color an adequate specification of quality. Monochromatic light can vary only in intensity. The eye may synthesize several different mixtures of wavelengths so that they appear as one color, yet photocells or a photographic film may behave differently. So color is an incomplete specification of quality for photography. However, if the wavelengths are spread out into a spectrum, the relative amounts of each become obvious and specification is unequivocal. The same specification can be given as a graph of the energy or the intensity versus wavelength as the abscissa.

SPECTRAL TYPES OF LIGHT SOURCES

From a practical viewpoint there are three types of spectrum: (a) discontinuous (or line), (b) continuous, and (c) mixtures of the two types.

When the light source is a gas or vapor, a discontinuous (line) spectrum is obtained, such as that of the mercury arc (Figure 9-28). Actually the vapors may come directly from a solid, as in the case of an arc or spark source. Moreover, the lines of the spectrum may be so close together, because of many wavelengths, that the total effect is that of a continuous source of white light (e.g., from the vapor of an iron arc). However, sources of line spectra are most useful in photomicrography because of the ease with which a single wavelength may be isolated (e.g., $\lambda\,546\,\text{m}\mu$ from the mercury arc).

A continuous spectrum is obtained when a solid is heated to incandes-
ence. Since most natural light sources, including sunlight and daylight,
fall in this category, this is the most familiar quality of illumination and
therefore is desired for reproduction by color photography.

Sources of the mixed spectral type (Type c) are usually a problem for
color photography. This type can include use of a gas or vapor light source
under very high pressure. This type cannot be disregarded, however,
because it includes some common and otherwise excellent light sources,
such as the concentrated arcs and fluorescent lamps.

INCANDESCENT SOURCES

Color Temperature

A very convenient method for specifying illumination quality as a single
number is available when the latter originates from an incandescent
solid. This is because the spectral quality of the light coming from any
such glowing solid is always the same when it is heated to the same temperature, irrespective of the particular chemical nature of the solid, *as
long as it was black when cool.* Moreover, if it is not a blackbody (e.g.,
if it is shiny like tungsten) although its spectral quality will not be the same
as that of an incandescent blackbody at the same temperature, it will
resemble that from a blackbody at a somewhat higher temperature
[3. Rutgers and DeVos (1954)]. Therefore, as pointed out by Lord
Kelvin, **the quality of the light is specified by giving the temperature of the
blackbody that would be furnishing light of this quality**. It is a complete
quality specification because the energy-wavelength distribution curve is
given by a mathematical formula, the Planck equation, which has the
temperature as the independent variable. The temperature of the corresponding blackbody, which is the **color temperature** of any incandescent
body, is given in degrees Kelvin (°K). This **absolute temperature scale**
is obtained by adding 273.18° to a centigrade temperature, thus shifting
the zero by that amount (usually 273 is added). Actually the relative
energy at any wavelength in the spectrum of an incandescent blackbody
can be obtained from tables (3. Pivovonsky and Nagel 1961). The
photomicrographer is rarely concerned with such detailed specification
although the spectral curves themselves are both instructive and useful.
On the other hand, it is frequently desirable to convert the quality of
light actually obtained from a light source to that which would be received
at another temperature. This can be done with **photometric filters**,
discussed later in this chapter.

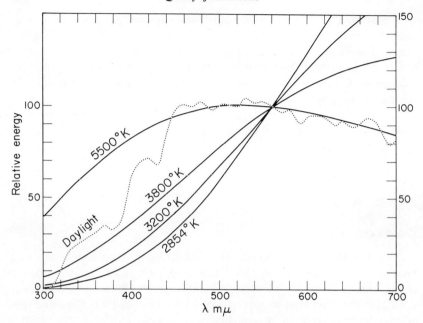

Figure 9-1 Relative spectral quality of some light sources.

The complete curves of spectral quality that represent two color temperatures are shown in Figure 8-15; Figure 9-1 shows the spectral quality of some light sources within the usual region of photomicrographic interest. The curves have been adjusted to represent the same light intensity for all sources at the wavelength of 560 mμ. (This is merely equivalent to an adjustment of the intensity of illumination by variation of the distance of the lamp or the number of lamps.) The four curves with solid lines represent four color temperatures similar to what might be obtained from incandescent tungsten at the specified color temperatures; that for the photographically important 3400°K would naturally be represented by an intermediate curve between those of 3200 and 3800°K. That for 2854°K represents an average household lamp. The reason for the different appearance of colored objects, including areas of microscopic fields, in illuminations of considerably different color temperature will be obvious if the fact is noted that **with a greater blue component there is also less red and conversely.**

Since any color temperature can be represented by one of a family of curves of the same basic equation, **specification of a color temperature merely becomes the selection of the proper curve within the family.** A particular curve and quality within such a group can be selected by merely specifying the relative height of two ordinates, that is, **the ratio of the brightnesses at two wavelengths** or colors, such as the brightness through red and blue filters. However, **if the quality of the illumination cannot be expressed by such a "blackbody" curve, the problem of its specification is not so simple** and probably more than the brightnesses at two wavelengths will have to be used. Unfortunately, individuals have been careless in expressing illuminations differing markedly from the blackbody in terms of color temperature. It is certainly a convenient method of specification, and sometimes within the tolerance of acceptability the equivalent blackbody quality will be acceptable.

Daylight and **sunlight** are examples of this custom of using color temperature to specify the quality of this type of light source. The bold curve of Figure 9-1 represents "daylight" as determined by a number of independent, trained observers and will probably become an ASA standard [3. Judd, MacAdam, and Wysecki, 1964]. Most photographers would call this quality "sunlight," since the area whose illuminance was measured was in full sunlight but also exposed to the surrounding blue sky. It will be seen that the quality of this daylight matches that of the color temperature, 5500°K, quite well, except in the far blue and ultra-violet region (below λ440 mμ). In fact, the correcting filter required to convert the incandescent lamp quality of 5500°K to daylight would be less dense than the Kodak Sky Filter, Kodak Wratten No. 1A (Kodak Publication B-3). Moreover, the daylight quality of light within a camera, which has come through the average glass lens, will have somewhat less ultraviolet than is shown in Figure 9-1. The same effect is undoubtedly obtained with the glass lenses of a photomicrographic system.

The spectral quality of "daylight" is much affected by the relative proportion of sun and blue sky illuminating a plane but is quite reproducible around the world. The daylight quality, represented by Curve D of Figure 9-11, was that from a horizontal white plane under open sky and is well matched (above λ 430 mμ) by the curve for 5600°K. As more sky and less sun illuminates the observed plane, the equivalent color temperature rises. The same is rather true for the presence of haze, clouds, and finally overcast. A color temperature of 6100°K has been mentioned as an approximation for such daylight with a diluted sun and is the "equivalent color temperature" of the famous Taylor-Kerr "standard daylight spectral distribution" used for so many years but now displaced by the curve on Figure 9-1. Finally, there is the quality obtained on a plane facing a blue sky directly opposite the sun. This light is often chosen for observing colored fabrics. A color temperature of 10,000°K has been mentioned as being representative; it may be slightly too high.

Although the photomicrographer may not use outside daylight (except in some photomacrography), he is likely to use color film that is balanced both in manufacture and processing for photographers who do use daylight.

Photometric Filters

A photometric filter is one that can be used to transform the illumination of the quality of one color temperature to that of another. We would have such a color filter if its **transmission corresponded, wavelength by wavelength, to the ratio between the transmissions of two color-temperature curves.** With a color filter we can, of course, only subtract light (by absortion) from one illumination quality to convert it to another. This is why photometric filters may have relatively high exposure factors, the bluish photometrics that raise the color temperature being worse in this respect than the orange type that reduces the effective color temperature.

If we had a perfect photometric filter, with its transmission having the above specification, it would make no difference whether we burned a lamp directly at a color temperature corresponding to one of the curves or at another color temperature and used it through this ideal photometric filter, except of course for an exposure factor. Unfortunately, there are no ideal photometric filters in practice, especially if the entire photographic spectrum is considered. In Figure 9-2 the solid curve represents the transmission of a perfect photometric filter, whereas the dashed curve

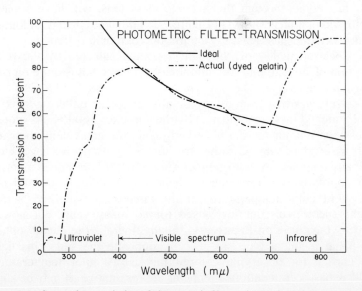

Figure 9-2 Spectral transmission of photometric filter to lower the color temperature.

represents that of a yellowish commercial filter actually available and considered quite good.

Within the wavelength range of quality in which we are interested, the spectral distribution of the illuminance may be expressed by a simpler equation of Wien rather than of Planck. H. P. Gage (3. 1933) pointed out that by utilizing the latter expression the optical density of an ideal photometric filter at each wavelength would vary as the reciprocal of the wavelength. Therefore **by plotting the optical density of the filter against the reciprocal of the wavelength, the ideal filter would be represented by a straight line**, which makes an excellent test for the perfection of any given practical photometric filter. Figure 9-3 shows the graph for a photometric filter which converts the quality of illumination from the color temperature 2360°K to that of 2850°K. **A filter line sloping up to the right would represent a bluish photometric filter that would raise the effective color temperature; that sloping the other way would represent a yellowish filter that would lower the color temperature.** The dotted curve represents the practical filter; note that it shows the same departure from the ideal at the two ends of the visual spectrum.

Improvement of Photometric Filters. Yellowish photometric filters can be of very good quality. Bluish photometric filters are more of a problem, yet they are especially useful, since they allow a lamp to be burned at a lower color temperature, hence for a longer life, than corresponds to the light quality to be used. It should be noted from both Figures 9-2 and 9-3 that when a photometric filter changes the quality of illumination from a lower to a higher color temperature the bluish filter should continue to absorb, increasingly, the unwanted excess illumination of the longer and longer wavelengths. However, nearly all organic dyes useful for making such filters transmit the infrared; the otherwise reasonably accurate dyed gelatin filters all have this defect.

Such filters are quite satisfactory visually, since we do not see the infrared. They are quite satisfactory photographically *if* the color-temperature correction is fairly small. However, **when a photoelectric cell is to be used** for any measurements, if it has red sensitivity, it will not have a sharp cut-off, as has the photographic film, and the correspondence between **photocell and film will not be the same for the directly burned illuminant and that obtained by means of a photometric filter.**

The solution of the problem is, of course, to **add the infrared absorption** and at just the rate of increase to correspond to that required by the curve of the ideal photometric filter. A solution of copper sulfate is an excellent infrared absorber, cheap, and perfectly practical to use on a bench in a water cell. **Table 9-1 shows the concentration of copper sulfate recommended**

Density

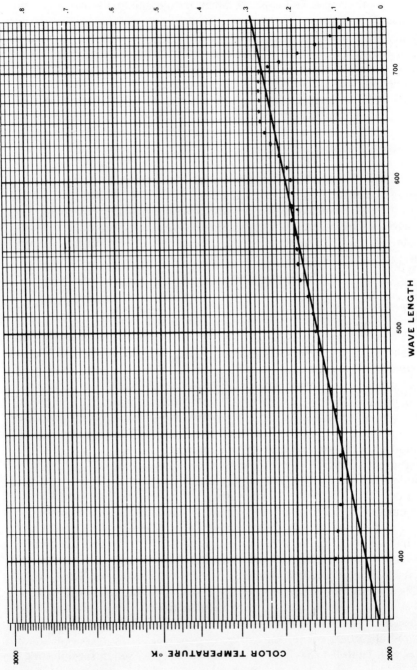

Figure 9-3 Graphical linearization of photometric filter curve.

for a 1-cm layer with the several photometric filters in the bluish Kodak Light Balancing Filter series; rounding off the concentration specifications allows the same solution to be used for several filters. These solutions also improve the photographic performance of the filters as photometric converters. **An interference heat cut-off filter could also be employed** (see Chapter 7).

Conversion Power. A great many photometric filters could be made, going from one color temperature to another, up or down the color-temperature scale, that is, from T_1 to T_2, through all the possible pairs of color-temperature curves. For years photometric filters were made to go from one specified color temperature to another. There is a more convenient method.

For each photometric filter the interval between one color temperature, T_1, and that which is desired, T_2, is a constant *if* this interval is expressed as the difference of the reciprocals $1/T_2 - 1/T_1$. In practice the constant is multiplied by a million to remove the decimal point. This constant, which is a characteristic of a given filter, is called the **conversion power*** Δ and is expressed in microreciprocal degrees, written as "mired."

$$\Delta = \frac{10^6}{T_2} - \frac{10^6}{T_1} = \text{conversion power.} \qquad (9\text{-}1)$$

A little arithmetic shows that a positive number is obtained when color temperature is lowered, that is, with yellowish photometric filters, and a negative number with the bluish photometric filters in which the color temperature is raised effectively by the filter. Moreover, **the conversion power of the filter is proportional to the slope of the straight line on the graph paper illustrated in Figure 9-3**; in this case, $\Delta = 72.8$.

Use of Conversion Power. With the above background, this important concept can be put to practical use. There are three factors: T_1, T_2, and Δ (conversion power), and any one may be found from the other two. For example, in the case of the ribbon-filament lamp it may be burning at 2848°K, whereas we wish a quality of 3200°K. Which filter should be chosen? Use of the simple equation definition of conversion power shows that a bluish filter of conversion power -38.6 is wanted. Reference to Table 9-1 shows that photometric filter 82B has a slightly lower conversion power. On the whole, it has been found preferable to use a slightly

---

Note: Unfortunately, the sign convention for conversion power was inverted in photographic literature from the more conventional notations of physics, as used by Gage and by Weaver (3. 1936), who wrote $1/T_1 - 1/T_2$. This may be confusing to those interested in the derivation and more general use of this notation.

lower color temperature in photomicrographs when departing from the exact one specified. The question might be: at what color temperature should the lamp be burned to give exactly 3200°K with the 82B filter. Although this is given in Table 9-1 (2900°K), it could also be calculated directly from the simple equation.

It is important to note that **conversion powers are additive**; for instance, if we had two photometric filters of 20 and 40 mireds, respectively, we should also have the equivalent of 60 mireds conversion power by using the two together. Our problem, therefore, is usually to find the value of T_1 to give us the desired T_2 color temperature, with the stepped values of the photometrics available, to make a lamp burn at the desired color temperature, with reasonable life but sufficient brightness.

The nomograph of Figure 9-4 is very convenient when the conversion constants of photometric filters are being used to determine the correct color temperature at which a lamp should be burned. For instance, if we wish to obtain a color temperature of 3200°K from a lamp for which this is too high for direct burning and have a bluish photometric filter of -20 mireds conversion power, we can use the nomograph as follows: a straight line is laid (a black thread is ideal for this) from 3200° on the right (T_2) through the value -20 to reach the scale of T_1, the original color temperature of the source at which it is burned (3010°K). This same value is also easily found, of course, by direct calculation from the equation (9-1) itself. A simple table of reciprocals makes even this easy.

If a color-temperature meter is available, the whole problem may be solved empirically. Suppose again that we wish to burn a ribbon-filament

Table 9-1   Kodak Photometric filters

Bluish*: To Raise Color Temperature

Wratten Number	Recommended Copper Sulfate (CuSO₄·5H₂O) % per cm	Mired Shift	To obtain 3200°K from	To obtain 3400°K from	Exposure Factor	Special Function
82	0.3	-10.0	3101°K	3288°K	1.25	
82A	0.3	-18.0	3026°K	3204°K	1.25	
82B	0.3	-32.0	2903°K	3067°K	1.6	
82C	0.35	-45.0	2797°K	2949°K	1.6	
82+82C	0.4	-55.0	2721°K	2865°K	2.0	
82A+82C	0.4	-63.0	2663°K	2800°K	2.0	
82B+82C	0.4	-77.0	2567°K	2695°K	2.5	
82C+82C	0.45	-90.0	2484°K	2603°K	2.5	
80C	0.5	-80.0	...	...	2.5	3800°K (Type F) to Daylight
80B	0.55	-112.0	...	...	4.0	3400°K (Type A) to Daylight

Yellowish†: To Lower Color Temperature

Wratten Number	Recommended Copper Sulfate (CuSO₄·5H₂O) % per cm	Mired Shift	To obtain 3200°K from	To obtain 3400°K from	Exposure Factor	Special Function
81	...	10.0	3306°K	3520°K	1.25	
81A	...	18.0	3396°K	3622°K	1.25	
81B	...	27.0	3503°K	3744°K	1.25	
81C	...	35.0	3604°K	3860°K	1.25	
81D	...	42.0	3697°K	3967°K	1.6	
81EF	...	53.0	3854°K	4148°K	1.6	
85C	...	80.0	...	...	1.25	Daylight to 3800°K (Type F)
85	...	112.0	...	...	1.6	Daylight to 3400°K (Type A)
85B	...	149.0	...	...	1.6	Daylight to 3200°K (Type B)

*To convert light quality at lower color temperatures than 3400°K to that of daylight add the correct Light Balancing Filter to the 80B filter, also adding in the corresponding copper sulfate concentration minus 0.25%.

The Corning Glass Works offer two bluish photometric filters of very good spectral quality.

Designation	Mired Shift	To obtain 3200°K from	To obtain 3400°K from	Exposure Factor	Special Function
CSl-71	−196.5	1965°K	2038°K	2.5	2854°K to 6500°K
CSl-72	−64.7	2651°K	2787°K	1.4	2854°K to 3500°K

†The following experimental yellowish photometric filters are available from the Eastman Kodak Company on special order. Their spectral quality, as positive blackbody-energy-converting filters, is excellent.

Designation	Mixed Shift	To obtain 3200°K from	To obtain 3400°K from	Exposure Factor	Special Function
8033	24.7	3475°K	3712°K	1.3	3066°K to 2850°K
8046	66.5	4065°K	4394°K	1.6	3516°K to 2850°K
8039	108.0	4890°K	5373°K	1.9	4117°K to 2850°K
8040	243.0	...	...	2.3	9268°K to 2850°K

lamp below the 3200°K that should be used for photography. **The color temperature of the lamp is simply determined through the photometric filter** while the current is varied until the readings show the effective temperature to be 3200°K. Note that we can use the concept of conversion power by utilizing any bluish photometric filter, and it may not be the one originally designed to bring the color temperature to 3200°K. If the

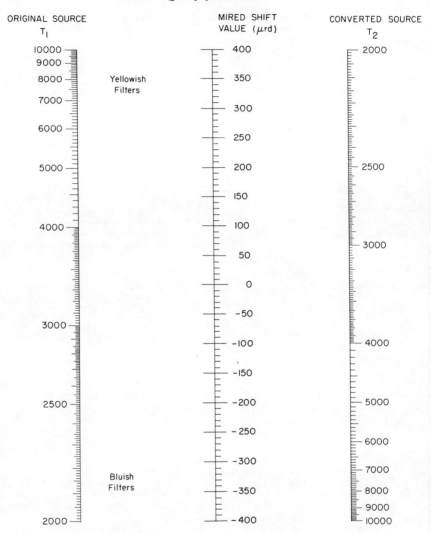

Figure 9-4 Nomograph for color-temperature shift with photometric filters.

color-temperature meter utilizes a photoelectric cell, it is important to remember the discussion of the effect of the excessive infrared transmission of the bluish gelatin photometric filters on a photoelectric meter. **For accuracy it is definitely important to use the auxiliary infrared-absorbing component, such as copper sulfate,** with the photometric filter. Although direct determination of the color temperature through the photometric filter plus infrared absorber is convenient and usually satisfactory, with practical (nonideal) photometric filters, a somewhat more accurate determination results when the color temperature of the lamp is measured directly and the filter requirement calculated by the formula.

MEASUREMENT OF COLOR TEMPERATURE

Color-Temperature Meters

A color-temperature meter is a device for selecting the correct color temperature by obtaining the ratio of the light output of a lamp at two wavelengths. This method can be used with either visual or photoelectric meters but only the latter are considered. Although visual meters are cheaper, they have different and serious problems.

The more sensitive the photocell photometer, the more sensitive and useful the color-temperature meter that can be made from it. If a relatively sensitive photometer, such as a Photovolt Model 200A or a Densichron, is available, it can be converted into a very satisfactory color-temperature meter. A device must be constructed for sliding first a blue then a red filter in front of the photocell when the photocell is held in the light beam from the lamp. The color temperature is obtained from the ratio of the readings taken through the two filters, as obtained by a slide rule, or, **if the reading through the blue filter is always brought to the same value, the reading with the red filter alone will give the color temperature after proper calibration.** Because of photocell fatigue considerations, the photometer should be used at the same general light level as specified in the calibration. Moreover, **the order, red-blue or blue-red, must be the same in use as in calibration.**

The device for doing this in the author's laboratory utilizes a photometer equivalent to the Model 200A and is shown in Figure 9-5. It was found to be **absolutely necessary to hold the photocell in a clamp and not in the hand**, so that a more compact device was of little advantage. The method is simple. **With the blue filter in front of the photocell**, the reading on **the meter is brought to some constant illuminance**, for example, 4.0 footcandles by varying the distance from the lamp or by closing the iris diaphragm in front of the photocell. **Then the red filter is slid in** and another meter reading taken. Utilizing this value and referring to a calibration graph (see Figure 9-6), the color temperature can be read off to ± 10–$15°K$.

Filters. Obviously, for determining the color-temperature curve from the family that applies, the two wavelengths selected should be well separated, and monochromatic filters would be best; in fact, they are

Figure 9-5 Color-temperature meter with photoelectric photometer.

theoretically necessary. Two narrow-band interference filters gave very reliable determinations. On the whole, however, they used up too much precious photocell sensitivity with their low total transmission. The photocell should be sensitive in both the blue and the red, but *now*, when the transmission band of the color filters is widened, the difference in the infrared limit of sensitivity between the photocell and the color film, previously discussed, becomes a factor and **an infrared absorber is needed** for best results. For the blue filter the following combination is recommended:

Kodak Wratten Filter No. 47 + Corning 9780 Glass Filter ($\frac{1}{2}$ standard thickness).
The glass-filter component is very important, for it is an absorber of the infrared transmission that nearly all blue dye filters exhibit.

There are two alternatives for a red filter according to their availability:

1. Corning 2403 Glass Filter (standard thickness) + Pittsburg Plate 2043 Heat-Absorbing Glass (7 mm thick).
2. Kodak Wratten No. 70 Gelatin Filter film.

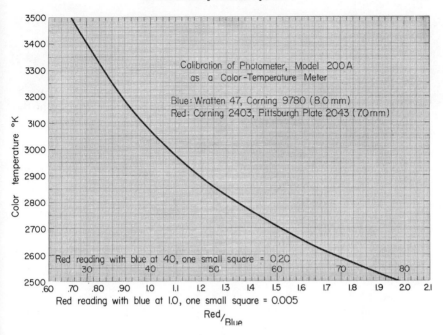

Figure 9-6 Calibration curve for color-temperature meter.

The second filter is optically good and reasonably stable, except that it is open to the infrared. The heat-absorbing glass filter was discussed in Chapter 8.

Calibration of Color-Temperature Meter. In setting up one's own color-temperature meter, there is the problem of calibration. Actually, this is the principal problem, since **it involves the reproduction of a specified color temperature with an incandescent lamp.** Obviously a low-voltage lamp should be avoided. Calibrated lamps may be obtained from the Bureau of Standards. However, the simpler and less expensive method of utilizing a commercial 500-W studio lamp for 3200°K is probably satisfactory for this purpose. A new lamp should be obtained and then aged for a few hours at below-normal voltage. The lamp should give close to 3200°K at its rated voltage, such as 115 V, although the author found them 10° to 15° higher even then. The best method of determining electrical conditions for other color temperatures will depend on whether direct current is available.

If direct current is available, a linear relation can be obtained through the 3200°K point by plotting the logarithm of the color temperature against the logarithm of the ratio of the voltage and amperage of the lamp.

A number of bulbs could be represented graphically by the following equation:

$$\log CT = 1.042 \log \left(\frac{V}{C}\right) + 2.025 \cdot \qquad (9\text{-}2)$$

If only alternating current is available, the best relation is the empirical plot of the logarithm of the color temperature against the log of the voltage, *taken directly across the lamp.* The equation is

$$\log CT = 0.4025 \log V + 2.684. \qquad (9\text{-}3)$$

When several lamps rated at 115 V were used, the curve passed directly through the following points:

log CT	3.501	3.445	3.395
voltage	110	80	60

Care must be taken to alter the current, hence the temperature of the lamp, with a rheostat and not an autotransformer, since the latter complicates the relations.

To establish the calibration curve of the color-temperature meter **it is convenient to plot it first as a straight line.** Thus only two or three points need be determined, that is, two or three cases in which a red/blue ratio must be determined for a lamp burning at a known color temperature. This can be done because **a straight line is obtained when the reciprocal of the color temperature is plotted against the logarithm of the ratio of the meter readings** (see Figure 9-7). It is worthwhile, however, to

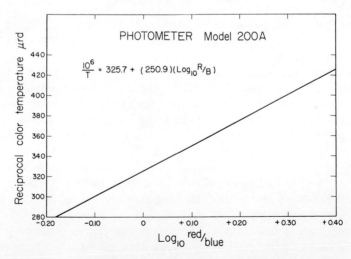

Figure 9-7 Intermediate calibration curve for color-temperature meter.

replot this graph onto arithmetic paper for the calibration curve that will be used for reference so that only the value of the red reading is needed and not its logarithm, as shown in Figure 9-6.

Commercial Color-Temperature Meters. If extensive color photomicrography is done with incandescent lamps, a color-temperature meter is most worthwhile. Some fairly accurate commercial ones are available, but they have been designed for studio use in the motion-picture industry, are relatively expensive, and usually not particularly adaptable to photomicrography. Since they do not also function as a simple photometer, and a simple and sensitive photometer is in itself very worthwhile in photomicrography, it would seem to be much better to buy such a photometer and convert it as described.

Less expensive, though less sensitive and accurate, commercial color-temperature meters are also on the market. In fact, they seemed hopelessly crude and insensitive at first. It was found, however, that the errors were random and that the average of three to five readings gave a smooth calibration curve, usually with a deviation within 50°K. Any instrument can be used as a calibration instrument to an accuracy limited by its reproducibility. Several of these instruments were found to be much more reproducible, if proper precautions were taken, than directly reading their scales would allow. This often included elimination of parallax in reading the position of the galvanometer needle and holding the meter in a frame and clamp during use. With these precautions, independent operators, taking an average of several readings on one such meter and utilizing the calibration curve shown in Figure 9-8, agreed with each other and a more elaborate color-temperature meter within ± 10°K and within the range of the calibration graph.

Color Temperatures of Lamps Typical for Use in Photomicrography

Many requests were received by the author's laboratory for general specification of the correcting filters that should be used for all the incandescent lamps sold for use in photomicrography. Unfortunately, **incandescent filament lamps cannot be depended on to give a specified color temperatue accurately unless set up by the manufacturer to do so or unless a calibration is provided.** Low-voltage lamps, particularly the ribbon filament, are especially nonreproducible from lamp to lamp.

A word must be said concerning the life and behavior of incandescent lamps with respect to the quality of their light, expressed as color temperature. A graph representing this temperature relation is shown as Figure 9-9 and other graphs expressing it will be found in the literature (3. Judd, 1936; 3. Blevin, Brown, and Sarma, 1955). An accurate relation

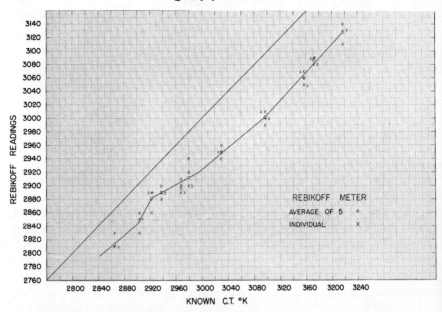

Figure 9-8 Typical calibration curve of a commercial color-temperature meter.

can be quite complicated; it is important to know which characteristic is held constant during such tests, for the shape of the curve representing lamp life can be quite different between them. Here we are interested in stability of quality, not light output as discussed in Chapter 8. The quality will change more slowly in the normal light-producing range.

Note from Figure 9-9 that only lamps of high wattage, as used in studios, can be expected to give a practical life at color temperatures directly suitable to color films for artificial lamps. **The lower wattage incandescent lamps, likely to be used in photomicrography, should be used at a lower color temperature with a photometric filter.** It must be remembered that the life of a lamp as normally quoted by a manufacturer or in this graph is for complete failure; the acceptable life at a stable color temperature may be a much shorter period. Actually, the final change in light quality is chiefly due to the color-filtering action of an evaporated tungsten film on the bulb and makes the light more greenish, departing from the blackbody distribution. With voltage held constant, there is at first a rapid rise of color temperature with time and a gradual and linear drift to lower color temperatures with time of burning. **Therefore a new lamp that is to be used for critical color photomicrography should first be burned at about 85 percent of rated voltage for about six hours before determining its color temperature.**

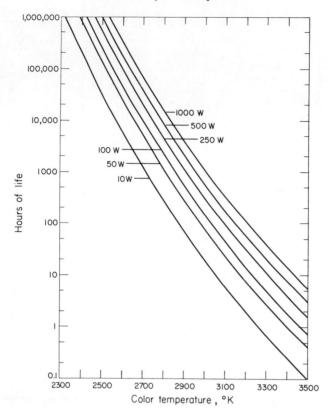

Figure 9-9 "Normal" hours of life of incandescent coil filament lamps versus color temperature.

When using a lamp for critical work, both the voltage, *directly across the lamp,* and the current in amperes should be measured. It will be found that two lamps cannot usually be brought to identical values of color temperature, voltage, amperage, and *lumen* output. Moreover, when the voltage-amperage relation starts to change, the end of stable quality with time is near.

Several manufacturers have stated that they felt it necessary to give some kind of advice to the buyers of these lamps. Table 9-2 is based on measurements in the author's laboratory of their color temperature. It is still considered preferable to determine the color temperature as discussed above. The "coil-filament lamp" of the table is the 6-Volt, 18-A, concentrated-filament lamp (CPR) of the same tube dimensions as the ribbon filament. At least 10 lamps were measured when figures regarding the variation among the lamps are given. **The voltage was taken**

Table 9-2 Color Temperature of Lamps for Photomicrography

	Lamp-Bulb					Color Temp. °K			Filter Recommendations[c]		
Code (ASA, etc.)	Type[a]	Specification	Used in (lamps)	Voltage[b]	Amperes	Average	Min	Max	Type B 3200	Type A 3400	Type F 3800
CMV	p	PH/300T8½-2SC	Brownie 300 movie projector	110	2.65	3053			82A	82B	82A+82C
CMV	p			120	2.75	3170			None	82+82	82+82C
	p	100G·16½/29SC120	AO 370A	110	0.85	2769±4	2762	2794	82C	82A+82A+82B	80C
		100G·16½/29SC120		120	0.89	2870±5	2856	2909	82A+82A	82+82C	82C+82C
CDS	p			103	0.84	2861			82A	82B	82B+82C
CDS	p		AO 735C	110	0.867	2936±4	2918	2958	82+82A	82C	82B+82C
CDS	p		AO 370A	120	0.908	3046±4	3022	3082	82	82B	82A+82C
CPR	p	Coil filament	Can replace ribbon filament	4.77	16.42	3000			82A	82+82B	82B+82C
CPR				5.50[d]	17.76	3163±10	3143	3183	None	82A	82A+82B
CPR				5.50[d]	18.05	3200			None	82A	82A+82B
		Horizontal ribbon filament	B & L lamps	5.50[d]	17.50	2890±15	2857	2929	82B	82A+82B	82C+82C
				6.07	18.46	3000			82A	82+82B	82A+82A+82B
#55	Min		B & L Pan focal, etc.	5.5[g]		2765			82A+82B	82A+82C	80C
#55	Min			6.5		2970			82+82A	82+82B	82B+82C
#55	Min			8.5		3040			82A	82A+82A	82C+82C
#88	Min		AO Microstar	6.75		2980			82+82	82+82B	82B+82C
#88	Min			7.00		3040			82A	82A+82A	82A+82A+82B
#88	Min			8.00		3200			None	82A	82A+82B
1130	Min		AO Microstar	6.4		3000			82A	82+82B	82A+82A+82B
1130	Min			7.0		3100			82	82+82A	82A+82C
1130	Min			7.6		3200			None	82A	82A+82B
1493	Min		AO vertical illuminator	6.0	2.76	2874			82A+82A	82+82C	82B+82C
1493			AO universal #353	6.5	2.88	2967			82+82A	82+82B	82B+82C
1493				7.0	3.00	3066			82	82B	82A+82C
1497	Min		AO 355	6.0	2.71	2854			82A+82A	82+82C	82C+82C
1497				6.5	2.84	2958			82+82A	82C	82B+82C
1497				7.0	2.95	3038			82A	82A+82A	82A+82C

384

Code	Type[a]	Lamp	Illuminator	Voltage[b]		Color temp.			Filters		
1630	Min		B & L vertical illuminator	6.0	2.65	2840			82+82B	82+82C	82C+82C
			B & L Trivert	6.5	2.77	2938			82+82A	82C	82B+82C
			B & L Nicholas	7.0	2.88	3028			82A	82B	82A+82C
38-01-77	Min	6 V, 15 W	Low voltage illuminator								
			Zeiss lamps	6.0	2.52	2840			82B	82+82B	82C+82C
			G.F.L WL universal	8.0	2.95	3160			None	82A	82+82C
LINID	Min		Leitz Monla								
			Leitz Aristophot.	6.6	6.0[c]	3000±15	2964	3024	82A	82+82B	82A+82A+82B
LINOP	Min	Osram 70259	Leitz Labolux	5.50	2.40	2592±6	2578	2607			
				6.00	2.50	2678±3	2668	2685			
				6.50	2.62	2766±3	2759	2773	82+82C	82A+82C	80C
				7.00	2.72	2840±6	2824	2857	82+82B	82+82C	82C+82C
LINOP	Min	Phillips 13347	Leitz Labolux	5.50	2.47	2782±10	2759	2808			
				6.00	2.58	2891±9	2865	2908			
				6.50	2.69	2996±13	2967	3033	82A	82+82B	82A+82A+82B
				7.00	2.81	3104±11	3082	3135	82	82+82A	82A+82C
#419	Min			7.50	0.700	2655					
				8.00	0.726	2712					
				8.50	0.754	2780			82A+82B	82A+82A+82B	80C
#870	Min			7.50	0.96	2852±5	2832	2873	82+82B	82+82C	82C+82C
				8.00	0.99	2922±3	2909	2937	82B	82C	82B+82C
				8.50	1.03	2995±4	2978	3017	82A	82+82B	82A+82A+82B
	Arc	Zirconium (100 W)		118 (line)	1.8[d]	3230			2B	2B+82A	2B+82A+82B
	Arc	Carbon dc		4.5		3645[f]			2B+81A	2B+81	2B+82
	Arc	Carbon dc		10		3820[f]			2B+81EF	2B+81B	2B

[a] Type: P for projector, min for miniature bulb.

[b] Voltage at bulb. Chosen as representing an average frequently encountered when no ammeter or voltmeter is used.

[c] For the concentration of copper sulfate solution to be used on the bench with these filters, see Table 9-1.

[d] Value on commercial power unit supplied with lamp.

[e] The standard power supply made by the George W. Gates Co. was used.

[f] Excessive ultraviolet (near).

With most of the incandescent lamps, the relation between the color temperature and voltage is sufficiently linear to allow interpolation or even *short* graphical extrapolation.

across the bulb. As discussed in Chapter 7 (See Index, *connections*) 6 V almost never exists at the lamp bulb as it is usually used, except perhaps in the large selfcontained outfits, in which the leads are small. They have not been measured. The value 5.5 V was chosen as typical. As mentioned in Chapter 8, an ammeter is definitely preferable to a voltmeter when only one is used. The amperage also gave more reproducible color tempera- iures among the lamps that were measured within the departments of these laboratories. The 6-V coil-filament lamps were all from the same batch, since local stocks were exhausted.

The probable error given (such as ±15°K for the ribbon-filament lamp) means that half the lamps had a bigger variation than that.

ARC LAMPS

The Carbon Arc

As a Source of Reproducible Quality. Arc lamps have very different electrical characteristics from incandescent sources. **The spectral quality of the light is much more reproducible in any laboratory and with any make of lamp than is that of an incandescent tungsten lamp.** The direct-current carbon arc has been recommended and is used as a standard of color temperature for pyrometry. Therefore, unlike most incandescent tungsten lamps, a carbon arc may be set up in any laboratory to meet a published quality specification if the same kind of carbon electrodes is used and at the same volt and ampere specification. Unfortunately, **the ultraviolet component of the arc complicates specification for color photomicrography.** Moreover, there is usually some variation in quality across the face of the anode, which is focused on the substage diaphragm, so that the image size of the arc versus the aperture size makes a difference.

When the arc is burned at a high current density, that is, with a high current per square millimeter of crater surface, small variations of amperage and voltage produce only slight variations of quality that may be negligible. Under these conditions also the arc is much less affected by slight cooling drafts, hence by the extent to which it is enclosed. Although the 4½- or 5-A dc arc with an 8 mm anode is the most common of the metallographic arcs, **decreasing the anode diameter** for this same current to 6 mm or **increasing the current** for the 8 mm diameter to 10 A, both of which have been done commercially, **makes an appreciable improvement in stability of quality** and also in intensity. For instance, a change of 1 A in the 10-A arc produces less than half the difference in color temperature produced by the 5-A arc and the same 8 mm carbon anode.

Both an ammeter and a voltmeter should be used to obtain this reproducibility. The distance between the carbons controls the voltage. This carbon arc used as a standard of color temperature for pyrometry utilizes spectrographic graphite electrodes which are impractical for photomicrography because of insufficient arc steadiness. The electrodes that are usually supplied, and should be chosen, have a core of amorphous carbon that does hold the incandescent spot in the anode. "White-flame" arcs, cored with rare-earth salts, have a complex spectrum and normally should be avoided for color photomicrography.

The Spectral Quality of the Carbon Arc. **The arc cored only with carbon** gives an illumination that **has excellent blackbody distribution except for the excess of near ultraviolet** caused by the superposition of the cyanogen formed by the arc. This can be seen from the graph of Figure 9-10 originally published by H. C. MacPherson (3. 1940). When the spectral distribution is plotted on Weaver's special paper (Figure 9-3), the spectral emission of the carbon arc is a straight line from the ultraviolet into the infrared, except for the interruption just below λ 400 mμ of the cyanogen lines.

In metallography the incandescent carbon anode is imaged in the objective field or aperture. Although the excess radiation below λ 400 mμ emanates from the arc stream, the core must be viewed through it, making the core image, therefore, a bluer white.

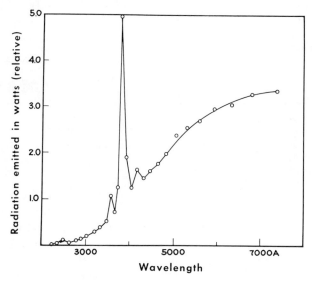

Figure 9-10 Spectral quality of light from carbon arc.

No perfect color filter for absorbing the correct amount of the excess ultraviolet in the wavelengths of the cyanogen line is known, although a fairly good one has been developed. It was found also that a 10 mm filter layer of 20 % (by volume) sodium nitrite solution just removes the *excess* ultraviolet from the arc lamp, as measured by the results on Kodachrome film, leaving the normal amount corresponding to the blackbody spectrum. However, the spectral distribution of the residual ultraviolet is not the same as that of a blackbody of that temperature. On the whole **it is preferable to remove all ultraviolet and recompensate with blue.**

Therefore the specification of the color temperature of the arc lamp for photographic application is rendered complex by the presence of these cyanogen lines, although they do not appreciably affect the visual appearance. It was found that for application to color film, surprisingly enough, the color temperature of the unfiltered arc determined visually was a more useful measure of its quality than that determined by the spectral curve. The values of the color temperature of the carbon arcs most frequently used in metallography are given in Table 9-3. Although they must be considered only as somewhat approximative to the effective color temperature to the film, they do represent, under proper conditions, exceedingly reproducible quality. These values were measured by using a colorless achromatic condenser to image the crater of the electrodes which were cored with amorphous carbon, as usually supplied for metallography. The carbon core has a color temperature very slightly lower than that of the crater as a whole.

Table 9-3

Amperes	Anode Diameter (mm)	Color Temperature (°K)
$4\frac{1}{2}$ dc.	8	3645
5 dc.	8	3680
5 dc.	6	3750
10 dc.	8	3820
10 ac.	6.4, 6.4	3475

Simple Correction of Carbon-Arc Illumination for Color Photography.
Even when the excess of ultraviolet radiation is just removed by such a filter as sodium nitrite solution, it is apparent from Table 9-3 that the color temperatures of the metallographic carbon arcs are not correct for direct use with any of the standard monopack color films, especially the professional sheet films that might be used for metallography in

color. For reasons that are discussed later in this book, as well as for the complication of the exact removal of the *excess* ultraviolet, often excellent color reproduction is not obtained by simple use of the color temperatures of Table 9-3 with the corresponding light-balancing filters, although the results may be within the tolerance for the color photomicrography at hand. The following recommendations were found preferable:

For a Type-A color film with a 4.5-A, dc carbon arc as the illuminant at medium and high magnifications, **use a 1-cm layer of 4% solution** (40 g/l) **of sodium nitrite** ($NaNO_2$). No further filter is necessary.

For photomicrography with a Type-B color film and a 4.5-A, dc carbon arc add a Kodak Light Balancing Filter No. 81A to this nitrite filter.

The 10-A arc is used principally in metallography. In a number of reports in the literature it has been used directly with color films for daylight, but, if this is satisfactory, it means that the excess ultraviolet, which registers as blue on color film, has fortuitously been about the right amount to offset the lower color temperature of the arc. This is dangerous because of the wide range of ultraviolet reflectivity of the metals. If color film for daylight is to be used, it would be preferable to remove the excess ultraviolet, as with a 2% solution of sodium nitrite, and then use the Corning glass filter for altering color temperature, 1-62, polished to the thickness to give a conversion power of 100 mireds. Assuming the color temperature of 3820°K from Table 9-3, a Kodak Light Balancing Filter No. 81EF should be used with Type-B color film, and for Type-A film, the correct balancing filter, 81C. However, for metallography, except with polarized light, some color compensation is usually necessary for the optics of the bench. This is discussed under that subject later in this book.

The Xenon Arc

The quality of the illumination from the xenon arc, XBO 150, as specified by its spectral distribution, is illustrated in the graph of Figure 9-11. It shows why colored film for daylight is usually considered suitable for use with the xenon-arc lamp but has nothing to do with any corrections introduced in photomicrography by the optical system or specimen slides discussed later in this book.

It may be wondered why xenon gas should show this continuous spectrum, except for the superposed lines, rather than the pure line spectrum of a gas. This is because xenon-arc sources have become important only after the very high intensity sources were developed when the gas is at an exceedingly high pressure, sometimes even dangerously so.

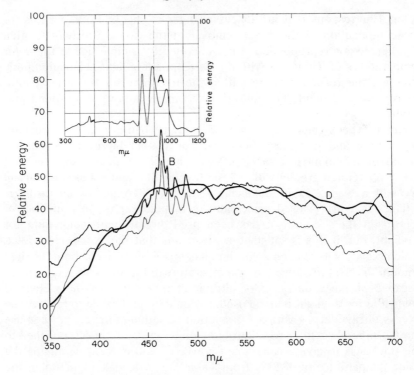

Figure 9-11 Spectral quality of light from Xenon Arc, XBO 150:
(A) range from near ultraviolet through near infrared (courtesy Wild);
(B) visual range, bare lamp;
(C) same as (B) but with Leitz heat-absorbing filter inserted;
(D) daylight that gives good color balance with Kodachrome Film for daylight.

There is an extremely strong spectral emission from the xenon arc in the infrared region from λ 800 mμ to about 1020 mμ, so that the energy peaks are more than three times as high as that at λ 462 mμ. Although this is far beyond the spectral sensitivity of the films, the optics, the eyes, and the specimen should always be protected from this intense radiant heat. The Leitz lamp carries a glass heat-absorbing filter, KG1. The effect of this filter on the illumination quality is also shown in Figure 9-11, Curve C.

The Metal-Halide Concentrated Arc Lamp CSI-250W

As mentioned in Chapter 8, the CSI-250W vapor arc lamp was developed for color photomicrography with color films for use with artificial light. It is said to have "a color temperature equivalent to 3800°K." If this is so, obviously it can be efficiently converted to a quality equivalent to 3400°K with a photometric filter of $+50$ mireds. The first case is probably best approximated by Kodak Light Balancing Filter 81C, the second case by Kodak Light Balancing Filter 81EF. See Table 9-1. Zeiss furnishes Lifa photometric filters for the same purpose.

To obtain this quality the ultraviolet and blue portion of the mercury spectrum is suppressed or diminished; the λ 546 mμ and longer spectral lines are most evident and much continuum is present (see Figure 9-12 and compare with Figure 9-29.) As with other lamps of mixed spectra, although a neutral gray will undoubtedly be well reproduced, the superposed line spectra could give poor reproduction to biological stains of narrow spectral cut if these absorptions occurred at critical spectral locations. It might be noted that in this lamp there is an excess of light near λ 600 mμ, where a slight deficiency occurs with the quartz-iodine lamp (see Chapter 8) and a deficiency about λ 560 mμ. The dilute neodymium filter might prove helpful with some magenta stains.

The Zirconium (Concentrated) Arc Lamp

The usefulness of the zirconium arc lamp for color photomicrography may seem to be a contradiction to the warning of light sources of mixed

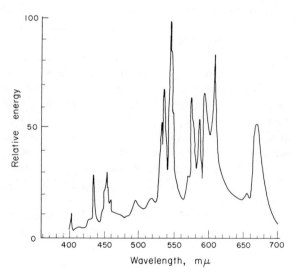

Figure 9-12 Spectrum of CSI 250-W metal halide short-arc mercury lamp (courtesy Carl Zeiss Inc.).

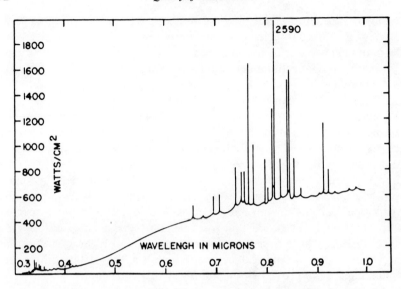

Figure 9-13   Spectrum of zirconium 100-W concentrated-arc lamp.

quality, specifically of those with spectral lines superimposed on a spectrum from an incandescent body (see Figure 9-13). However, it will be noticed that the spectral lines of appreciable intensity lie outside the visual spectrum, whereas the light within that region comes almost entirely from the incandescent zirconium oxide cathode, although the manufacturers state that there is a thin layer of incandescent zirconium and zirconium oxide vapors. Its color temperature is 3200°K. The discrepancy from this blackbody temperature is less than that caused by the factors of the photomicrographic system discussed in Chapter 19, but because it is an arc, the color temperature of 3200°K is less affected by small variations of voltage than in the case of tungsten lamps. Therefore **it is a most suitable source for use with Type-B films**. It can be used with Type-A films by utilizing Kodak Light Balancing Filter 82A, which has an exposure factor of only $1\frac{1}{4}$, which may be negligible. A heat filter for the infrared is normally advisable.

The high intensity of the source will usually allow the fractional second exposures required of most color films for standard camera use.

The Tungsten Arcs

From Table 7-2 we can see that the color temperature of the tungsten arc, as typified by the Pointolite, is 2800°K. This is blackbody, as represented by incandescent tungsten, and quite reproducible. The color temperature is unusually low. The author has not experimented with other currents through this arc since it comes* with its own power supply with its circuit in a box.

The newer G-E Tungs-Arc supplied by Bausch & Lomb has been discontinued as a commercially available source, except that many are in use and the lamps themselves are still available. Since mercury vapor is enclosed in the bulb "to start the arc," the spectrum represents that abomination in color photography, a mixed source. However, it was found that a neutral-gray balance could be obtained with Type-B film if a Wratten 2B Filter were used with a 2 mm thickness of Corning Didymium Glass Filter No. 5120. A Kodak Light Balancing Filter No. 82A should be added to this combination if a Type A color film is to be used.

CONTROL OF LIGHT QUALITY, COLOR FILTERS

The choice of the proper lamp may only provide the source from which the required light quality may be obtained. If the source is incandescent tungsten, a gamut of quality can be obtained within the blackbody specifications and the limits of color temperature just discussed. Desired wavelengths cannot be added, but unwanted wavelengths can be subtracted by absorption, that is, by use of color filters to give nearly any quality. However, unless the lamp quality is almost correct, the process may be very inefficient; for example, there may be a filter factor of several hundred. However, a color filter with the appropriate light source remains the most effective means of controlling the spectral quality of the illumination. Use of a prism or grating to disperse the light, from which a small portion is physically selected, that is, with a monochromator, is usually considered an inconvenient necessity when it must be used.

The use of color filters is usually best discussed in combination with the particular function such as for the control of contrast, for which they are applied. On the whole this is done in this book; in fact, it has already been necessary to discuss the use of photometric filters for altering color temperature in this chapter. At this point only their types are described.

---

*Information has been received that this lamp is no longer available. There are, however, many of these lamps in use, especially in England.

Color filters exist as the following types, according to their functions:

Type	Discussed
1. Photometric (light balancing)	Chapter 9
2. Color correcting	Chapter 19
3. Color compensating	Chapter 18, 19 and 21
4. Tricolor	Chapter 15, 19
5. Contrast control	Chapter 10
6. Cooling (heat absorption or reflection)	Chapter 7
7. Modulating (neutral)	Chapter 9
8. Monochromatic	Chapter 9

Filters may also be classified* according to their nature and composition.

1. Liquid filters.
2. Glass color filters.
3. Dyed gelatin filters.
4. Dyed plastic filters.
5. Physical filters (interference).
6. Monochromators.

Combinations of these filters are used.

The spectral characteristics of a color filter are best ascertained from a graph that shows the transmission against the wavelength of the light. Normally it may be given in two forms:

Transmittance versus wavelength, that is, T versus λ.
Optical Density versus wavelength, that is, D versus λ,
where $D = \log_{10} 1/T = -\log_{10} T$.

The graph of the transmittance might seem the simpler to an individual not accustomed to these graphs and in some cases it is. However, the density absorption graph is often the more useful and is more frequently used, at least in technical circles. There are a number of reasons for this. When it is important to exclude light from some portion of the spectrum, a graph line below 1% transmittance can give a false sense of security.

---

*Another method of classification of color filters (by spectral type) is given by P. T. Scharf [2. Kingslake, (1965) Vol. I, Chapter 3]. He also adapts an USA specification for the Selective Transmission of Lenses (PH3.37–1961) to the spectral specification of photometric and compensating filters.

On the other hand, a density graph usually is carried to 3·0. The following equivalent values are useful:

$D = 0.3$, $T = 50\%$; $D = 1.0$, $T = 10\%$; $D = 2.0$, $T = 1.0\%$; $D = 3.0$, $T = 0.1\%$. The light on the front side of a filter may be intense and a photographic film especially sensitive to the portion of the spectrum supposedly removed; therefore even a density of 2.0 may be insufficient protection.

The principal advantage of the density graphs is in combining filters. With the transmission graph the transmittances T of the two filters for each wavelength must be multiplied, wavelength by wavelength, to obtain T for the combination. On the other hand, the densities of the two component filters can simply be added, wavelength by wavelength. This is neglecting a 4% surface reflection.

Liquid Filters

This oldest type of filter, which is often the cheapest and most quickly available, has some very desirable characteristics for use in a horizontal illumination beam. It has been the writer's experience, however, that the question of their apparent desirability for such use depends chiefly on the individual, his background and facilities. For chemists who are accustomed to making up solutions, have the glassware for it, and if the chemicals or dyes are easily available, liquid filters are not only simple but their advantages are quickly appreciated. This is especially true of compensating filters, discussed later. The writer has received thanks for information that was successfully used by some of these people. On the other hand, for many people, unused to making solutions, especially if the facilities are not right at hand, the advantages do not seem worth the trouble. Of course, once a stock solution is made up, its use, possibly including dilution, would seem a simple matter.

Liquid filters have the following **advantages**.

1. Availability for individuals in a laboratory, especially if not geographically convenient to other supplies.

2. Much greater stability from fading in intense light than solid dyed filters is often available. If not, the inexpensive batch of filter can be discarded and the filter cell refilled frequently and cheaply.

3. Sometimes preferable spectral absorption is available from a liquid-solution filter. This may be due, especially with blue filters, to lack of color in the vehicle or to more absorption in the far red, where dyes often fail.

4. Better accuracy and more specifically chosen composition and

concentration for the particular job. This is especially important with compensating filters, as discussed under that subject.

An example of the application of some of these advantages is the use of metal salts, especially copper and ferrous salts as heat filters. No dyed filter is so good for this purpose, although these metal ions have been incorporated in glass filters with somewhat different absorption.

Liquid filters are quite useful with arc lamps, especially metal-salt solutions; dryed filters, for instance, often fade in the beam of a 10-A arc.

Filters for far ultraviolet are chiefly liquid, and the choice for this region is quite limited. These filters are discussed in Chapter 14.

The **disadvantages** of liquid filters are the following:

1. They are rarely useful for pictures by reflected light. They can be used in glass pans in a vertical beam or in a closed cell such as that of Leitz.

2. The nuisance involved for many individuals who dislike making stock solutions.

3. The thickness of the filter in its cell is sometimes more than inconvenient if the only portion of the illumination beam available is highly convergent. Insertion of a filter cell may require a little refocusing of some optical components. However, a water cooling cell is often present and can be utilized.

Liquid filters have the following characteristics:

1. The color saturation and the absorption density at any wavelength within the absorption band increase directly with the concentration of the metal salt or dye. This law may fail somewhat at higher concentrations.

2. The color saturation and the absorption density increase with the thickness of the filter layer.

Filter Cells.   Commercial photomicrographic benches once invariably carried a water cell in front of the lamp to absorb heat; this cell is now often replaced by a heat-absorbing glass filter. However, a water cell is most suitable for use with liquid filters, particularly the heat-absorbing type that uses acidulated copper sulfate (see Figure 7-13).

If liquid filters are to be used extensively, it is worthwhile to obtain a number of the 1-cm-thick cells used in colorimetry. They will be considerably better optically and their standard thickness of 10.0 mm makes metering of the liquid filter an easier task. The cells sold for the Klett Colorimeter by laboratory supply houses are satisfactory; they have $90 \times 40 \times 10.0$ mm   inside   dimensions   and   are   the   least   expensive commercial cell among those known to the writer. With thicker cells the concentration is merely decreased proportionally to the thickness in centimeters. When the smaller glass or quartz cells are used, it is wise to have a small holder for a pair of them so that two liquid filters can be used simultaneously.

For a single cell for a liquid filter the cells sold by the microscope manufacturers for cooling cells for photomicrographic lamps are suitable. The fused Pyrex cell sold by Bausch & Lomb (Cat. No. 42-47-28) is most suitable. The liquid thickness is 32 mm with a cross section in the beam of about $2\frac{3}{4} \times 3$ in. The LUNOM water cell of Leitz is unique in that it is designed to lie flat in a vertical light beam or to be used in a horizontal beam. It is a flat-ended cylinder, 2 in. in diameter with $\frac{1}{2}$-in. inside thickness. A small tubular neck that can be corked extends from the side of the cell.

Actually the best cells for the purpose are those made in the laboratory. This type can be completely disassembled for cleaning, an almost necessary feature for some filters for the far ultraviolet and usually preferable for most filter solutions. The cell consists of two glass or quartz disks held apart by a separator consisting of a ring cut from a tube of at least 2 in. inside diameter; the diameter of the disks is the same as the outside diameter of the ring or very slightly greater. If the disks are 3 mm plate glass and the ends of the rings are ground flat, no leakage will occur when the disks are pressed directly against the ring ends and held tightly in place by a metal washer of larger diameter pulled together by a circle of outside bolts around the cell. A neoprene gasket should be used between the metal and the glass. If the quartz or glass end disks are not flat, neoprene washers will also be needed between the end plates and the separator. The tube from which the separator ring is cut can be well-leached phenolic tubing for many filters but is not satisfactory for work in the far ultraviolet. If a diamond saw is available, a glass ring separator is relatively simple to make; it can also be purchased. Two holes in the top are used for filling and emptying.

A filter cell with simple glass plates, selected for flatness, can be easily made and is good for most aqueous solutions. A square piece of board, 1-cm thick (or other selected thickness), with a side length of, say, 4 in., is obtained and a 3 in. hole is cut centrally. The board is preferably a piece of phenolic composition, such as Synthane, or paraffin-impregnated wood. Its outside is coated with an adhesive wax, such as a beeswax-rosin-paraffin composition. The cell is assembled in a laboratory oven heated just above the melting point of the wax. The glass plates are pressed against the separator and cool to form a neat leakproof cell, which can also be disassembled with the aid of the oven. The only care is to avoid smearing the central aperture internally with wax. The assembled cell must not remain long in the oven.

Examples. There is very little limitation in the variety of liquid filters that can be used, both aqueous and nonaqueous, except the discovery of suitable absorption spectra and knowledge of suitable concentrations. Some may be found in the literature. Many laboratories have automatic spectrophotometers available by which the spectral absorption of a solution is determined in minutes. This is a prolific source of new filters as well as a means of determining the desirable concentration.

Some liquid filters found useful in the Kodak Research Laboratories are specified below. Others are given elsewhere in this book.

Ultraviolet-Transmitting and Selecting. These filters are specified in Chapter 14 on ultraviolet photomicrography.

Ultraviolet- and Blue- Absorbing. 1. *Sodium nitrite solutions* make a most useful, stable filter for removal of near ultraviolet (see Figure 9-14). They have been mentioned in this chapter for use with the carbon arc. Nonchemists must beware of confusing them with sodium nitrate, which is useless for this purpose but more widely known. This series of filters, varying in concentration only, has the characteristic that the long wavelength edge of the absorption band seems to advance proportionally with the logarithm of the salt concentration, being stopped by solubility limits at about $\lambda\,412\,m\mu$ (density $= 1.0$). This makes the exact concentration of the salt a less critical matter.

2. *Sodium salt of dinitrobenzene sulfonic acid* (Eastman Chemical T4276) (see Figure 9-15). This is useful as a filter whose upper edge of eliminated wavelengths advances from the ultraviolet ($c\,\lambda\,300\,m\mu$) into the visible as a yellow filter. This is quite useful since photographically the ultraviolet-blue region is a spectral unit domain. Since the concentrations used with this filter are much more dilute than the nitrite, the absorption edge advances more rapidly with concentration in the ultraviolet than does the nitrite and at high concentrations is a fairly deep yellow (limit about $\lambda\,490\,m\mu$ at $D = 1.0$).

It is important for this purpose to obtain the technical rather than the pure grade of this chemical because the pure grade has a far ultraviolet transmission band. The free sulfonic acid is unstable, the sulfonic acid group hydrolyzing off. Therefore the solution should be made very slightly alkaline with sodium hydroxide. Moreover, this hydrolysis is hastened by heat so that appreciable warming of the salt-water mixture to promote solution should be avoided. If the solution is allowed to stand at least overnight, a flocculant brown coagulant, which can be filtered

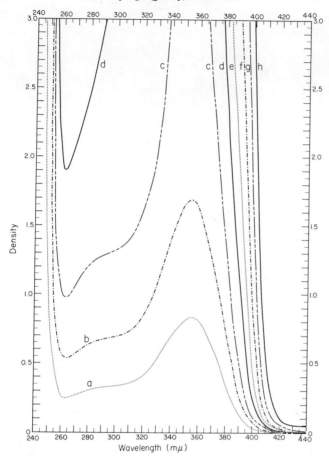

Figure 9-14 Sodium nitrite solutions as ultraviolet absorbing filters (1 cm); (*a*) 0.25%; (*b*) 0.5%; (*c*) 1.0%; (*d*) 2%; (*e*) 4%; (*f*) 8%; (*g*) 16%; (*h*) 32%.

off quite easily, will act as a clarifier to leave a sparkling-clear solution. The stock solution should be kept in a brown bottle. This filter solution is compatible with copper sulfate solution, with no precipitate. The mixture might not be so stable on long standing.

Many other ultraviolet-absorbing filters are available. Many times there is a transmission window in the farther ultraviolet, expecially with yellow filters, that is not caught in routine spectrophotometry.

3. A solution of **quinine bisulfate** in 1% concentration cuts very steeply at almost exactly λ 400 mμ to eliminate ultraviolet. The disadvantage of this old standby is its fluorescence, which may not be important if the cell is far from the substage aperture.

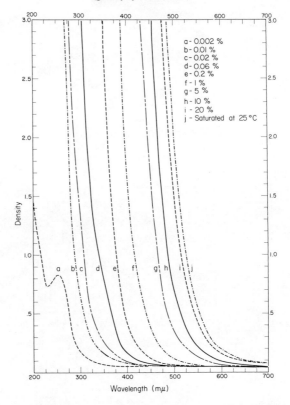

Figure 9-15 Sodium dinitrobenzene sulfonate solutions as ultraviolet and blue absorbing filters.

4. **Picric acid**, neutralized with sodium hydroxide, cuts out all below λ 490 mμ in 1% solution ($D = 1.5$) and with a sharper cut than the dinitrobenzene. Below 0.1% solution a transmission band appears at λ 285 mμ.

5. **Potassium or sodium chromate and dichromate** can be used to make stable deep-yellow filters, although the "toe" of its absorption band is not so sharp as that of some of the yellow dyes used in filters (see Figure 9-16). However, sodium dichromate is extremely soluble and can be made into an orange filter with increasing concentration. Figure 9-16 shows the spectral absorption of a series of sodium dichromate, commercial grade, with increasing concentration of the solutions. Any intermediate concentration can, of course, also be used. Where its absorption edge will cut can be predetermined quite closely by making from these graphs a curve of the absorption at one density, say 1.0, of the salt concentration against wavelength. It should be noted that it is the sodium dichromate

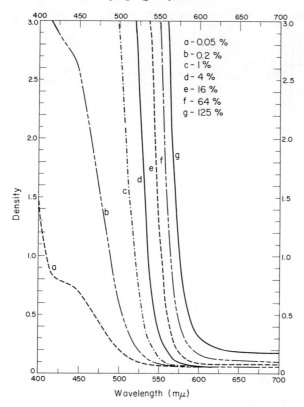

a - 0.05 %
b - 0.2 %
c - 1 %
d - 4 %
e - 16 %
f - 64 %
g - 125 %

Figure 9-16 Sodium dichromate solutions as variable yellow and orange filters.

that has the extreme solubility that is useful here. The potassium dichromate is less soluble.

So far the only color-filter solutions considered have been those with clear transmission at all wavelengths longer than the absorption at the short-wave end, that is, open-end filters. To carry the absorption farther along the spectrum than is shown for the dichromate filters would give red filters. These filters can best be obtained with dyes if liquid filters are desired.

Cyan, Blue, and Green Filters. *Cyan Filter.* As noted also in Chapter 7, copper salt solutions, most conveniently copper sulfate, are the most efficient absorbent of the far red and near infrared. Therefore to make a cyan (minus-red) filter one merely increases the concentration beyond that used for heat filtration (see Figure 9-17). A 25% solution is about the limit of solubility, but the depth of the containing cell can be increased

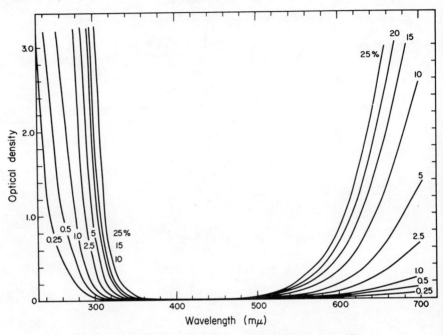

Figure 9-17 Concentrated copper sulfate solutions as cyan filters.

to increase the absorption. The absorption density at any one wavelength may be considered proportional to the thickness of the solution to the light beam. The absorption of the near red (just above λ 600 mμ) is very much increased by the addition of *a very little* methylene blue dye and the resulting mixture is light stable. The copper sulfate should be dissolved in an acidulated solution of methylene blue (instead of water).

A concentration of 10% copper sulfate (0.4 molar) in methylene blue (1 : 100,000 of water) solution has an absorption curve about equal to 25% copper sulfate but with more absorption below λ 630 mμ to give a better minus red. This combination is useful as a cyan compensating filter (see p. 408).

Blue Filter. A well-known deep blue filter, with a transmission maximum at λ 411 mμ, is that of the ammonia-copper complex. Its spectral density is shown by Figure 9-18. It is made by redissolving the original precipitate formed when ammonium hydroxide is added to copper sulfate solution. A satisfactory formulation is

copper sulfate ($CuSO_4 . 5H_2O$) – 25% solution   80 ml,
ammonium hydroxide – 28% strength   60 ml.

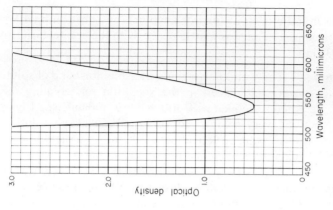

Figure 9-18 Curpric ammonium solution as a narrow-cut deep blue filter.

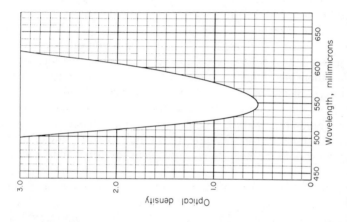

Figure 9-19 Inorganic, narrow-cut, green liquid filter for achromats.

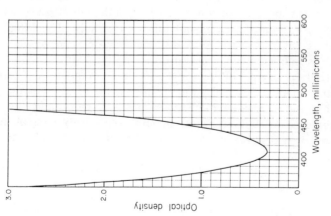

Figure 9-20 Organic dye, narrow-cut, green liquid filter for achromats.

Pyridine may be used instead of ammonia by substituting 40 ml. The spectral transmission band is a little wider, with the greatest transmission about λ 408 mμ.

Green Filter. The well-known green transmission band of solutions of nickel salts is not well placed in the spectrum for isolating its green portion if the green is defined as λ 400 mμ to λ 600 mμ (see Figure 1-2). Although they are very soluble, concentrated nickel-salt solutions (i.e., 50%) have a transmission peak at about λ 500 mμ. A more dilute solution, that is 10%, backed up for red and infrared absorption by a 20% solution (0.4 molar) of hydrated copper sulfate (see Figure 9-17), leaves more green transmittance. To cut this close to λ 500 mμ requires a separate yellow filter. One of sodium dinitrobenzene sulfonate, selected from Figure 9-15 for example, 5%, is suitable. The resulting green filter would be very stable and would withstand even stronger light.

A narrow-cut green filter is probably the most useful of all the liquid filters, since it is just such a filter that gives optimum quality with achromatic objectives and colorless specimens. Therefore this filter is almost routine in much metallography when an intense carbon arc is used as a light source. This is because achromatic lenses, including achromatic microscope objectives, are only truly corrected for a single wavelength in the apple-green region of the spectrum, about λ 550 mμ (see Chapter 1). It will also be noticed from Chapter 1 that inclusion of some of the longer wavelengths should be less deleterious than an equal bandwidth short of λ 550 mμ.

In order to obtain sufficient concentration of copper sulfate, it is necessary to use a cell such as the Klett that is at least 2.0 cm thick. The common and satisfactory 32 mm cell of Bausch & Lomb could also be used by decreasing the concentration to $\frac{2}{3}$, although at these high concentrations the dilution law is not followed exactly. The filter, whose density-wavelength curve is shown in Figure 9-19, and which uses the Klett cell, is equivalent in spectral transmission to the Kodak Wratten 58 + 15 (B + G) filter combination, which is standard for use with achromatic objectives (see Figure 9-24d). However, the gelatin filter is not stable in the intense beam of the carbon arc. **This liquid filter is preferable (because of narrower cut) to most supposedly equivalent green glass filters** provided for the purpose, although not so efficient as an interference filter.

The formulation of the liquid filter solution, whose spectral transmission is shown in Figure 9-19, is as follows:

copper sulfate ($CuSO_4 . 5H_2O$)	25.0 gm
sodium dichromate ($Na_2Cr_2O_7 . 2H_2O$)	0.50 gm

water to make   100.0 ml.
cell thickness: 20 mm

The width of the spectral cut of this combination can be adjusted within wide limits by varying the concentration of the two components with Figures 9-16 and 17 as a guide.

In fact, with a liquid filter its dilution, or even the relative concentration of the components, can be adjusted to suit the particular occasion.

Dye Solutions as Liquids Filters. It will be noted that the liquid filters described above consist of inorganic salts rather than dye solutions. It is true that solutions of inorganic salts are most easily reproducible and usually stable (except for evaporation). Principally, they are easily available to any individual in the quantities needed. Unfortunately, this in not true of most dyes; the dye industry is not accustomed to specifying its dyes chemically and furnishing them in very small quantities. The great and fortunate exception consists of the group of dyes designated as biological stains, which are adequately and well described in the book *Biological Stains* (13. Biol. Stain Commission 1961). A large number of microscopists will be connected with some industry from which textile and other dyes may be obtained in small amounts and specifically identified; for example, by their Color Index Number or Schultz Number. These two compendiums of dyes (13. Society of Dyers and Colourists, 1956, and 13. Schultz, 1931–1932) are invaluable. Atlases of absorption spectra exist. The biological stains and some other dyes may be obtained from: Eastman Organic Chemicals Department, Eastman Kodak Co., Rochester 50, New York; Hartman-Leddon Co., Philadelphia 3, Pennsylvania; Roboz Surgical Instruments Co., Washington 7, D. C. (importer of Chroma Stains). **It is wise to specify the dyes by their Color Index Number** when it is known. Small amounts of other types of dye, such as textile dyes, of specific identification are less easily obtainable in small quantity. Bachmeier and Co. of 154 Chambers Street, New York 7, lists some dyes used for photographic purposes.

Use of a liquid filter consisting of a dye solution may be very practical. Very little dye is usually required, even for a concentrated stock solution from which secondary stock solutions are made up by simple dilution with water. The water would be poured directly into the filter cell which could be emptied frequently if alteration by light or evaporation were even suspected.

Most of the liquid dye solutions as filters will be specified with the application for which they are useful.

Narrow Green Filter for Achromats. Again the narrow green filter is probably the one of greatest importance in photomicrography.

The spectral absorption of an all-dye liquid filter is shown in Figure 9-20. Its components are

0.50 gm   Tartrazine (C.I. 19140)
9.025 gm   Filter Blue-Green (Pina)
0.025 gm   Thymol
to   100.00 ml with water

The thymol is merely a preservative. The Filter Blue-Green Pina is a unique dye with high far red and near infrared absorption. It is made by Hoechst for use in color filters, a very limited purpose. It can be obtained from Carbic-Hoechst Corp., Mountain Side, New Jersey.

A liquid green filter of almost identical spectral absorption can be made from dyes that are common on the shelves of biologists but requires a copper sulfate solution in a separate cell to remove the high red transmission of the green dye. This filter is made as follows:

Solution A   0.50 gm Tartrazine (C.I. 19140)
              0.003 gm Fast Green FCF (C.I. 42053)
              0.025 gm Thymol
to   100.00 ml with water
Solution B.   20.0 gm copper sulfate as $CuSO_4 . 5H_2O$
to   100.00 ml of solution with water

Liquid Color–Compensating Filters.   Compensating color filters serve a different purpose in color photomicrography than the contrast filters discussed so far. In general, they should have broad absorptions, but **it greatly simplifies their use if a given filter affects only the one or two sensitive layers** of the film it is desired to affect. Moreover, the variable is the concentration or thickness of the filters, which are usually very dilute. Thus liquid filters are especially suitable and often perfectly practical on a photomicrographic bench.

The use of these compensating filters is discussed under the appropriate sections of color photomicrography. However, if good reproducibility is desired in the use of the compensating filters, including the ability to utilize suggestions from others, the compensating dye filter solutions must be made up quite accurately, originally with an analytical balance. Since the liquid compensating filters themselves are so dilute, it is **most practical to make a primary stock solution**, which, if well kept, will last one individual indefinitely, and from it make a secondary *working stock* solution 100X more dilute than the primary stock solution, that is, 10.0 ml of primary stock, measured by pipet, diluted to 1 liter in a volumetric flask. From this solution further dilutions are made, volumetrically with distilled water, to make the liquid filters themselves. The stock solutions, especially the primary stock, should be stored out of direct sunlight and preferably in a cool place. Evaporation of water is the greatest danger, assuming reasonable accuracy of dilution and precautions against contamination. Glassware must be exceedingly well rinsed after cleansing with a detergent because of the tendency of the detergent to adsorb to the surfaces. **Evaporation of water occurs with extended time from bottles with screw caps and ordinary cap liners** because of the irregularity of the top edge of a blown bottle neck. Evaporation can be prevented from such a bottle by wrapping the neck with Parafilm or other good sealant.

Preparation of Working Stock Solution.   Bring the temperature of the bottle of primary stock solution to $20 \pm 2°C$ $(68 \pm 3°F)$ or to the temperature at which the available volumetric glassware is calibrated for use. Shake the bottle if there is danger that some water may have distilled and condensed on the upper sides of the bottle. Also bring an adequate supply of distilled water within the same temperature range, making it neutral to litmus (or to pH 6.0 to 8.0) with a little sodium carbonate if necessary. Use a pipet and volumetric flask to make the following dilution:

>liquid compensating filter, concentrated stock   10.0 ml,
>water, distilled to   1.0 liter.

This working stock must also be protected from sunlight (long term) and evaporation.

Preparation of the Liquid Filter Solutions.   These solutions are made up in essentially the same manner, except that a variable number of milliliters are pipetted into a flask according to the requirements of the time. Several sizes of volumetric flasks (100, 250, 300 ml) are convenient for this purpose. The dilutions required are discussed with specific applications.

The principal liquid compensating filters for Kodachrome and Ektachrome are

yellow: sodium 2,4-dinitrobenzene sulfonate
         stock solution: 20% in water (see Figure 9-15)
magenta: acid rhodamine B (Color Index No. 45100)
                                      0.300 gm
        thymol   0.025 gm
        water (distilled) to make   100.0 ml

This stock solution should be kept in brown bottles out of the light.

This filter is an exception to the general rule that compensating filters should have absorption bands wide enough to affect all the wavelengths contributing one color. In this case some correction for false rendition of some magenta biological stains is included (discussed later). Thus this filter, designed specifically for color photomicrography, will differ from the magenta compensating filter of general color photography.

The spectral transmission of the magenta liquid filter diluted *10×* *from the working stock*, 10 ml ws : 100 ml with water, is shown in Figure 9-21. Note that this is the specular transmission, taken from a distance. This is because the filter is used this way and does show some fluorescence. It is quite light stable in solution, much more so than when used in a dyed filter film.

Cyan: copper sulfate (as $CuSO_4 \cdot 5H_2O$)   5.00 gm
        water (distilled) to make   100.0 ml
        sulfuric acid
        methylene blue (Color Index No. 52015)   0.435 mg

The concentration of the biological stain, methylene blue, is 1 : 230,000 in the copper sulfate solution. It may be more convenient to dissolve the copper sulfate in the methylene blue solution made on a larger scale.

Surprisingly, this solution is light stable. Its absorption is reversibly temperature dependent, however, like all copper sulfate solutions, and increases somewhat with temperature.

This liquid filter has the merit of being a good cyan. Figure 9-22, Curve 1, shows the spectral absorption of the stock solution quoted above. It has about three times the color compensating power of the Kodak Cyan Color Compensating Filter CC50C. It should be diluted with acidulated water (1 drop of concentrated sulfuric acid per 100 ml) to the desired compensating power. Curve 2 of Figure 9-22 gives the spectral transmission of this liquid cyan compensating filter diluted 10X from the stock solution.

Glass Color Filters

Except for breakage, glass color filters should offer the convenience of solid filters with quite a large selection of spectral absorption specifications. Like liquid filters, their color saturation, or spectral density, increases proportionally with their thickness and the concentration of the coloring matter. Therefore glass filters can usually be ordered ground and polished to a variety of thicknesses or may even be furnished as molded blanks to be finished by the purchaser. However, their color density per millimeter thickness may vary somewhat from melt to melt. The Corning Glass Works took a big step forward when they

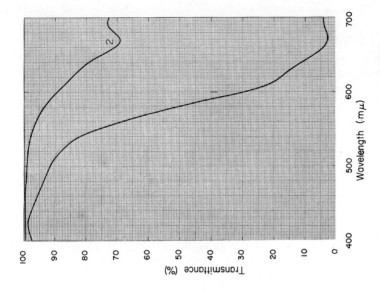
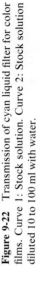

Figure 9-22 Transmission of cyan liquid filter for color films. Curve 1: Stock solution. Curve 2: Stock solution diluted 10 to 100 ml with water.

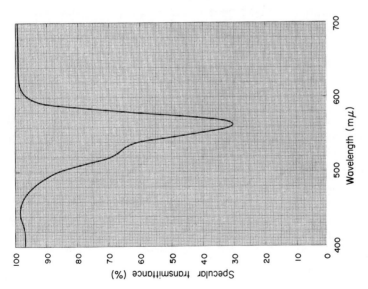

Figure 9-21 Transmission of magenta liquid filter for color films; working stock diluted, 10 to 100 ml with water.

established standard spectral curves for their glass color filters and offered them in whatever thickness corresponded to this standard. Filters of one-half standard thickness, etc., can also be ordered.

To readers familiar with some of the older literature and specifications of glass filters the new and really improved nomenclature of the Corning Glass Filters may be confusing. The old four-digit numbers still specify a given glass and spectral absorption but neither its concentration nor thickness. The number 9780 still specifies a valuable minus-red filter glass, but the designation 4-76 specifies a spectral curve given in their book of filters which will describe a filter of glass 9780 that can be obtained reproducibly.

Also, like liquid filters containing the metal ions of either copper or ferrous iron, glass filters can be made with any degree of far red and infrared absorption, which is not easy to obtain with organic dye filters. This is a special advantage of bluish photometric filters, and quite good heat-absorbing glass filters are available (see Figure 7-13).

Unfortunately, in obtaining a variety of desirable spectral absorptions, the manufacturers of glass filters have sometimes had to give up desirable physical and optical properties. Some glass filters do not even contain silica. As a result, they may be structurally weak, have a high coefficient of expansion so that they break with absorbed heat, or contain seeds or striae that distort a focused light beam. For instance, the unique and valuable Corning Glass Filter, Glass No. 9863 (or Filter 7-54), which transmits ultraviolet down to $\lambda\,240\,m\mu$, yet removes the visible light, is easily broken, either mechanically or by heat, and damaged by prolonged immersion in water.

On the other hand, there are glass filters that with a little care will stand up well in the beam of a focused, high-intensity carbon or xenon arc to which an organic filter should not be subjected, at least without very special protection (q.v.). Therefore they are valuable in metallography and most metallographic benches have a glass filter in the beam. They are also invaluable with the high-intensity sources used for illumination for fluorescence microscopy.

The bulletins furnished by the manufacturers* of color filters contain the spectral absorption and transmittance characteristics of their filters and are therefore most useful. The filters can be combined; by neglecting the 4% reflectance at the surfaces the densities at any one wavelength

---

*The three largest and most famous of the makers of glass color filters are Corning Glass Works, Optical Sales Department, Corning, New York: Jena Glass Werk, Schott & Gen. (Fish-Schurman Corp., New Rochelle, New York, American agent). Chance-Pilkington Optical Works, England (Alfa American Corp., W. 42nd St., New York 36, New York, American agent).

can be considered as potentially additive in the combination. For instance, a yellow-green isolation filter for achromats may be made by combining Corning Glass Filter 4-96 with 3-69 or 3-68, according to the spectral cut desired.

Dyed Gelatin Filters

Most organic dyes are utilized in water solution and an enormous variety of aqueous dyes are available that will dye gelatin when mixed and coated with it. The number is greatly reduced when selection is made for light stability, sharpness of cut, etc. They share with liquid filters the practicality of obtaining the absorbing agent in precise and reproducible concentrations. When spectrally sharp-cutting filters are needed, and they often are, the absorption edges of dye filters will usually be sharper than those of glass, except for interference filters. The gelatin filters are also relatively inexpensive. The result is that gelatin-dye filters are often used in microscopy and photomicrography, and in this country they are likely to be Kodak Wratten filters. **The Kodak Scientific and Technical Data Book, B-3,** which lists these filters and gives the graphs of the spectral absorption of all of them, has become a common handbook among professional photomicrographers.

Kodak Wratten filters can be obtained as gelatin filter film, which can be trimmed to shape on a cutting board or with shears, or as film mounted between glass. In recent years **gelatin filter film** has been protected by a lacquer that makes this type quite practical to use with reasonable care and **has excellent optical quality.** In fact, if kept clean, as it should be, it can be used in the image beam which is sometimes desirable; for example, laying a piece over a lens that is pointed down toward a field illuminated complexly by reflected light. Therefore a photomicrographer can probably best keep a main stock of Kodak Wratten filters as 2- or 3-in. squares of gelatin filter film, reserving those mounted in glass for the one or two used regularly in the beam and that might require frequent cleaning. These filters are normally mounted in "B" quality glass for photomicrography.

Care of Gelatin-Dye Filters. The **central area** of a gelatin-film filter **should never be directly touched with the fingers,** but it can be picked up by a corner or the edges. It is most practical to hold it in a negative envelope with a large hole cut in its center or in the filter holder for 2- or 3-in. square films sold by the Eastman Kodak Company. Then, **when it is not in use, it should be stored in a labeled protective envelope.** The protective lacquer on its surface will allow it to be dusted with a camel's-hair brush. It should be kept flat and dry. When it is to be cut, it should be held between wrinkle-free sheets of paper and the whole sandwich trimmed or cut with sharp shears.

Cemented filters should be treated with the care given to fine lenses; they should be kept in their cases when not in active use. No water or other solvent must ever come into contact with their edges. Water will, of course, cause the gelatin to swell, thus straining the glass mounting, frequently allowing air to enter, and nearly ruining the filter.

The light source should never be focused on the filter; normally it is kept nearer the lamp condenser and protected by a heat-absorbing filter such as a water cell. The various color filters vary greatly in their light stability. The relative stability of the Kodak Wratten filters is given in Data Book B-3, previously mentioned. The light stability of a dye filter may be enormously increased by protecting it with *both* a short-wavelength-absorbing filter such as the 2A or even a yellow filter *and* a heat-absorbing filter such as a copper sulfate solution. For instance, the life of a Kodak Wratten No. 58 (B) Filter is not satisfactorily long when used in the beam of a 10-A carbon arc, as set for metallography. Its life is increased so that it may be practical to use it thus *when* protected from each end of the spectrum.

The Kodak Wratten M Filters. The Eastman Kodak Company, paying attention to transmission, sharpness of absorption, and stability, has selected a group of filters especially adaptable to photomicrography and can be used in pairs to divide the spectrum into narrow segments when approximately monochromatic light would be advantageous. This feature is discussed further in Chapter 10 and the filter combinations of narrow spectral transmission are included in Table 10-1. This group has become famous in microscopical literature, in which the letter designations are generally used. In fact, competing companies employ the same alphabetical designation for filters of similar spectral cut. However, when ordering M filters from the Eastman Kodak Company, **the numerical designation should be used.**

Table 9-4

Number	Name	Visual Color	Spectral Transmission
25	A	Orange-red	From 590 mμ to red end
58	B	Green	From 480 to 620 mμ
47	C5	Blue-violet	From 370 to 510 mμ
35	D	Purple	From 320 to 470 mμ and from 650 mμ to red end
22	E	Orange	From 550 mμ to red end
29	F	Pure red	From 610 mμ to red end
15	G	Strong yellow	From 510 mμ to red end
45	H	Blue	From 430 to 540 mμ
11	X1	Pale green	For correct tone reproduction with tungsten light

Color-Compensating Film Filters. The gelatin film color-compensating filters, which are normally made in the six additive and subtractive primary colors of increasing depth of color but still comparatively dilute, will be discussed when their application to color photomicrography is considered.

Dyed Plastic Filters

Color filters are also made with nonaqueous dyes of such plastic materials as cellulose acetate or a methacrylate. The latter would be used as a cement between glass. Thus the filter would not be susceptible to water-vapor damage, as might an unprotected gelatin filter. Unless the edge of the glass-mounted filter is protected, one must be careful when cleaning them with solvents such as xylene, which are common to microscopists. There is not the huge number of commercially available dyes from which to select, that there is of aqueous dyes for gelatin filters, and this limits the precision or variety of color filters that can be made to specification.

The Tiffin Marketing Co. sells Photar and Hi-Trans Color Filters with the coloring material in the cement, which is "thermobonded" to hold two glass plates together as a cover.

SPECIAL COLOR FILTER SETS

Filters for Visual Microscopy

The Kodak Wratten Visual M Filters are primarily intended for visual work with the microscope. The nine filters, listed in Table 9-5, alone or in combination, are most generally useful.

Wratten Visual M Filters are supplied in 33 mm circles, thin enough so that several can be placed in the substage-condenser ring. The glass is the same quality as that of cover glasses. Extreme contrast may be secured by combining the two filters that will transmit a very limited region of the spectrum.

The control of color contrast obtainable with these filters is indicated in Table 9-6, which gives the filters, and in three cases pairs of filters, arranged in the order of their dominant wavelengths.

Rheinberg Differential Color Filters

Julius Rheinberg devised a system of filters that often gives a dramatic appearance to the field in the microscope, especially with motile organ-

Table 9-5

Number	Color	Function
78	Blue	A photometric filter was made to convert the illumination quality of light from incandescent tungsten lamps of the common type to that which is visually equivalent to daylight. This filter is often employed for viewing colored specimens with their commonly accepted standard daylight appearance.
38A	Blue	A filter for increasing the apparent contrast in faintly stained yellow or orange preparations. Helps in the resolution of fine detail.
45A	Blue-green	Especially useful when the highest resolving power visually possible is required, as in the study of diatom structure. It has no red transmission and its dominant wavelength is at about 470 mμ.
66	Light green	A contrast filter for use with pink and red-stained preparations. Preferred by some workers for general use in place of No. 78.
58	Green	A contrast filter for use with faintly stained pink or red preparations.
15 22	Yellow} Orange}	For increasing the contrast in blue preparations and for helping in the observation of detail in insect mounts by reducing the contrast between the preparation and the background.
25	Red	Contrast filter for use with preparations stained with methylene blue, methyl green, etc.
96	Neutral tint	A filter for moderating the intensity of the illumination. The density supplied transmits about one-tenth of the incident light.

isms. The specimen and its background can be given contrasting color, with wide choice possible; that is, brilliant orange-red paramecia could be seen swimming in a green field. With the advent of cinephotomicrography in color there has been a revival of interest in this method. Since it really utilizes the principles of darkfield illumination, these filters are discussed in Chapter 10 under that heading.

Table 9-6

Filter	Dominant WL	Filter	Dominant WL
45A	484	66	512
38A	489	38A with 15	557
38A with 66	497	15	586
38A with 58	528	22	599
58	538	25	617

Neutral Filters

Neutral gray optical filters can be most useful and even important in photomicrography. With the introduction of color photography, their importance became much greater and the specifications they should meet much stricter, especially their neutrality throughout the *photographic spectrum*.

One use for "neutral filters" is that of modulating the intense light required for some photomicrography when the system is to be used visually, since reducing the illumination aperture is an unacceptable way of reducing the illumination; this is usually the case. Another case is modulating the light in adjusting photographic exposure, when variation of time of exposure is unavailable, as in flash illumination, or unacceptable because of effects on color quality, as discussed in other chapters.

In the cases in which modulation of the light is only for visual use of the microscope, only visual neutrality is required and there is no problem. With high-intensity sources, all-glass filters, such as those supplied by Bausch & Lomb, are most satisfactory because the absorbed energy may be great, with detriment to dyed filters.

A simple and often satisfactory solution to reduce light intensity for visual use of a microscope is a variable-density filter made of two disks of Polaroid light-polarizing filter, mounted so that one is held firmly as the other is rotated. With this device maximum transmission of about 50% is obtainable with both polarizers parallel, to a practical minimum transmission of 5%, as limited by a stop. These disks are not a good neutral, since such a polarizer fails at both extremes of the visual spectrum. With a green filter, so often used in photomicrography, they might be satisfactory for photographic use. A heat filter should always be present in the illumination beam, since the polarizing ability of these films fails when they become too warm. It should be possible to flip them out of the way when not wanted. A variable-density filter, obtained from Harry Ross, New York, 10007, is shown in Figure 13-6.

Neutral Filters for Photographic Use. Some fundamental characteristics of neutral filters should be known, if they are to be employed photographically, with an understanding why, so far, there is no one type that embodies all of them for any purpose.

First, for the criterion, the transmission-wavelength plot from a spectrophotometric analysis and, of course, the density-wavelength plot for such **an ideal neutral filter would be a rigidly straight horizontal line across a graph paper** that would include wavelengths from the short ultraviolet through the visual region into the far infrared. If it consists of a wavy line through the visual region, as some dye neutral filters can show, it can be made to appear neutral in one quality of illumination but fail in another. This shows why the designation of blue-green-red factors can never be a complete specification.

The specification of neutral filters is by their density, defined on p. 394 and further discussed in Chapter 15. Even more common and troublesome than the spectral defect is that of scattering some of the transmitted light so that the **effective density** of the filter (a measure of the fraction of the originally incident light prevented from going through an aperture) **depends on both the absorption and the scatter of the filter**. The effect of the scatter depends on the distance of the filter from the aperture. Moreover, since the shorter waves are scattered more, the filter will seem yellow, the more scattered light misses an aperture. When used in the illumination system of a photomicrographic camera, the light attenuation will more nearly correspond to the specular density, whereas the specification of the density is usually the lesser value of the diffuse (ASA) density (see Chapter 15). A medium that scatters considerable light is said to have a high Q factor (see Chapter 15).

The spectral curve for the Kodak Wratten Filter, No. 96, at a density of 1.0 is shown as Figure 9-23 and illustrates the above discussion. That the increase of density is not due to the gelatin vehicle can be seen by reference to the spectral curve* of Kodak Wratten Filter, No. 0, which consists of the clear vehicle with negligible absorption above λ 300 mμ and only a

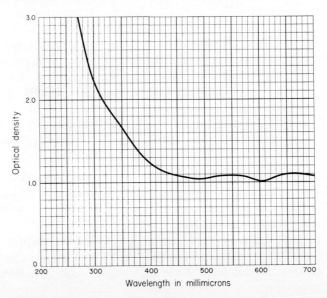

Figure 9-23   Kodak Wratten Filter No. 96. Visually neutral density of 1.0.

*Refer to Kodak Data Book B-3, "Kodak Wratten Filters".

density of 1.0 at λ 280 mμ. The Kodak Wratten Neutral Density Filter, No. 96, is supplied in 13 standard densities, constituting steps of 0.1 up to 1.0 and also of 1.0, 2.0, 3.0, and 4.0. It can be a most useful set.

Obviously, if the illumination is confined to the region above λ 450 mμ, this type of filter will behave as a true neutral filter of the specified density. This is often the case in black-and-white photomicrography when yellow or green filters are used. It is in color photomicrography that these defects are really troublesome, since the case often arises in which the illumination should be modulated, often by big factors, without affecting its color temperature. Neutral filters are necessary for flash photomicrography.

At the present time the **use of dispersed particles**, such as carbon or silver possibly augmented with dyes, has displaced pure organic dyes as the only absorbent. **As the particle size grows larger, the scattering action decreases**, but if this factor is allowed to improve the scattering defect appreciably the neutral filter will become horribly grainy in appearance as it becomes more neutral.

However, it is easy to forget that if the screen is placed in a plane conjugate to an aperture, or in an aperture, it may not be noticeable in the image plane and **very bad graininess in that plane may be acceptable.** The limitation is depth of focus; with reasonably high apertures this method, which includes the use of mechanical screens, is most successful. The secret is to stop a ray of light or to let it pass unhindered. The writer had two screens made, each with a density of 1.0, by graphic-arts techniques with ultraviolet-transmitting glass. Each screen consisted of relatively coarse parallel lines, each absolutely black with not even any vehicle lying on the clear linear interstices. The result was an absolutely photographically neutral filter of exact density 1.0 and zero Q factor. The two are additive by crossing the gratings perpendicularly to give a good neutral density of 2.0. No hint of the pattern can be seen in the image plane of a compound microscope. These screens are most satisfactory for color photomicrography. Unfortunately they are not simple to produce to a specified density on a production basis.

Reflection Neutral Filters. During World War II a method was developed for evaporating Inconel, a stainless-steel alloy, as a film on a glass surface **to give a density that was truly spectrally neutral**, provided the glass substrate was neutral. This film can be put on a quartz plate. It can be avaporated to a specified density, such as 1.0. Its Q factor is zero.

Inconel neutral filters can be made so that interposition of one of them in an illumination beam will not affect the color balance of color film, except as this might be affected by the illumination level itself.

The weakness of these metallic filters is as follows: because they are mirrors that reflect but do not absorb the incident light, the light is still around to do harm by rereflection unless some care is taken. Obviously, **if stacks of these filters are made, the density is not additive**, as with absorption filters. However, at least two filters are sometimes usefully combined if an independent way of measuring the illumination exists (e.g. with a photometer or other exposure meter).

The Inconel neutral-density filters are an important advance in color photomicrography. They can be obtained, made to specification, on special order from several optical manufacturers. Bausch & Lomb sell a set of four as 2-in. square filters on ordinary glass in a density series of 0.3, 0.6, 0.9, and 1.2. The writer found it worthwhile to obtain a special Inconel density of 0.15, which allows a series of illumination levels to be obtained with step ratios of $\sqrt{2}$.

Use of Color Filters

In color films color filters are employed as photometric (light-balancing) filters and as correcting or compensating filters. In black-and-white films they are used to increase the definition of the image with achromatic optics or to control contrast of the microscopical image. These uses are discussed in the appropriate places, as given on p. 394.

If the image is colorless, or relatively so, the microscopist is free to use wavelength (color) control to improve the definition of his image, and it is for further definition that magnification is being used in the first place. All objectives have some color aberrations with the exception of mirror (catoptric) objectives, which have other limitations. Therefore definition will improve with restriction of the spectrum of the illumination, **all objectives giving their best definition with monochromatic light.** Achromatic objectives are most commonly used and improvement of definition usually shows quickly with a restricted spectrum. As noted in Chapter 1, (see Figure 1-20) an achromat gives the best image in the green at about λ 550 mμ, but the change with wavelength is not symmetrical, so that a band from λ 530 to λ 590 mμ might seem quite acceptable. The acceptable narrowness of the transmission band of the filter will depend on the available illumination among other factors.

Green Filter Series of Narrowing Cut. One of the following Kodak Wratten filters, or filter combinations, may be chosen, when working with achromats, to improve the definition (given in the order of *decreasing breadth* of transmission band): X1(11), X2(13), 40, B(58), 61, B + G(15), 74, 44 + 15, 44 + 16. Finally, of course, employment of the truly monochromatic wavelength 546 mμ, is very practically available. (See Figure 9-24).

MONOCHROMATIC ILLUMINATION

Obviously chromatic aberration is entirely eliminated when the illumination is confined to a single wavelength, that is, monochromatic light. Indeed, it can be shown (4. Trivelli, 1930) that **there is improvement in image quality when monochromatic illumination is substituted for light through a narrow-band color filter** in photomicrography, even with apochromatic objectives. In the author's laboratory nearly all work with non-colored specimens including particle-size photomicrography is done with monochromatic light; see Figure 14-3*a*, which represents a picture of the grains of a photographic emulsion X2500 in monochromatic λ 436 mμ radiation with a 2 mm apochromatic objective, NA 1.30.

Even though limited range achromats are now available, even for the far ultraviolet, the definition is especially superior with monochromatic illumination. However, this special case is discussed under the heading "Ultraviolet Photomicrography."

Monochromators

When the particular wavelength to be chosen is specified by the absorption of the specimen and the required wavelengths may be anywhere in the spectrum, it is usually essential to use a monochromator for isolation of the wavelength required from a continuous source, although sometimes, and preferably, a discontinuous source can furnish a handy isolated wavelength. This combination of a monochromator and a discontinuous source, such as a mercury-vapor arc or a spark from metal electrodes, is really an excellent combination and is much used. In actual practice, in order to obtain adequate exposure, the slit of the monochromator is often widened so that the best obtained is a narrow band of wavelengths. With monochromatic objectives the definition then degrades correspondingly unless an isolated line can be used.

Most of the optical companies furnishing photomicrographic apparatus also make monochromators. Bausch & Lomb and E. Leitz both have good instruments, that of Bausch & Lomb specializing in the ultraviolet. The optics, including the grating, should be chosen for the region of the spectrum to be used. Then there is the famous but quite expensive Leiss monochromator. On the whole individuals obtaining such equipment will be prepared, with the help of the manufacturer, to set them up. In the author's experience it takes great care to ensure stability to allow them to stay aligned for weeks and months.

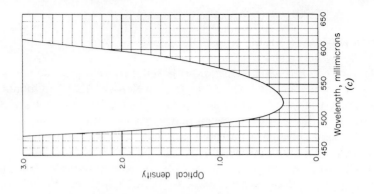

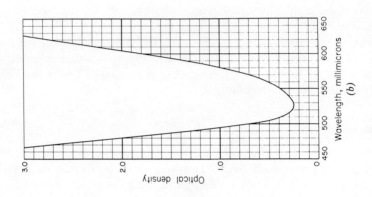

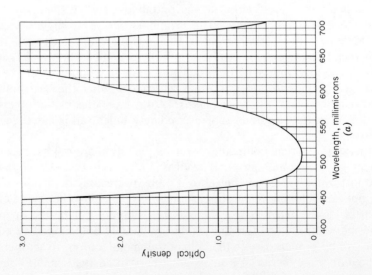

420

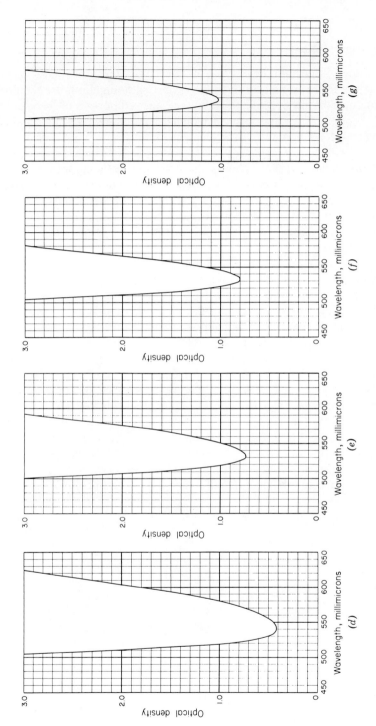

Figure 9-24 Kodak Wratten Green Filters, useful with achromatic objectives. (*a*) No. 40; (*b*) No. 58; (*c*) No. 61; (*d*) No. 58 + No. 15; (*e*) No. 74; (*f*) No. 44 + No. 16.

Interference Filters

Another important new type of filter has appeared on the market. Actually there can be said to be *two types*, although both use the phenomenon of light interference to isolate the desired portion of the spectrum. The phenomenon of interference, which is a fundamental property of light, is discussed in Chapter 1.

1. Transmission Filters. This is the common designation, but no good single word seems to exist to distinguish this type, since the other type may also be used by transmission and this type reflects most of the light not transmitted, more than 50% being reflected. Since they are older and more common, some manufacturers merely refer to them simply as "interference filters."

This type, developed by the Schott Glass Works (Jena) in 1939, was used during the war and became commercially available soon after. The filter is made by evaporating a partially reflecting film of silver on one surface of each of two optically flat glass plates that face each other in parallel. Thus they make a Fabry-Perot interferometer. Instead of air, however, a layer of a transparent dielectric, such as cryolite or magnesium fluoride, also by evaporation, cements the two layers firmly and permanently together (see Figure 9-25). This sandwich will have high transmission only for light whose wavelength is twice the optical distance between the reflecting layers or for multiples of this half wavelength. The selectivity may be very sharp.

Interference filters are characterized by three constants: (a) the **wavelength position** λ of the peak, (b) the transmittance T at the peak of the transmitted band, and (c) the **half-width** HW of the band at the transmittance equal to half that of the peak (see Figure 9-26).

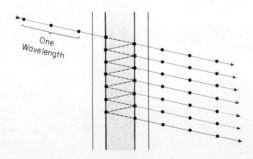

One Wavelength

Figure 9-25

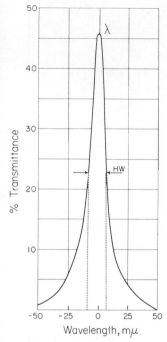

Figure 9-26 Interference filter, specification terms.

An interference filter is called a **first-order filter** when the optical path between the reflecting films is one-half the wavelength; it is a **second order filter** when this optical distance is two half-wavelengths, and so on. For equal transmittance a second-order filter will have a narrower half-width than will the corresponding first-order filter. Second-order filters are therefore usually chosen when monochromatic light must be isolated.

First-order filters usually have the higher transmittance, which, with the somewhat wider half-width, gains considerable transmitted energy with a continuous source.

It is obvious that **a given filter will have several pass bands corresponding to the multiples of the half-wavelengths.** They may not fall within the sensitivity range of the eye or photographic film. Second-order filters, however, may have one on each side within observable limits (see Figure 9-27). The unwanted pass bands must be eliminated with a secondary absorption filter; the cover glass may be colored for this purpose.

Interference filters should not be allowed to become too warm; Bausch & Lomb gives 80°C as a limit.

The light path between reflecting films is obviously dependent on the angle of incidence of the light on the filter, with the perpendicular giving the minimum path. **The transmitted wavelength is shortened if the filter is tilted.** This applies, of course, to convergent or divergent light, although it is practical to use them in focused light of not too high convergence. By 20° the transmitted bands split in two and each beam is polarized perpendicular to the other. Since the majority of the incident light is reflected by the filter, it may be necessary to tilt the filter very slightly, not to be troubled by this rejected light.

Bausch & Lomb offers the following list of 2-in. square transmission first-order interference filters.

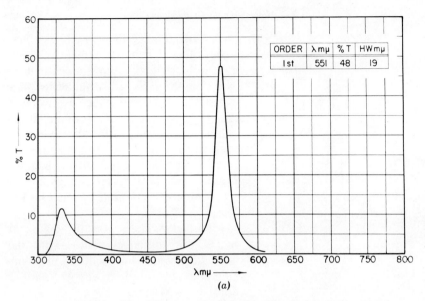

Figure 9-27a   Representative first-order interference filter.

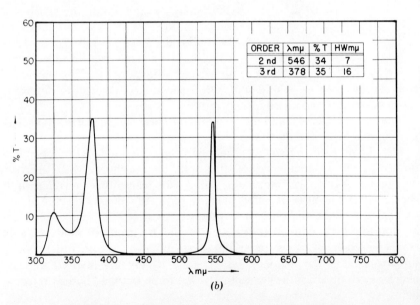

Figure 9-27b   Representative second-order interference filter.

Table 9-7   Photomicrographic Series — First-Order Interference

Catalog Number	Wavelength (mμ)	Peak Transmittance (%)	Half-Width Maximum (mμ)
42-47-55	450	45	20
42-47-56	500	45	20
42-47-57	550	45	20
42-47-58	600	45	20
42-47-59	650	45	20

They also offer a list of second-order filters for the isolation of spectral lines.

Table 9-8   Selected Line Series — Second-Order Interference

33-79-40	405 mμ (Hg)	35%	12 mμ
33-79-43	436 mμ (Hg)	35%	9 mμ
33-79-48	486 mμ (H)	35%	8 mμ
33-79-51	517 mμ (Mg)	35%	8 mμ
33-79-54	546 mμ (Hg)	35%	8 mμ
33-79-57	578 mμ (Hg)	35%	8 mμ
33-79-58	589 mμ (Na)	35%	8 mμ
33-79-64	644 mμ (Cd)	35%	8 mμ
33-79-65	656 mμ (H)	35%	9 mμ
33-79-76	768 mμ (K)	35%	10 mμ

This type of interference filter is made and sold in this country by Baird Atomic, Inc., Bausch & Lomb, Inc., and Farrand Optical Company. Surplus filters of the less commonly ordered wavelengths are often offered. On the whole, this type of filter may be 10 times more expensive than the absorption-type filter, but several of the latter may be needed to isolate a spectral line.

2. Multilayer, All-Transparent Dielectric Filters. The previously described type of interference filter, although qualitatively excellent, has poor light efficiency, some light being absorbed. A filter of very high light efficiency can be made by depositing alternately, by evaporation, a series of layers of high- and low-refractive-index material of carefully controlled thickness as a $\lambda/4$ optical path. Alternate layers of zinc sulfide ($n_D = 2.3$) and cryolite ($n_D = 1.35$) have been used. In this case all the light not transmitted is reflected and is of the strictly complementary color. Wide wavelength-band transmissions with very sharp absorption limits (cut-offs) can be made. Since **the same filter can be used both as a reflection and transmission filter**, they are useful in split-beam tricolor cameras. They are made for both 90 and 45° incidence.

As colors filters this type of interference filter is primarily custom made and primarily for instruments.

However, the multifilm filters are much more stable, especially toward heat, than the silver-film type. Therefore **their greatest use has proved to be as infrared filters**, both for selecting and for rejecting that portion of the spectrum with great sharpness of cut-off. For photomicrographers the heat-rejecting filters are most interesting. This type has already been discussed in Chapter 7 and the absorption curve of one of them is shown (Figure 7-14). Relatively narrow band-pass filters of high efficiency compared with absorption filters, and for any portion of the spectrum, can be made. **For extremely narrow transmission bands one would choose the reflecting silver filters** unless relatively high temperatures were encountered.

Quality of Mercury Arcs

So-called "spectral lamps", can be obtained which emit the discontinuous spectra of many of the suitable gases and vapors so that a variety of monochromatic lines may be isolated. These lamps are especially made by Osram of Germany but are sold in this country. **The best spectral purity** desired by spectroscopists, however, **is obtained at low pressures**. At increasing pressures many spectral lines broaden and at higher pressures the background for continuous spectra appears. Since these spectral lamps are marketed especially for spectroscopists, most of them are not intense enough for photomicrography at appreciable magnification after a single line has been isolated. On the other hand, **mercury-arc lamps are available in many types** and over the range of pressures so that a suitable one for the purpose may be chosen.

From some microscopical literature the impression can be gained that the spectral quality of mercury arcs is a specific and fixed quality. It is true that, like all arcs, their quality is not so susceptible to voltage and current variation as are incandescent tungsten lamps and also that almost invariable quality is obtained by isolating a spectral wavelength or group of wavelengths. The qualification "almost" comes from the variation of continuum present and which increases according to the pressure within the tube or bulb of the mercury arc; some of this is quite likely to be included by most practical means of isolating the spectral wavelength "lines." In fact, the mercury arc forms a particularly excellent example of the effect of pressure on the spectrum of a vapor arc. As discussed in Chapter 14, the General Electric **Germicidal Lamp** represents one commercial extreme; with exceedingly low vapor pressure, most of the energy of the lamp is radiated at λ 254 mμ, the far ultraviolet fundamental wavelength. Cooling the arc reduces this pressure further and increases the purity of the λ 254 mμ radiation, although the total output decreases. Increase of pressure, removing the cooling, and increasing the voltage gradually increase the proportion of other wavelengths above λ 254 mμ, such as those at λ 265 mμ. The effect of pressure on the visible spectrum of the mercury arc is well illustrated by Figure 9–28. A General Electric mercury lamp, UA-2, was first cooled to reduce the pressure and then allowed to operate under ambient conditions. The first spectrum represents a "low pressure arc," that of Figure 9–28b a higher pressure arc, neither case being an extreme. The increase of the relative radiation at the longer wavelengths λ 546 mμ and λ 577–79 mμ is obvious. However, note the increase in the continuum of "white light" but *remember that the height of the ordinates of Figure 9–28b should be increased* twice to offset the scale difference. The spectrum of the commonly used **HBO 200 mercury arc** (supplied by Osram) is shown in Figure 9-29, and that of the English

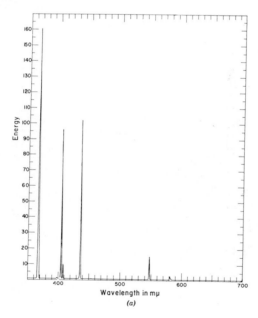

Figure 9-28a   Spectrum of a low-pressure mercury arc. G.E. UA-2 Arc cooled, 120 V, standard conditions.

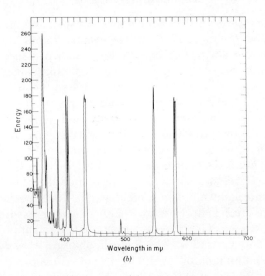

Figure 9-28b   Spectrum of a mercury arc at higher pressure. G.E. UA-2 Arc, normal conditions, 120 V.

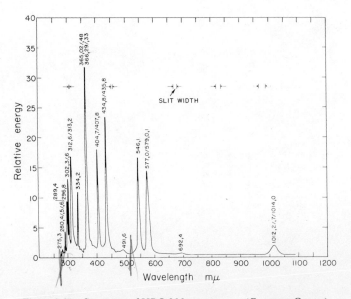

Figure 9-29   Spectrum of HBO 200 mercury arc (Courtesy Osram).

428

Table 9-9

Color	Wavelengths (mμ)	Filter(s) for Isolation	Transmission (%)	Practically Useful Filter	Transmission (%)	Principal Impurity	Transmission
Yellow	577–579			22	19	612–672	Total
Green	546	77A	68	77	72	577–579	0.5%
						612,672	66%,83%
Blue	436	2B + 34A	49.5	...	...	612	None
						672	79%
Violet	405–407	2B + Corning 5970 (C.S. 7–51)	4	...	...	None	None
Ultraviolet	365	18A	38	...	...	None	None

429

BTH compact arc is very similar. The continuum is higher yet, but still not so high that it will impair the use of these arcs for "monochromatic" photomicrography by isolation with narrow-cut filters and corrected objectives, as discussed in this chapter and Chapter 14. The cooled H6 mercury arc, discussed in Chapter 7, does represent an extreme. In this case the continuum is very great, with the spectral lines showing up in the graph like peaks in a high plain.

The quality of the mercury arc is well suited to most photomicrography not done with color. For black-and-white photography of noncolored specimens the ultraviolet, the blue, and the green constitute the gamut that is suitable to achromatic objectives. Fortunately, these three important wavelengths are easily isolated by relatively inexpensive Kodak Wratten or Corning color filters. They are listed in Table 9–9 together with the yellow doublet that is sometimes useful. The green line, λ 546 mμ, is near the peak of visual brightness and also close to the wavelength for which achromatic objectives are corrected, both achromatically and spherically. It is undoubtedly the most used and useful monochromatic wavelength in the available spectra for photomicrography. On the other hand, the blue line, λ 436 mμ, will give the extra resolving power of the shorter wavelength and is sufficiently strong that photomicrographs at 2500X magnification are made in routine with this wavelength in the author's laboratory. Ease in seeing this line will somewhat depend on the pigmentation of the eye. The doublet lines, λ 405 to 407 mμ, are not very strong and so less useful; they lie at the limit of visibility.

All the filters chosen except the last two freely transmit the red lines of mercury which are too weak to be directly useful. Although they do not interfer badly, **noncolor-sensitive or orthochromatic plates or films should be used with the lines in monochromatic photomicrography with the mercury arc**, in which case they cannot interfere at all. If only panchromatic material is available, these lines may be eliminated by use of an auxiliary filter. For the green line the Kodak Wratten No. 57 or No. 58 filter should be combined with the 77A. For a blue line the Kodak Wratten 38A filter should be placed over the 2B + 34A or even better a copper sulfate solution (15% of $CuSO_4 . 5H_2O$ per 1-cm layer) may be used. The red radiation of the mercury arc does bother a little visually, especially in examining the aperture, so that it has been the custom in this laboratory to employ the 38A filter during visual examination and then withdraw it before making the exposure with the Kodak Metallographic Plate.

The mercury-arc lamps themselves have been discussed in Chapter 7.

Chapter 10

Image Contrast

One of the most important factors in photomicrography is the contrast, or the brightness differences, in the image and in the corresponding tones of the photograph. Actually, contrast is specified as the *ratio* of the brightnesses in the image, but in the photograph we usually speak of the differences in the densities, since density is already a logarithmic term. We can have the contrast of a given area of the image against the background or the brightness or density contrasts within details of the specimen among themselves. The factors in the specimen producing contrast differences, with which we see the image at all, are (a) refractive-index differences, which represent a bending of light across a boundary, or (b) light absorption, which may be in an area. The absorption may be neutral across the spectrum, giving gray or black detail, or differential with respect to the wavelength, giving the familiar colors of microscopical images. Often the photomicrographer will wish to change the original contrast of the image. Usually, but not always, he will wish to increase it. In photomicrography the degree of contrast of the picture is usually under the control of the operator, almost always so in making black-and-white photomicrographs.

OPTICAL FACTORS THAT DEGRADE IMAGE CONTRAST

In discussing the technique and characteristics of illumination in Chapter 7, it was found that anything that produced *flare light* in the optical system would reduce the contrast of the image. This discussion was fairly thorough and very pertinent and, unless well remembered, should be reviewed here. Even so, since we are now considering it from the standpoint of the image, some of these factors are mentioned again in this summary.

431

1. Any diffusing surface in the optical system, including just dirty surfaces, lowers image contrast. As we have seen, if such diffusion is in a plane conjugate to an image of the light source or imaged in the field (hence lies in front of the field), it is in the position to do the least harm.

2. Adjacent air-glass surfaces produce flare by interreflection; the closer they are, the worse the effect. A horrible example is a cover glass over a specimen on a slide with no intervening homogeneous immersion. Even air-glass surfaces within an objective, or course, have this effect and there can be too many of them. Modern antireflection coatings on optical surfaces have been a major advance in reducing the effects of this factor.

3. Any extraneous light from outside the system is deleterious and may be disastrous. This may come from the light source itself, when inadequately housed, or from the room. Sometimes it will be found that a large highly reflecting surface near the microscope is diverting light into the system. When using a condenser or objective of long working distance, this factor must be watched. Many people have to work in rooms common with others who need light. This is perfectly practical if proper provisions are taken. Often simple shields around the equipment are enough. Shillaber (4. 1944) recommends a cardboard or black-paper tube to surround a dry objective as a shield. In an optical bench setup a cardboard shield of considerable size, between the microscope and a powerful light source, is often worthwhile.

4. Improper focusing of the elements of the illumination system will have the effect of introducing flare. In practice the same effect is obtained if the condensers of the illumination system, including the lamp condenser, are inadequately corrected for their purpose. A single high-aperture lamp-condenser lens with no correction for spherical aberration and coma, not even simply splitting it into weaker elements, will always give some flare due to the poor image of the light source in the aperture plane. This has been discussed in preceding chapters.

5. An excessively large image of the source in the aperture gives flare. This calls for adequate aperture diaphragms sufficiently early in the system. Moreover, too large an image of the source in the aperture plane makes a smaller illuminated field than is necessary and wastes light (see Chapter 2).

6. The size of the illuminated field in the object plane may be much larger than the area that is used in the image plane, thus giving flare (see Chapters 2 and 8). This calls for an adequate field diaphragm sufficiently early in the system.

7. An improper tube length or setting of the correction ring on a 4 mm objective will cause flare (see Chapter 2).

8. Internal reflections within the apparatus can easily go undetected unless sought for. The image space should be scanned with a small hole, but not a pinhole, in front of the eyes at different distances from the microscope, including the image plane and over an appreciable area, to look for gleams of light as rings, or otherwise, that may be present. The gleams need not be very bright, as they would be when coming from a large surface in the bellows. Do not forget that *a photographic material reflects back into the camera more than 80% of the light incident on it.* This can come back into low-density areas.

9. Finally, flare from the photographic material and its support may occur as "halation," which is a variation of Item 2 of this list. Since photomicrographers often take pictures of fine linear detail directly against a light source, that is, ordinary brightfield illumination, they may well encounter degradation from halation with photographic materials that are perfectly satisfactory in their ordinary photographic application. For instance, one must take care when ordering the special plates that are available and so very useful for special purposes. If they are to be used in the ultraviolet, especially the far ultraviolet, usually one can consider that the light will be absorbed by the photographic emulsion before practical halation effects are obtained from reflection from the rear surface. Otherwise, if the plates are to be used in brightfield microscopy, in which excellent detail is expected, they should be ordered with an antihalation backing that may require special comment in ordering. The plates sold especially for photomicrography, such as the Kodak "M" Plates and the Kodak Metallographic Plates, are already adequately backed. A glass plate with an adequate antihalation backing can actually be preferable to a film that has none, although normally relatively little trouble is experienced with antihalation on film, since most films carry an antihalation layer.

METHODS OF CONTROL OF IMAGE CONTRAST

None of the items mentioned above consider the original specimen on the microscope slide; this is principally outside the province of this book. However, if it is a biological material, it may be a thin slice of protoplasm or cellulose, chiefly water, colorless, and with low refractive-index and few thickness differences. If it is an industrial material, it can be many things, often plastic, often colorless, and with little visible internal detail. Therefore the usual desire (but not always) is to obtain contrast enhancement. The following methods of contrast control exist:

1. Use of hue contrast; that is, in full color. If the specimen is originally

colorless, we impart a color to it with dyes. Photomicrography by this method requires color films or plates.

2. Use of color filters. This gives complete control with colored specimens on black-and-white films and plates.

3. Enhancement of refractive index or thickness differences.

    (a) Darkfield.

    (b) Phase microscopy.

    (c) Interference microscopy.

4. Use of anisotropy (polarized light).

5. Photographic means during recording of the image (change of gamma) are discussed in a later chapter.

Use of Visual Color Contrast

Adequate visual contrast in full color may exist in the original specimen. However, as already observed, the thin layers of biological samples are usually very transparent, although there may be some details that absorb. This is illustrated in Figure 10-1, which shows the appearance, familiar to any histologist, of a bit of tissue, in this case a calf hide, as it will look if merely cut and placed under the microscope. The contrast on reproduction is unity, but the substage iris had to be reduced somewhat in order to see it. This, of course, reduces the resolution of detail. Staining such tissue to give differentiation of detail with color is an old but very effective method. In fact, it is the chief method used in microbiology (see Figure 1-30 in color insert). For our present purpose we can consider that we are looking at the original specimen through the microscope instead of at its reproduction. The modified Mallory stain that was used colored the general connective tissue blue, the other living tissue, magenta, and the tissue that was dead when the sample was fixed, yellow. In a color photomicrograph, although the dyes are now the standard dyes of the three layers of the picture, this differentiation carries over into the record.

Black-and-white photomicrographs, however, are still required for most publications. Sometimes color photomicrographs are otherwise unavailable.

Use of Color Filters with a Colored Specimen

For black-and-white photomicrography a colored specimen is still most useful and staining is utilized, since image contrast is under complete control by the use of color filters. What if two differentially colored areas are reproduced by the same shade of gray when photographing with white light? With a color filter either area can be lightened or darkened with respect to the other.

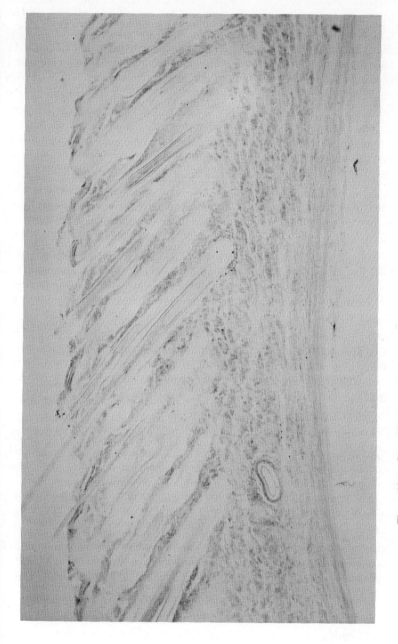

Figure 10-1 Steer hide from plate, X25.
Frozen, 20 μ cross section, unstrained, mounted in clarite. Kodak Wratten No. 58 Filter.

To understand the principles of the method the discussion of the nature of color, given in Chapter 1, must be kept clearly in mind. Here, it will be noted, **it is normally the absorptions of the colored materials** we see **that determine** the light sent back to the eye, hence **their color** (see Chapter 1). The use of this principle can be illustrated by Figure 10-2 (see color insert), which is a color reproduction of a specimen originally stained with safranin and fast green FCF. First, we must make the assumption that the absorptions of the inks used in this book for the illustration approximate those of the stains. Second, we must assume that holding the filter in front of the eye is equivalent to putting it in the illumination beam or over the camera. There are two extreme conditions: (a) the illumination is by light entirely absorbed; that is, the spectral distribution of the light lies entirely within the absorption band of the area. For this condition use a red Kodak Wratten Filter No. 25 and notice the green detail; with a green filter, notice the red detail. Maximum contrast is obtained for the red areas if a combination of Kodak Wratten Filter No. 58 plus No. 45 is used. (b) The illumination may be by light not absorbed by an area. For this case examine the red detail with the red filter and the green detail with the green filter. Here the detail is washed out; that is, we obtain minimum contrast. Note that some areas (e.g., around the periphery) are stained by both dyes and take on more contrast with both filters.

Obviously, if we used a wide distribution of light between that transmitted by the maximum-contrast filter and white light or if the transmission of the filter were displaced with respect to the absorption bands of the light, we would obtain intermediate contrast, hence complete control. Note that neutral gray or black components of the specimen and image are not amenable to this method of contrast control.

Rule 1. **To render areas (details) as dark as possible and with the highest contrast against the background use illumination that is completely absorbed by these areas, that is, which lie entirely within the absorption band.**

Rule 2. **To render detail as light as possible with minimum contrast against the background use illumination that is entirely outside its absorption band.**

Rule 3. **To control the contrast against the background for the detail of one color select a filter of appropriately wider transmission or with a displaced transmission with respect to the absorption of the specimen area. It is wise not to block detail with too much contrast.**

To enhance the contrast within the specimen and simultaneously to diminish it with respect to its background is sometimes a difficult problem. On the whole, we utilize Rule 2 and also increase the photographic contrast (gamma). This is illustrated by Figure 10-4, which is a photomicrograph of the head of a stable fly. Since it is almost entirely yellowish (brownish) and absorbs all of the shorter wavelengths, including the ultraviolet and blue and even into the green, no differentiation of detail can be obtained with a filter transmitting this light; Figure 10-4*a* was made with the blue Kodak Wratten Filter No. 47. If the converse technique is used and the wavelengths that would be absorbed by the specimen are first absorbed by the filter, the light transmitted by the filter will also be largely transmitted by the specimen and its internal detail will appear. Figure 10-4*b* was made with the deep red Kodak Wratten Filter No. 29. Note that this same principle is used when photomicrographs are made of brown chitinous beetles with infrared-transmitting filters. As a matter of fact, the same general principle is employed by the physician who makes use of x-rays of the human body. Figure 10-5*a* and *b* shows the application of the same principle but in which the two ends of the spectrum are used in opposite manner. Since these crystals are an intense blue, to which the eye is not very sensitive, they are rather opaque when viewed

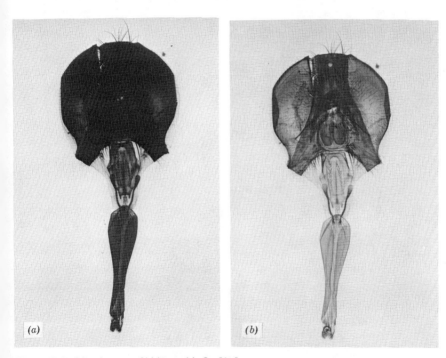

(a) (b)

Figure 10-4 Mouth parts of biting stable fly, X15.
(*a*) Photographed through blue filter; Kodak Wratten No. 47. (*b*) photographed through deep red filter, Kodak Wratten No. 29.

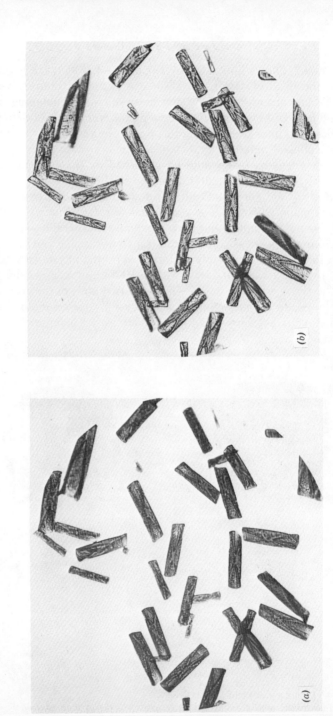

Figure 10-5 Kodak "M" Plate, ribbon filament. Blue cobalt mercuric thiocyanate crystals. X125:

(a) with Kodak Wratten No. 11 Filter; (b) with Kodak Wratten No. 47 Filter.

by yellow tungsten light and are reproduced as opaque with the Kodak Wratten Filter No. 11 which renders the tones as they would appear visually. With a blue Kodak Wratten Filter No. 47, they become transparent and show their internal detail.

The photomicrographs made with the filters that transmitted light totally absorbed by the specimen certainly gave maximum contrast; in this case, also, all detail was blocked up. Most inquiries to the author, however, have asked to increase the contrast when stains or colors are very weak. Here we use Rule 1. The visual criterion is to use the complementary color. Because the nature and extent of the absorption bands may not be obvious from the color, especially if the staining is weak, this general statement may be an inadequate guide when maximum contrast is wanted. If a good selection of filters is available, actually placing proposed filters in the illumination beam and observing the effect on the contrast is usually a most satisfactory method. **If the identification of the biological stains is known, a comparison of their known absorption bands with the spectral transmission of available filters** (e.g., as given in the Kodak Data Book B-3 on Kodak Wratten Filters) **will usually allow an efficient choice to be made.** Combinations of filters that are even more efficient can be calculated if it is remembered that the **densities at each wavelength of two filters are additive if they are used together.** The percent transmissions are not additive; they must be multiplied together.

The following may be used as a general guide **for obtaining maximum image contrast.**

For blue-stained areas use an orange to red filter.
For green-stained areas use a magenta filter.
For red-stained areas use a cyan filter.
For magenta-stained areas use a green filter.
For yellow-stained areas use a blue filter.
For brown-stained areas use a blue filter.
For purple-stained areas use a green filter.
For violet-stained areas use a yellow filter.

Most frequently the so-called red biological stains, such as fuchsin, eosin, and safranin, are really magenta, absorbing in the green, and therefore a green filter (transmission) will give areas stained with these dyes their maximum contrast.

Of course, the narrower the spectral width of the filter transmission band, the more likely it is to achieve the maximum contrast of an area if it is selected so that this lies at the peak of the absorption of the speci-

men area. Combinations of Kodak Wratten filters can be used to give
a series of narrow transmission bands across the spectrum. Such bands
are said to have a **dominant wavelength;** this is the wavelength that
gives the same visual color as the whole transmission band; in a narrow
transmission band this wavelength probably lies near its center. At both
ends of the spectrum, however, the apparent color will be that nearer the
edge of greater *visual* brightness. Therefore, although this term is in frequent use, even for this purpose, and should be understood, another admittedly arbitrary criterion is more advantageous when a color filter is
being selected for use with a photographic material sensitive over the
whole visual plus ultraviolet spectrum (panchromatic); it can be considered as having a reasonably flat response across the transmission band
of a narrowly transmitting color filter. The significant wavelength is now
called the **mean wavelength.** This wavelength is **found by averaging the
two wavelengths on each side of that of maximum transmission, whose
transmission is one-half that of the peak transmission or the minimum
density.** In a density-wavelength graph this is equivalent to adding a
density of 0.3 to the minimum density. The narrowness of the spectral
transmission, however, is best judged from the difference in wavelengths
at a transmission that is one-tenth that of the peak, which is equivalent to
adding a density of one.

Table 10-1 shows some of the especially useful combinations of Kodak
Wratten filters, plus several single filters of unusually narrow spectral
transmission. The width of the transmission band is given in the last two
columns by the 10% peak value discussed above. Sometimes the difference between the use of two combinations that are almost alike will
be quite marked if only one of the two lies entirely within an absorption
band of a stain.

Interference filters, which were discussed in the Chapter 9, are even
more efficient in giving a very narrow transmission band of low peak
density. As observed there, they are affected by the angle of the illumination beam, including its aperture. All of these filters, having such narrow
transmission bands, will have very high exposure factors. The use of a
filter combination of narrow cut to fit the absorption band of a biological
stain is illustrated in Figure 10-6, which shows how the transmission
of the B + E filters fits into the absorption band of aniline blue.

Obviously, using a photographic material with only blue sensitivity
is equivalent to using a blue filter with a panchromatic material, and using
an orthochromatic plate with no red sensitivity is equivalent to using
a deep cyan filter. **Orthochromatic plates and films are sometimes most
useful to obtain a sharp long wavelength limitation in photographic
effect.**

Table 10-1

No.	Filter A	Filter B	Mean λ (mμ)	Peak λ (mμ)	T (peak)	Spread (mμ)	λ at 10% of Peak (mm)
1.	18A	···	355	358	61.6	67	319–386
2.	*36†	···	415	415	48	96	355–451
3.	*35	2B	422	421	40.5	64	393–45/
4.	34	38A	425	426	38.8	111	361–472
5.	*35	2A	430	428	36.2	49	409–458
6.	*34	2B	438	434	52.5	66	410–476
7.	34A	38A	441	435	42.2	70	412–482
8.	98	2E	444	440	34.2	60	420–480
9.	35	45	448	447	2.0	40	430–470
10.	94	···	454	455	7.6	56	428–484
11.	47	45	468	461	12.6	70	436–506
12.	*39	4	474	473	6.6	52	454–506
13.	38A	45	476	473	18.6	86	437–523
14.	75	···	489	487	16.7	64	461–525
15.	45	8	499	500	18.7	61	469–530
16.	45	58	508	508	7.3	49	483–532
17.	65A	9	512	513	20.4	75	477–552
18.	45	12	518	517	4.3	38	500–538
19.	65	9	519	522	17.6	82	479–561
20.	44	12	528	526	24.4	57	504–561
21.	44	15	534	532	15.5	52	512–564
22.	74	···	534	532	16.4	60	509–569
23.	44	16	539	537	9.0	50	517–567
24.	93	···	541	541	6.9	56	517–573
25.	64	16	544	542	24.4	67	517–584
26.	58	15	547	541	38.6	75	516–591
27.	99	···	548	546	24.0	66	521–587
28.	99	16	553	551	17.4	61	527–588
29.	64	21	558	556	10.1	54	536–590
30.	58	21	563	560	17.8	56	539–595
31.	57	21	566	562	25.1	64	540–604
32.	58	22	573	571	10.5	49	553–602
33.	90	6.47 mm 2043 glass	578	575	29.2	72	555–627
34.	72B	···	608	605	7.2	51	588–639
35.	25	···	···	··	···	···	590–
36.	29	···	···	···	···	···	608–
37.	92	···	···	···	···	···	625–
38.	70	···	···	···	···	···	655–

λ = wavelength, T = percent transmission

*Those filters marked with an asterisk are cases in which there is appreciable transmission of red light below 700 mμ. If used with blue-sensitive or orthochromatic films, this transmission will not bother. When panchromatic materials must be used, it may be removed with the inclusion of a Kodak Wratten 38A filter in the set at the cost of some lowered transmission. The red filters in the table are open ended to the longer wavelengths, depending for the long-wavelength limitation on the ending of either visual or photographic sensitivity.

†The wavelength spread of the transmission of Filter No. 36 can be substantially decreased by inclusion of a 1-cm layer of 0.15% sodium dinitrobenzene sulfonate in water solution, or its equivalent (see Figure 9-13).

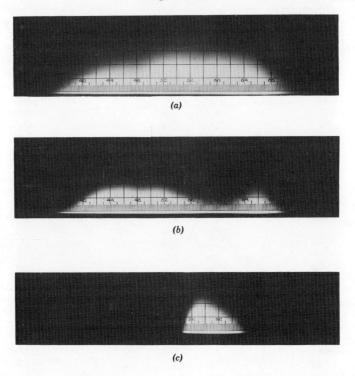

(a)

(b)

(c)

Figure 10-6 Choice of filters whose transmission lies within the absorption band of a biological stain to obtain maximum contrast:
(*a*) spectrum of panchromatic plate; (*b*) Absorption spectrum of aniline blue; (*c*) transmission band of B and E filters.

Microscopical Stains and Suitable Filters. The use of color filters to control contrast can be further illustrated with a specimen stained magenta with acid fuchsin, as represented by Figure 10-7. This has an asymmetric absorption band with a peak of about λ 550 mμ but transmits freely beyond λ 580 mμ. Therefore the red Kodak Wratten 25 (A) Filter, also transmitting this region, almost washes out the image. The use of a very narrow green filter, with transmission within the dye absorption, gives maximum contrast but blocks up the detail in some cells. The Kodak Wratten 62 used is almost identical with the 74 filter. The wider transmission of the Kodak Wratten 58 (B) Filter, which allows some light through that is not absorbed by the dye, gives better detail yet adequate contrast.

Usually there is more than one stain. The most common stain-counterstain combination in biology is hematoxylin-eosin, the visual appearance

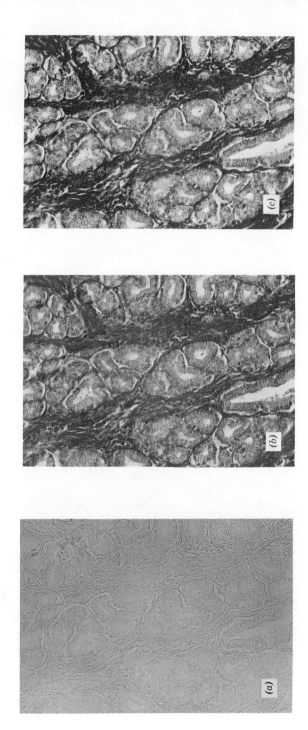

Figure 10-7 Section of cat duodenum, stained with acid fuchsin, X325: (*a*) photographed through filters having the same color as the specimen; (*b*) photographed through filters having transmission of wider range than the absorption band of the stain; (*c*) photographed through filters whose transmission was within the absorption band of the stain.

of which is illustrated by the color in Figure 10-3. Most black-and-white photomicrographs of specimens so stained are probably made with the No. 58 Filter, which gives high contrast to the areas stained with eosin but not the maximum contrast. The contrast of the nuclei stained with hematoxylin usually does not need enhancement. If the areas taking the eosin are well stained and contain much detail, a green filter of even wider cut than the 58 is preferable. Figure 10-8 shows a case in which a Kodak Wratten 13 (X2) Filter was used with this stain combination. A list of green filters, starting with a wide-open transmission and becoming the narrow-cut combination 44 + 16, is given on p. 418.

A polychrome staining of several dyes may pose a problem in which choice is made to enhance the contrast of specific areas of interest or the best general representation. Again, using Figures 10-1 and 1-30 as a base, the use of stains and color filters to control image contrast can be illustrated. Here the trichrome Mallory's stain certainly enhanced the visual contrast, as noted before. In representing this control in monochrome, which the unstained specimen can be considered to be, the large areas of the blue corium will be noticed first. If the illumination were confined to that giving maximum contrast of the aniline blue, as shown in Figure 10-6, these large areas would certainly block up. Going to the complementary case, when a blue filter (the 47), is used, the blue areas are washed down in contrast but blue-absorbing (yellow) areas stained by the orange G are now darkened (see Figure 10-9). Here an important phenomenon is noted. **Some areas take more than one stain but may not**

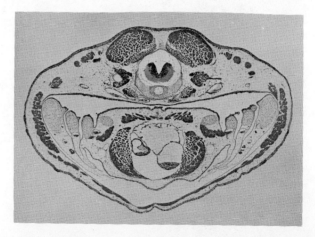

Figure 10-8   Section through head of larva of newt (*Triturus*), X10. Hematoxylin-Eosin, Kodak-Wratten No. 13 Filter, tungsten ribbon filament. 32 mm *f*/4.5 Micro Tessar Kodak "M" Plate.

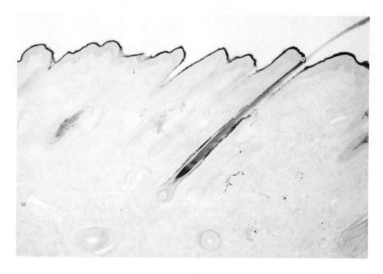

Figure 10-9   Calf skin from Foreleg, X35.
Frozen 20 μ, cross section. Stain: Mallory's triple. Kodak Wratten No. 47 (blue) Filter.

show the color of a less visually predominant absorption. The yellow stain converts the magenta of the fuchsin in the originally dying cells to red, and these areas are even darker with the blue filter. Probably the best general rendition is given when the 58 (B) filter is used (see Figure 10-10).

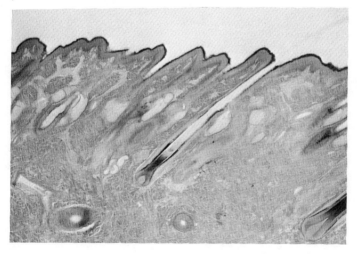

Figure 10-10   Calf skin from foreleg, X35.
Frozen, 20 μ, cross section. Stain: Mallory's triple. Kodak Wratten No. 58 (green) Filter.

Table 10-2 Biological Stains

Stain	Color Index (1st ed.)	Color Index (2nd ed.)	Spectral‡ Absorption	Maximum Contrast Filters	Band‡ Used
*Acid fuchsin	692	42685	530–560	99	521–587
*Aniline blue, W. S.	707	42755	550–620	22 + 58	553–602
Azocarmine G	828	50085	470–540	45 + 12	500–538
*Azure I	...	...	580–640	72B	588–640
*Basic fuchsin	...	42510	520–550	74	509–569
Bismarck brown Y	331	21000	General in blue and violet	47B	380–480
Brilliant cresyl blue	877	51010	570–640	72B	588–640
Brilliant green	662	42040	600–640	29 (F)	610–680
*Carmine	1239	75470	500–570	74	509–569
Chlorazol black E	581	30235	General with low maximum at 600	90 + 6.5 mm 2043 glass or 90 + 7% copper sulfate	555–640
Congo red	370	22120	480–520	45 + 58	483–532
Cresyl violet acetate	...	...	550–630	90 + 6.5 mm 2043 glass	555–627
*Crystal violet	681	42555	550–610	22 + 58	553–602
Eosin B	771	45400	480–550	58 + 45	483–532
*Eosin Y	768	45380	490–530	58 + 45	483–532
Erythrosin B	773	45430	510–536	12 + 45	500–538
Ethyl eosin	770	45386	490–540	12 + 45	500–538
Fast green FCF	...	42053	590–650	72B	588–640
Gallyocyanin	883	51030	560–686	25 + 3% copper sulfate	580–710
*Hematoxylin (Ehrlich)	1246	75290	Gradual through green	58 + 15	516–590
Hematoxylin (Heidenhain)	...	...	560–600	22 + 58	553–602
Indigo-carmine	1180	73015	560–650	22 + 58	553–602
Janus green B	133	11050	560–640	22 + 58	553–602
Light green SF yellowish	670	42095	600–660	29 (F)	610–680
Malachite green	657	42000	590–640	72B	588–639
Martius yellow	9	10315	380–450	2B + 35	393–457

Methyl blue	706	42780	550–635	90 + 6.5 mm 2043 glass or 90 + 7% copper sulfate	555–640
*Methyl green	684	42585	620–650	29 (F)	610–680
Methyl orange	142	13025	430–500	47+45	436–506
*Methyl violet 2B	680	42535	550–600	22+58	553–602
*Methylene blue	922	52015	600–620 and 650–680	92	625–
Methylene violet (Bernthsen)	…	…	500–650	12+57	500–630
Neutral red	825	50040	480–550	12+45	500–538
Nile blue A	913	51180	560–650	72B	588–640
Nigrosin, W. S.	865	50420	General with maximum at 580–600	22+58	553–602
Oil red O	…	26125	470–570	65A+9	477–552
*Orange II	151	15510	460–510	39+4	454–506
Orange G	27	16230	470–500	45+8	469–530
Orcein (for elastic tissue)	1242	…	460–570	65+9	479–561
Orcein (acetic-) (for chromosomes)	1242	…	460–570	65+9	479–561
Phloxine B	778	45410	510–550	45+15	505–558
Picric Acid	7	10305	445–open	36 + 10% copper sulfate	
Pyronin B	741	45010	540–570	93	517–573
Pyronin Y	739	45005	530–560	93	517–573
*Safranin O	841	50240	480–540	12+45	500–538
*Sudan III	248	26100	450–552	45+8	469–530
*Sudan IV	258	26105	452–554	65A+9	477–552
Thionin	920	52000	560–610	22+58	553–602
Toluidine blue O	925	52040	550–650	72B	588–639

*The most commonly used stains are marked with an asterisk. Gentian violet belongs to these stains, but it is not officially recognized by the Biological Stain Commission. It is a poorly defined mixture of violet rosanilins, nearly synonymous with methyl violet. The same filters recommended for methyl violet may be tried.

†The spectral absorption of the stained object may differ somewhat from that of the stain in solution from which these data were taken.

‡Bandwidth between 10% of peak transmission.

Note. Use of Kodak Wratten No. 44 or 44A filters, instead of No. 45 with the No. 12 Filter, will relax the exposure factor appreciably without great loss in contrast.

Sometimes the optimum filter may depend on the detail that is to be emphasized, as illustrated by Figure 10-11. The blood vessels have been injected with a blue liquid. The photomicrograph (Figure 10-11a) brings them out quite clearly. The other photomicrograph (Figure 10-11b), taken with the No. 58 Filter, shows more structural detail of the skin itself.

Table 10-2 lists many of the commonly used microscopical stains and indicates their spectral absorption and the Kodak Wratten filters that will give high contrast, often close to the maximum. Some stains, such as hematoxylin and nigrosin, have such a large gray component that color filters do not exercise the control they have over the more intense stains. Moreover, in multiple staining **the absorptions of all stains covering one area may affect the contrast filter.** In the Mallory triple stain the red areas have taken both the yellow and the magenta dyes. Bacteria differentially stained with fuchsin and methylene blue show this phenomenon in that the bacteria stained with fuchsin have also been stained with the methylene blue.

The *exposure factor* required by the Kodak Wratten filters with some photographic films and plates are given in Chapter 17, Vol. 2.

Darkfield Microscopy

Sometimes specimen color is unavailable as a separate factor for enhancement of contrast, even though the specimen has some color.

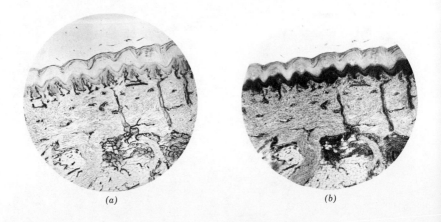

(a)   (b)

Figure 10-11   Skin of man, cross section, X15, with dye-injected blood vessels: (a) Kodak Wratten "M" Plate and F Filter; (b) Kodak "M" Plate and B Filter.

This is often the case with the many samples of colorless cells, especially if they are living, and with many industrial samples. If brightfield illumination gives utterly inadequate contrast, as in Figure 10-1 or 10-12, darkfield illumination may be utilized as the most effective method of contrast enhancement that magnifies sudden refractive-index differences, even though very slight. Such differences occur at surfaces and edges.

Darkfield illumination can be used at any magnification; its nature and technique are considered in the next chapter. Here it is important to recognise the method and realize its usefulness.

The method is most effective when the specimen is made up of separate bodies whose combined area is still small with respect to the whole field. The degree of contrast enhancement that is available is illustrated in Figures 10-13 and 10-14, which should be compared with Figures 10-12 and 10-1, respectively.

Just as diatoms have been used as a test object which all microscopists can use mutually to test the resolving power of objectives, so epithilial cells, scraped from the inside lining of the mouth, have come to be used as a general test object for contrast studies, etc., and are even more available. They need be covered only with a glass slip and mounted in saliva; the slip may be ringed with grease or wax to prevent evaporation. The resolving power, at the same objective aperture, of the darkfield should be better, since in brightfield the substage iris must usually be closed too much to gain sufficient contrast for seeing the specimen at all.

If the refractive-index differences are small, the great contrast enhancement that is obtained may deceive the observer in regard to the image brightness, if observed in a darkened room. The eye adapts to a night sky but the scattered intensity may be low. Therefore photographic exposures may be long; exposure factors of 1000X more than that for a stained specimen may be encountered. For cinemicrography at high magnification of living cells, such as bacteria, in which the exposures must be fractional seconds, the light sources must be powerful arcs (12. Loveland, 1933). The measure of the contrast enhancement is discussed in the comparison of the alternative methods near the end of the chapter.

Rheinberg Differential Color Illumination

In 1896 Julius Rheinberg realized that if the opaque stop in the condenser, illustrated in Figure 12-1 were replaced with a contrast color filter, such as a Kodak Wratten 61, which was opaque to only part of the spectrum, the background would be colored; for example, green. Now, if the annulus around the central stop, which gives rise to the illumination cone that may be scattered by the specimen, is covered with a contrasting color filter, for example, the red Kodak Wratten 22, the specimen

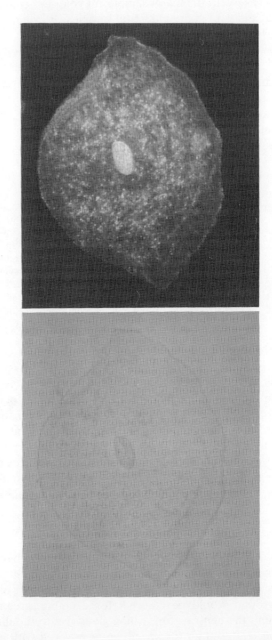

Figure 10-12 Oral epithelial cell, X750. Brightfield 4 mm fluorite objective (oil) NA 1.0, λ546 mμ (green from mercury arc).

Figure 10-13 Oral epithelial cell, X750. Darkfield cardioid condenser, 4 mm fluorite objective (oil) NA 1.0, λ546 mμ (green from mercury arc).

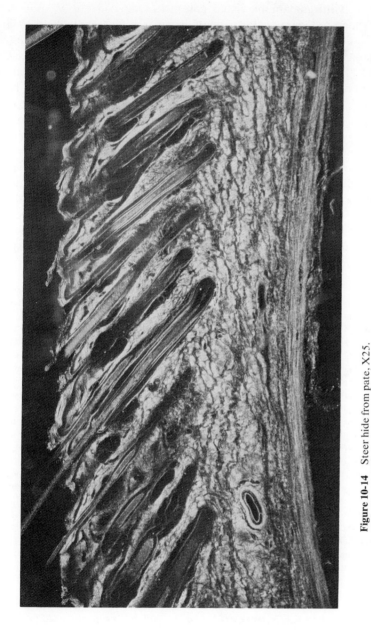

Figure 10-14 Steer hide from pate, X25.
Frozen, 20 μ, cross section. Unstained: Mounted in Clarite. Darkfield: Simple microscope (Case 7) Enlarging Ektar at f/5.6, stop in external aperture plane. Kodak Wratten No. 58 (green) Filter.

would show with maximum visual contrast. One can easily show a brilliant orange-red paramecium swimming in a bluish-green field or a blue-yellow combination could be used. The Eastman Kodak Company sold "Rheinberg Filters" for many years. Before the war Carl Zeiss marketed a "Polychromar Condenser" which did the same thing optically and allowed a variety of color combinations. Now that the effect can be used as a subject for color photomicrography, including cinemicrography with Kodachrome II Film, there is more reason to use the method. Probably no more information is gained, however, than with the black-and-white darkfield.

Not only can anyone make up his own Rheinberg color filter set, but there is often an advantage over any commercial set or device. The commercial set has to assume an objective aperture in setting the diameter of the central stop. The Kodak filters assumed a 10X (16 mm) 0.25 NA objective and condenser of average focal length. When making a filter, of one's own, the filter set can be made around the actual objective-condenser combination in use.

To make a filter it may be wise to **prepare a prototype dummy** of a piece of fixed-out film or any stiff but thin plastic. The proper dimensions of the two filter components must first be determined. For easy reference in the following directions let us assume that the outside annular ring will be a red filter and the inner disk, a green one. The outside diameter of the red disk should just fit into the filter holder at the iris diaphragm of the substage condenser or into the depression made by the iris of the condenser. A too sloppy fit might allow the colors in the wrong aperture. The inside boundary diameter is the projection of the objective aperture down onto this filter plane. It is easily determined by closing the substage iris while looking down the microscope tube, preferably through a pinhole eyepiece, until it is *just* at the edge of the objective, as visually determined. The hole in the middle of the red disk should be a trifle smaller than this in diameter; the green disk should be a bit larger to make about 1 to 2 mm overlap.

The final composite filter can be made from Kodak Wratten gelatin filter film. To avoid bad fingermarks it can be handled between two sheets of unwrinkled tissue paper cut with it. This type of filter film is very brittle and will harden further with age; cracks may start in the sheet if it is cut with scissors that are even slightly dull. This brittleness may become extreme if the filter film has lain around for a long time. The outside diameter of the red filter is easily cut with the corner of a razor blade by tracing around the outside edge of a metal disk of the correct diameter pressed onto the sheet of film. The inside hole of the red filter and the outer diameter of the green filter may be too small for this method. With a proper punch this operation becomes simple. With one condenser of short focal length a standard paper punch (round hole) had just the right diameter. Ordinary brass cork borers will just shatter the filter. A good steel borer can be sharpened to be satisfactory. A punch with which both the hole and the central disk are usable, with clean edges, can be made in a machine shop so that a drill rod with a flat sharp end will just fit into a hole drilled in two plates between which the filter film is sandwiched. With the cork-borer type of punch the side toward the bevel may crack. With the drill-rod punch there must be auxiliary locating pins. With this punch the central green disk fits inside the hole of the red disk to made a single plane. A little lacquer, applied with a sharpened toothpick and preferably black, can be used to unite the two components as a cement. An aqueous cement will cause the film to buckle unless great care is taken, but glue can be used. With the lacquered gelatin filter film, it is unnecessary to mount the Rheinberg filter between glass disks, as done originally. The unmounted but cemented filters must be handled very carefully, preferably with tweezers, and stored between squares cut from cards. If round thin glass disks can be obtained, it is possible to cement them to a single glass disk rather than form a sandwich to keep the assembly thin. Lantern-slide cover glasses are a source of thin sheet glass. An optician's shop can cut the disks.

Phase Illumination

In Chapter 2 phase microscopy, a relatively new method of contrast enhancement, was discussed. Phase microscopy increases the contrast of edges, in which the refractive-index *difference* exists, rather than the differentiation of areas as does most color staining. This method has become very familiar to many microscopists. Its application, as a unique method of illumination, is considered in detail in Chapter 12B, Vol. 2.

Equipment is sold in terms of "dark contrast" and "bright contrast" and most discussions of the subject, including those in this book, are in those terms. Yet this distinction can become quite confusing. There is a reversal of phase and therefore of *relative* contrast with level of focus, which, except for very thin specimens, can lead to reversal of brightfield and darkfield appearance within the same field and the same mechanical level of focus. Reversal of the refractive index levels between specimen and the mounting medium can bring reversal of this contrast appearance. When many edges are scrambled closely together in a pattern, that is, close detail, phase illumination may be confusing, as it is in a concentrated colloidal system.

It is most instructive to make two test slides of broken cover slips, including some dust from the breakage. In one slide the fragments should lie just above the refractive index of the medium; in the other the solid should be just below the field medium which may be an immersion oil. The larger fragments will show up only at the edges; with dark phase and with higher index in the glass fragments, a bright halo lies outside the edges. The converse is true with bright phase or with reversal of index. On the other hand, small fragments, with inner variations of surface and thickness, are depicted beautifully throughout. Nearly all disappear with normal transillumination and reasonable aperture. It is most instructive to note the difference in appearance of those edges of large fragments, with plane interiors, in which the edges are wedges. The appearance, brightfield and darkfield, of those wedges with an upward slanting surface is opposite those of the downward slant and the appearance is reversed by the factors just mentioned.

A simple example of phase-illumination is provided by comparing Figures 10-12, 10-13, 10-15, and 10-16, which are of the same epithelial cell mounted in saliva.

Simple bright phase gives the same appearance as darkfield, yet note that unlike darkfield, in which the field itself should have no illumination, the background for A⁺(dark) phase and A⁻(bright) phase is the same, equivalent to a medium density in the photograph. Yet because the brightness of the image, and the densities of the photographic image can be both above and below the background density in bright phase, contrast differences approach those of darkfield.

The position of the edge halo between A⁺ and A⁻ can be noted too. In dark contrast the bright halo is outside. It can be much more marked, with

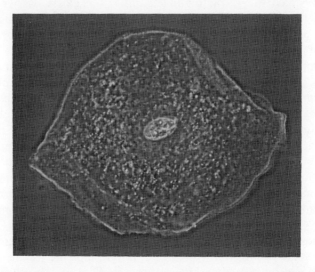

Figure 10-15   Oral epithelial cell, X750. Bright phase, saliva mount. A-O phase objective, A-, achromat, 1.8 mm NA 1.25. 5X compensating ocular, 546 mμ (green from mercury arc).

a larger difference in refractive index; with the fragmented glass it was most marked outside the sharp edge of a wedge of proper slant.

The calf-hide sections are examples of the other considerations discussed above. Here, the thickness of the sections, plus the packing of detail, akin to wedging, prevents a simple bright or dark contrast appearance. At first glance the appearance does not seem to be too different between A⁺ and A⁻ objectives or with reversal of index level with respect to the mounting media. Close examination of individual detail shows more differences. On the whole the less confusing appearance seems to be with the objective and conditions to give dark contrast with simpler specimens. The same objective was used for Figures 10-17 and 10-18. In Figure 10-17 the unstained calf hide was mounted in Clarite; the desiccated tissue in some regions had a refractive index very close to that of the medium. Figure 10-18 was made from a frozen section mounted in normal saline solution so that the index of the tissue was appreciably above that of the medium.

Comparison of Methods of Contrast Enhancement

The author made a study of the contrast relations of these different methods of illumination by the methods of photographic photometry. Photomicrographs similar to those shown here were made of (a) an

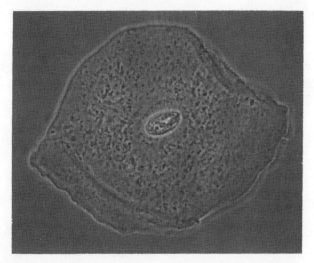

Figure 10-16 Oral epithelial cell, X750.
Dark Phase, saliva mount. A-O phase objective A+, achromat, 1.8 mm NA 1.25, 5X compensating ocular.

unstained calf skin in a resin mount at 100X magnification and (b) epithelial cells in aqueous mounts at 1000X magnification to represent this common type of living protoplasm. A quantitative step exposure wedge by the same quality of light and at the same general intensity level was impressed along a previously masked strip of the same negative. Then the final negatives were analyzed by a microdensitometer, for which the same significant areas of the same specimens taken with the different kinds of illumination were chosen. Thus the original brightnesses of the images could be calculated. The results are shown in Table 10-3. The exposure factors obtained when white light was used with Kodak Pana-

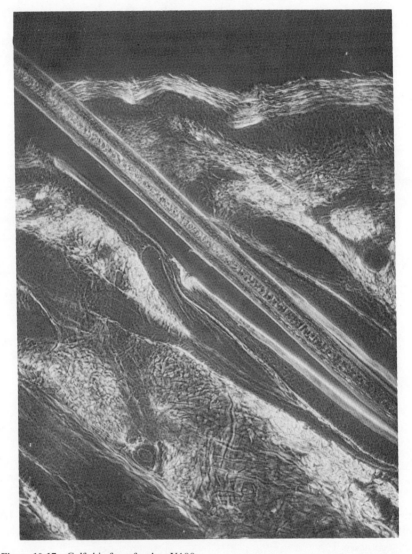

Figure 10-17 Calf skin from foreleg, X100.
Frozen 20 μ cross section. Unstained: mounted in clarite. Dark contrast $0.10 + 1/4\lambda$ achromat 16 mm 0.25 NA, $\lambda 546$ mμ (green from mercury arc).

tomic-X Film are included; this is a direct, though relative, measure of the brightnesses of the images.

The brightfield pictures made at 0.9 aperture, as if the specimens had been stained, were done to give the relation to ordinary brightfield, with

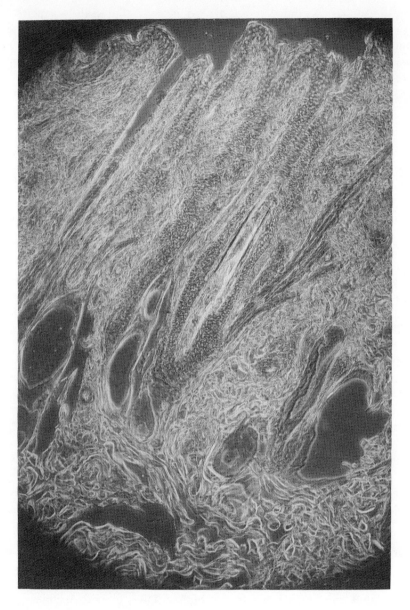

Figure 10-18 Calf skin from foreleg, X100.
Frozen 20 μ cross section. Unstained: aqueous mount. Dark Contrast: 0.10 + 1/4λ achromat 16 mm 0.25 NA, λ546 mμ (green from mercury arc).

its normal resolving power. The image had a contrast so low that focusing the image was difficult; no one would have chosen this aperture for visual examination. The aperture was then closed (column 3) to the minimum degree that could be allowed for practical contrast in the visual image and the photomicrograph; this is somewhat arbitrary but not greatly so.

Table 10-3

Illumination Type	Brightfield 0.9 Aperture	Brightfield Reduced Aperture	Phase Contrast A −	Phase Contrast A +	Darkfield
Exposure factor	1	2.25X	22.5X	22.5X	1200X
	0.44	1X	10X	10X	500X
Contrast enhancement					
Resin mount X100	···	1X	4.4X	···	24X
Aqueous mount X1000	···	1X	4.4X	11X	12X

Note that **bright phase does give more contrast enhancement than dark phase and almost as much as darkfield.** Since the brightness of darkfield images depends on (a) the refractive-index differences of the specimen and (b) the intensity of the light poured obliquely through the object field, certainly much greater contrast-enhancement factors can be obtained. The factor for phase illumination would increase also, possibly proportionally, but one suspects that a perfect darkfield is a more efficient factor. However, note the **exposure factors!** They are the same for dark and bright phase and **50 times better than for darkfield.** This is good news for cinemicrographers and exhibits the reason that phase microscopy is so largely employed for cinemicrography of the living unstained cells. The illumination requirements for darkfield may be 1200X greater than for normal brightfield at 0.9 aperture.

Use of Anisotropy (Polarized Light)

The use of polarized light as a quantitative measure of crystal structure, etc., is an important and broad subject that cannot be discussed in this book. More and more, most microscopists, certainly the industrial microscopist, are likely to need some knowledge of this field. Some good texts are mentioned in the bibliography, classified under petrography, which deals chiefly with transmitted light. Cameron (6. 1957, 1959, 1961) has done much to extend the effectiveness of the method to a field in which reflected light is used. Inoué and Hyde (4. 1957)

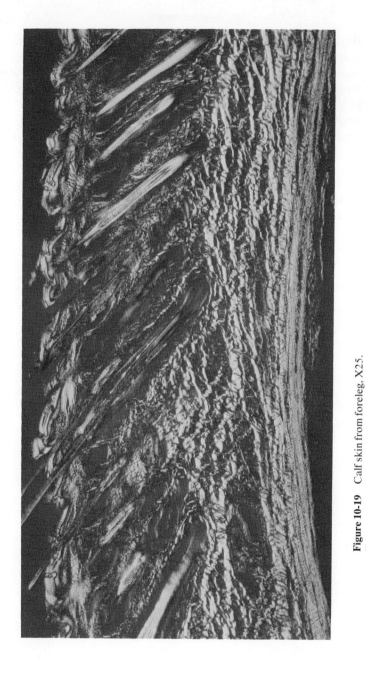

Figure 10-19 Calf skin from foreleg, X25.
Frozen, 20 μ cross section. Unstained: mounted in clarite. Polarized light. Simple microscope (Case 7) enlarging Ektar at $f/5.6$. Kodak Wratten No. 58 Filter.

have developed compensation for high apertures so that the relatively slight anisotropy of living cells can be effectively studied.

Polarized light, however, can be pragmatically useful also, as a method of contrast control, sometimes for specimens for which it might not be considered for its more fundamental employment. One application is the use of a polarizer in photomicrography by reflected light to remove unwanted specular reflections. This method of contrast control has already been discussed in Chapter 4 (p. 185).

It may come as a surprise that crossed Polaroid filters, that is, polarized light with an analyzer, were found to be useful in industrial microscopy for the quick examination and identification of unstained cross sections of animal hides. A photomicrograph at 25X of a cross section of the hide of a carabao (water buffalo) made by polarized light is shown in Figure 10-19. The section, of course, is not easily visible in high-aperture brightfield.

Chapter 11

Eyepiece and Roll-Film Cameras

The actual procedure to be followed in making a photomicrograph with a commercial eyepiece camera is given on p. 472. A procedure for making photomicrographs by utilizing almost any available hand camera is described under the title, "Simplified Photomicrography" (pp. 473 and 482). Some apparatus assembly is required.

There is no question of the convenience of the cameras that use 35 mm film when placed directly on the microscope. Even they, however, can be independently supported with advantage. They have been discussed in Chapter 3 and are considered again under "Simplified Photomicrography" later in this chapter. It must be possible to remove and replace them quickly if they are to be convenient at all.

The very convenience of such a photomicrographic camera, coupled with the inability to survey and analyze the image in the film plane when it is used, may lead to the practice of taking several exposures on the roll film with slight variation of conditions in the hope that one of the series will be excellent, as opposed to the deliberate effort to bring all factors to the optimum as practiced with the view camera and sheet photographic materials. This careless habit will surely lower the competence of the photomicrographer and the quality of the product.

On the other hand, the user of 35 mm film has a wide variety of film materials available. This is especially important for color photomicrography in which the economy of this size may be important when many photomicrographs are to be made. For only occasional color work, however, the sheet materials will be more economical.

One of the following methods will be used for viewing and focusing the image if commercial equipment is considered.

1. An aerial image is viewed by a telescope containing a reticle which lies in a plane conjugate to that on the camera film; that is, it is the same *optical* distance from the eyepoint of the microscope.

(a) This aerial image may be presented alternately with a camera image by a mirror or prism that moves out of the way to allow one *or* the other image to be formed.

(b) The image beam above the microscope may be split by a semi-reflecting mirror or prism, so that both images are present simultaneously.

2. The camera can be slid or rotated above the microscope so that either the camera or a matte plane is in place above the microscope; the latter allows the field to be viewed and focused.

Method 2 is definitely superior to Method 1 when a wide variety of photomicrography is to be done, varying in type of specimen, magnification, or illumination. This method allows inspection and evaluation of the quality of the image for each picture and simpler determination of the exposure for each case, whenever that is desirable. On the other hand, when many photomicrographs are to be taken with standardized brightfield illumination and the same type of specimen, Method 1 is faster and simpler; a change of exposure with change of magnification can usually be calculated. Either submethod (a) or (b) can be used; some devices such as the Leitz MIKAS micro attachment are equipped for use with either the moving or split prism. If there is movement in the specimen or a change with time that also needs observation up to the time of taking the picture (often flash), Method 1 is essential. In this case the use of roll film, the disadvantages of which are removed by Method 2, is a definite advantage.

The following discussion of the optical principles involved in use of an eyepiece camera, or a camera containing its lens, is important for anyone constructing his own apparatus and also for anyone who expects to use the method of simplified photomicrography described later in this chapter. It should also be useful to anyone who is choosing or preparing to use a commercial eyepiece camera.

OPTICAL PRINCIPLES OF EYEPIECE CAMERAS

As noted on page 57, the compound microscope with its objectives has been designed only for visual work with the corrected image at "infinity, " but a bellows distance as close as 10 in. can be used with negligible degradation of image. The film plane of an eyepiece camera lies much closer. On page 60 it was also noted that one method of correcting the condition is to send the image *optically* back to infinity by placing a lens above the microscope with the film in its focal plane; that is, a camera focused for infinity. Some practical aspects of this method should now be considered.

The optimum position of the lens is that which places its front surface (vertex focus) in the plane of the eyepoint. If the lens is lowered so that its front surface is below the eyepoint, the quality of the image degrades perceptibly and rather rapidly. Thus it is not best to let the camera lens rest on the top of the microscope, as has been advised. If the lens is raised above the eyepoint, the beam will also spread out conically to cover more area of the lens. Even in this case, however, the lens is being used at the aperture determined by the diameter of the eyepoint (or slightly less), since **each separate pencil of image beam occupies only that diameter**. The image also degrades slightly as the lens is raised but not so rapidly as when it is lowered, so that careful examination may be required to notice it.

Another effect is much more obvious and quickly noted. **With the lens at the eyepoint, the whole** illuminated **field appears** at the focal plane, assuming the microscope to have been focused visually. This gives a picture of the whole round field in a camera that has been focused at infinity. The center will be in focus, but this central area will be surrounded by an out-of-focus peripheral band because this is the condition seen *at a single position of focus of the microscope*, when one looks into it, usually much improved with the new flat field optics. In a photomicrograph it seems quite objectionable, since the chance to twiddle the periphery into focus with the fine adjustment is not available. The central portion of this negative, however, can be selected by enlarging in printing and cropping to the central area. This, in effect, is the situation that is normal with a large professional bench, since the microscope image is directly enlarged by the long bellows draw and only the central area is included within the camera frame. It is important to note that use of a camera focused for infinity above the microscope is an optically excellent method and that this central area of the image, which is in focus, has a definition as good as can be obtained with large professional equipment, assuming equally good optics in the microscope.

When the camera is raised so that the lens is increasingly higher above the eyepoint, the image seems to remain in focus (because of the enormous depth of image focus at these camera lens apertures); the magnification stays the same but the diameter of the illuminated field seen in the image or recorded on film becomes smaller as the periphery of the camera lens or the lens diaphragm seems to close down, acting as a field diaphragm. Thus at some height only the central area of good definition is included in the smaller imaged area. This effect should have been expected. Each separate beam of the rays issuing from a point in the field and passing through the tip of the illuminated image cone (the eyepoint) comes to a separate point in the image. The peripheral points of the field form the periphery of the image cone.

This situation produces a second noticeable effect. **Each pencil from each point** in both field and image **passes through a small spot of the camera lens.** Every spot, including marks or dirt on the lens, will be in apparent focus in the image and often with high contrast. In more usual use of a hand camera light focusing to any one point of the image has filled the whole open lens aperture. **The lens must be kept very clean for this use.**

Akin to this defect is another noticeable disadvantage. With many cameras, when the camera is placed at the eyepoint, an in-focus sharp image of the eyepoint is formed in the image plane by reflection from the lens surfaces; it is a "flare spot." It goes out of focus and seems to disappear if the camera and lens are raised appreciably. This defect will be at a minimum with a coated lens.

Effect of Lens Type

If the vertex focus of the camera (or correcting) lens is at the eyepoint, it has been noted that only an extremely small relative aperture of the lens is being used; it might be a 3 mm central portion of a lens of 6-in. focal length, or $f/50$, or a 2 mm lighted disk in a 2-in. lens of a 35 mm camera equal to $f/25$. Moreover, as the camera is raised, seemingly filling the whole lens with light, *the relative aperture remains the same,* since each image point is still fed by the same lens area. This corresponds very definitely to the paraxial use of a lens (see Chapter 1). A highly corrected lens is no advantage; in fact, thickness and a multiplication of lens surfaces is a definite disadvantage for this use. On the other hand, it is advantageous for this camera or correcting lens to be an achromat so that its focal length is not too different for all colors. The great depth of focus at the small aperture is a help here. A simple plano-convex cemented doublet is therefore ideal. A simple spectacle lens is surprisingly good, especially if color filters are used.

Focusing (Use of Accessory Telescope)

It was noted that *if* the observer's eye were relaxed the microscope image would be at infinity. Therefore a camera focused at infinity would have the image correctly focused on the film. It would also be in focus in a telescope focused at infinity. It is advantageous to have a reticle in the mutual plane of the objective and eyepiece to force the eye into its focus to tell where the image of a microscope has been optically placed or to force the focusing of the microscope to a specified image plane. The telescope need be focused only to this plane, possibly out a window for "infinity focus." This point is discussed further in a later section.

Magnification

The magnification unit distance for the projected image from a microscope is 10 in. (250 mm); the magnification for any other projection distance is obtained by multiplying the catalog magnification (objective × ocular) by the actual bellows distance and dividing by 10 in. (or 250 mm). As observed in Chapter 2, p. 60, the presence of a lens above the eyepiece does not change this relation. Therefore, with a 5-in. lens, the magnification is only one-half that seen by direct vision into the microscope; with a camera with a 25 mm lens the images on the film will be one-tenth that seen directly. In the first case the loss in magnification over what would have been obtained by direct viewing may be overcome by using an ocular with correspondingly higher magnification; but, mostly, with the compact projection system the assumption (and practice) is to enlarge the negative in printing.

COMMERCIAL EYEPIECE CAMERAS

Most commercial eyepiece photomicrographic cameras consist of a miniature camera for 35 mm film.

As we have seen the simplest arrangement, in principle, is merely to place a 35 mm film camera over the microscope with its lens focused for infinity. Some manufacturers have done just that, often including the lens that comes integrally with it. However, use is made of an observing telescope at right angles to the photographic axis, since, when focused at infinity, the image is simultaneously in focus in camera and telescope. A reticle must be at the focus of the objective to force the eye to focus the microscope image into the correct plane. Usually the split beam (or removable) prism occupies the plane of the eyepoint so that the camera lens must lie just above it (see Figure 11-1). Such a device is sold, for instance, for the Kodak Retina camera. The viewing system removes a serious disadvantage (otherwise) of roll film for photomicrography in that now the field for each frame can be specifically viewed and focused. The disadvantages of the use of a hand camera above the microscope, including the problem of the peripheral out-of-focus field, were discussed at the beginning of this chapter.

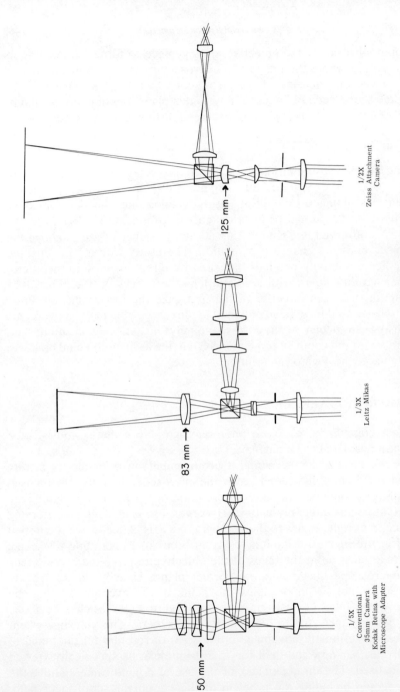

Figure 11.1 Split-beam-observation eyepieces for photomicrography.

50 mm

1/5X
Conventional
35mm Camera
Kodak Retina with
Microscope Adapter

83 mm

1/3X
Leitz Mikas

125 mm

1/2X
Zeiss Attachment
Camera

The disadvantage of the peripheral out-of-focus field can be reduced somewhat by increasing the focal length of the lens, yet retaining the same frame size on the film. The image is thus masked down more. A popular choice for focal length is about 83 mm because the photographed magnification (83/250) is restored to the catalog magnification if the negative is enlarged 3X, the most popular printing enlargement. Both the Bausch and Lomb Model N Eyepiece Camera and the Leitz MIKAS (formerly Micro Ibso) and their newer micro attachments use this focal length.

Simultaneously, a much simpler lens is chosen for the camera lens. It is cheaper and more suitable than an excellent but complex lens normal for a hand camera and is much less likely to give a flare spot on the film. In the Zeiss adapter the lens is bent to a meniscus, although it is still a positive achromat; it is not likely to produce a flare spot on the film.

Only in the Zeiss adapter is the lens at the eyepoint; here the prism has been relegated to the upper position above the eyepoint. This arrangement requires a larger prism but is optically somewhat superior. Note that the magnifications given by the Zeiss device are one-half the standard (catalog) values. This is done by choosing a lens of about 125 mm focal length ($\frac{1}{2}$ of 250 mm) with a projection distance this length. The device is not so compact, but a smaller central fraction of the visual field is chosen for photography.

Leitz has two other adapters for the MIKAS which can be used alternatively for less reduction of magnification from the apparent visual value; that is, $\frac{1}{2}$X and 1X.* A rather clever scheme was utilized to allow them to be used with the same telescope and simultaneously to keep the adapters more compact than would mere substitution of the corresponding 125 and 250 mm lenses for the 83 mm lens. This can be understood from the following discussion, assuming that the microscope is set up with the standard observation telescope in use to focus an image in the field. Let us put a positive achromat of about 56 mm focal length above the eyepiece. The image will come to focus at 56 mm above the lens and be almost one-fifth the size at 250 mm. Now, if a strong negative lens of at least equal power is placed against the positive lens, it will almost neutralize the positive lens, thus causing the image to focus again at a distance. On the other hand, if it is at the image plane of the positive lens, it will have negligible effect. When located above the positive lens, there will be some intermediate position at which the magnification on the film is exactly one-half the catalog value. The con-

*The latter is no longer available in the United States.

jugate image in the telescope will still be in focus but with a different size relation. At a reduction factor of $\frac{1}{2}$X the height of the device is still very little greater than for the $\frac{1}{3}$X adapter. The same principle is applied for the adapter with a 1X factor, but the focal lengths of both the positive and negative elements are somewhat greater. The projection distance for this device is only 128 instead of 250 mm.

Leitz has introduced a new model of the attachment camera (codeword: KAVAR). At this writing the MIKAS is still available, since the complete assembly of the new model is necessarily considerably more expensive. The KAVAR assembly fits on top of all the Leitz microscopes available at present, and, since the only fitting required is that of a standard body tube, it should be usable with any microscope unless there is no room for its larger outer dimensions.

A GF 10X widefield eyepiece is part of the KAVAR assembly; the imaging lens system is similar to that shown for the MIKAS in Figure 11-1, except that the prism is mounted on a plate so that with the pull of a lever the plate rotates to substitute another prism which sends the main beam to a photocell instead of to the camera. Release of the lever causes the camera prism to rotate back into position over the ocular. The eyepiece of the observing telescope should be focused on its reticle for the individual observer. A photocell exposure meter is part of the KAVAR assembly. An intermediate optical system (an $f/2$ lens) isolates a middle area that corresponds to about one-third of the entire 35 mm frame. This can be a distinct advantage in making the determination of exposure, but it is not small enough to allow determination of the brightness range of the image in most cases. The darkest areas may be very small (e.g., the nuclei of cells). The specimen can, however, be moved slightly for the exposure determination from the framing that will be used for the picture. It probably will still be preferable to utilize clear field brightness in most cases, (with Ec or Eg as discussed in Chapter 17. Vol. 2). The lack of blue sensitivity of the cadmium sulfide photocell (e.g. the Gossen Lunasix) is a considerable disadvantage in most photography, compared with the sensitivity similar to that of color films that is characteristic of selenium photovoltaic cells. The fact that with a filter it can be made efficiently to match the sensitivity of the eye is not pertinent; the measurement is made to prophecy the behavior of a photographic film. The effect of a color filter is usually quite different for this photocell than it is for a photographic film. The "filter factors" for the cell are given for some glass color filters supplied by Leitz. The exposure meter covers an enormous range of exposures, from a small fraction of a second to at least an hour; therefore the extra calibration requirements of this type of meter may be worth the disadvantage.

The fact that the KAVAR micro-attachment uses the Leica Series M Camera is an advantage; the image plane is available for examination when there is no film in the camera. The camera can be removed from the assembly when a roll of film is partially exposed, since the focal plane shutter protects the film. Most of the assembly, including the camera, is supported on rubber to separate it from the microscope and prevent the shutter action from degrading the definition of the image.

With any camera that sits on a microscope, it is recommended that a test for shutter vibration be made with black-and-white film and a sensitive specimen: first take a picture at a small fraction of a second using the shutter, and then on another film frame make an exposure with a suspended outside shutter or make a time exposure after reducing the light intensity with a neutral density of at least 2.0. In the case of a KAVAR or an Ortholux microscope no image degradation due to vibration was found at fractional second exposures, including the use of "high-dry" objectives. This was unexpected from experience with focal plane shutters.

A dummy camera can be made to allow examination of the image when the camera is loaded; it is merely a block of wood that has the thickness of the camera and a hole at least as large as a 35 mm frame. A piece of ground glass or, more conveniently, a sheet of matte film is laid on top to help locate the image.

The Kodak Instamatic Reflex Camera with its Microscope Adapter represents a system that mechanically and operationally is very easy to use. All lenses are removed so that the image beam goes directly to the film plane except for about one third that is directed by a split prism and side arm to the port of the photocell on the face of the camera. Focusing is done through the reflex finder (utilizing the mirror-curtain as usual). The "electronic shutter," governed by the photocell, sets the exposure time; the user is unaware of the actual duration of exposure.

The advantage of the device is obvious — the great simplicity of use. Its disadvantages are:

1. The shutter is not in an aperture plane. Vignetting and even shutter pattern occur at fast shutter speeds. This presents no difficulties at light levels for most built-in light sources.

2. Because the user does not know the exposure time, no provision can be made for reciprocity effects except by guess or by trial.

3. The integrated light from the field is used for the exposure criterion (see Index, exposure criteria).

A practical rule is to set the exposure scale for f/4 when the background of the field predominates and for f/2.8 when a normal section fills the field.

4. The film plane is close to the eyepiece with no optics intervening to send it optically a great distance. In my opinion, when best definition is important it is worthwhile to place a spectacle lens in the adapter (preferably a meniscus lens concave side down). Its exact power is unimportant, since the focus is independently set through the finder; a 5–10 diopter lens is suggested.

5. The focusing is done by viewing a ground glass. Normally this is not serious for detail and ensures proper image location.

Attachment Cameras Utilizing Primary Image

In Chapter 3 (p. 150) we discussed a type of attachment camera that was somewhat different in that it projected the primary image onto the film with a complex negative lens that reached below the plane, where this image from the objective normally would have been formed, and refocused it onto the film plane. Naturally the location of this plane and that of the imaging negative lens must be very exact, especially for low magnifications. With the positive lens, *if the focused distance of lens to film is exact*, as it is in any good camera, the position of the whole camera unit is not extremely critical.

Both the American Optical Co. and Bausch & Lomb utilize the negative lens. With the Bausch & Lomb 35 mm Camera Assembly (42-14-69-64), the magnification on the film is 2.5X greater than the labeled magnification of the objective (at 250 mm), except that the zoom lens of the Dynazoom gives a variable of 1 to 2X factor over this. With the camera assembly for the A-O MicroStar Series 10, there are two magnification factors available (i.e., 2.8 and 5X); each factor is applied to the stated magnification of the objective. Each uses the Kodak Colorsnap Camera. With each there is an available method for adjusting the focus on the film, as precalibrated visually with an open camera back. The special "Field of View Eyepiece" (No. 1054) of the American Optical Co. makes this simple and helps to fix the visual focus throughout its use.

The camera is merely a device to hold the film in the specified plane and to use the standard 35 mm film cassettes. The leaf shutter is in the support below the camera in both cases, as discussed in Chapter 3. In each case, also, there is a simple gate slide to protect the film in the camera. It is wise to keep this shut except during an exposure, not allowing it to stand long protected only by the leaf shutter. This device also permits the camera to be removed while the roll film is still in use. However, this gate protector is so easily moved that there is considerable danger that it may partly open while the camera is being handled and stored off the stand. A rubber-band protective device was found important by the writer.

Because of this facility for removing the camera and the relative cheapness of the camera, several of then may be used to make several types of roll film available at the same time, an advantage, usually, only of professional cameras that employ sheet film. However, the chief advantage of this facility is to provide easy access to the image plane at any time when using a roll film camera.

The Dummy Camera

When using a roll film camera, which is usually also an eyepiece camera, the author has normally found it worthwhile to make a *dummy camera* that would quickly allow direct observation and a brightness measurement of the projected microscope field at any time. With a very complete outfit, such as the Leitz KAVAR, this dummy may not be needed. It is relatively simple to make, being a block of wood, and great accuracy of dimensions is not required, since no photography is done with it. In general, there are two types of equipment:

1. The camera has a lens integral with it.
2. The camera, without lens, can be removed. The film is protected by a focal plane shutter, as in a Leica, or by a simple gate slide.

The first case is discussed in the section of this chapter devoted to Simplified Photomicrography.

In the second case the dummy camera is simply a block of wood whose total thickness is exactly that from the camera support to the film plane when the camera is in place. There is a hole through it to clear the image beam. Its top diameter may be just sufficient to clear the frame format of the camera (i.e., about $1\frac{3}{4}$ in. for a 35 mm camera) or it may be made large enough to rest the head of a photoelectric photometer on it (i.e., $2\frac{3}{16}$ in. for a Photovolt Model 200A). The bottom may be flat also or it may need to be indented to fit the support onto which the camera is fitted.

The following directions apply literally to making a dummy camera for the A-O MicroStar Photomicrographic Equipment to allow the use of the Model ID Photometer described in Chapter 17, Vol. 2. The change is obvious for a similar block for the B & L Dynoptic or Dynazoom Assembly. Other equipment can adopt quite similar plans; for a smaller top hole only a simple wooden block is necessary. The block should be painted a dull black.

A block of wood planed on both faces and large enough to accept a $2\frac{3}{16}$-in. hole is obtained. A $3\frac{1}{2}$- by 4-in. piece will do this easily. A sheet of $\frac{1}{8}$-in. Masonite or other comparable material is then obtained. A $1\frac{3}{8}$-in. hole through this sheet allows the image beam to clear to the edge of the top hole. This sheet is glued with uniform pressure to one face of the larger block. The total thickness of the assembly should now be 44.1 mm. Then, when the block is laid on the support over the microscope and a piece of stiff matte acetate sheet (or ground glass) is laid face down over it, the image will be in the same plane as in the camera. Locating pins, that is, small headless nails, are added to the center most easily by driving them into a small projecting piece of Masonite glued onto the ledge inside the block.

Taking the Picture (Commercial Adapters and Attachment Cameras)

It is assumed that the microscope is well set up with a specimen, as discussed in this book. Moreover, it is also assumed at this point that the operator is familiar with the specific camera and its assembly to the adapter, at least as given by the manufacturer's instructions. If the camera sits on a monocular tube that slips vertically for adjustment of tube length, it is probably wise to fix its position to prevent it from slipping, unnoticed during the work. This can be done most simply with a small rubber band. It is quite easy, if simple shop facilities are available, to make a ring clamp. Zeiss sells a ring clamp for this purpose.

There will be an observation tube, which is a telescope with a reticle, as previously discussed. **The eyepiece must first be focused on the reticle for the operator, individually.** This should be done with the color of light to be used for the photography. It is also preferable to do it with diffuse light coming through the telescope, but it is usually done with the illumination set up at the microscope. It is assumed that the accuracy of the position of the reticle, as conjugate to the image in the camera, is built in.

With the apparatus set up and in order, the first step is to **set the shutter for the exposure time** that is to be given. The focal-plane shutter of a camera should not be used for timing the photomicrograph unless it has been proved that it does not cause vibration when actuated.* If the camera has such a shutter, it must merely be kept open during the true exposure interval. Some adapters, such as those of Leitz and Zeiss, contain a separate shutter, whereas cameras with between-lens shutters used with an adapter, such as the Kodak Retina, use these shutters. If the exposure time is a fraction of a second, or even longer, the safest procedure for sharpest definition without vibration is to **open the protecting camera or adapter shutter and time the exposure with an auxiliary shutter** in the path of the illumination beam of the microscope. **For relatively long exposures** this method is the simplest anyway and can be done with **a simple cardboard.** Outside shutters are discussed in Chapter 3.

---

*However, refer to the discussion of the Leitz KAVAR micro-attachment, p. 468.

If a shutter must be cocked while on the camera or attachment, **the final focusing of the image onto the film should be made after this is done.** Placing the image correctly is crucial for good definition; fortunately, the higher the magnification, the more depth of image focus. Assuming the common Method 1 (see p. 462), the image must lie in the plane of the telescope reticle. The eyepiece has been prefocused, but even then **the eye should be swayed to the side slightly** to ensure that there is no motion of the image with respect to the reticle. If the two lie in the same plane, there cannot be relative motion between them.

The choice of photomicrographic film is best dealt with in a later chapter. The determination of the proper exposure when using roll film for photomicrography is probably the biggest problem of this method and is specifically discussed in Chapter 17. It can be stated here, however, that **with an attachment camera in use for the first time for photomicrography, at least some film should first be devoted to determining optimum exposure** by trial with typical illumination and specimen. Normally, after making what seems to be an underexposure, separate exposures are given to successive frames, each twice or even four times longer than the previous exposure time. A cheaper substitute photographic material for exposure trials is discussed in the chapter on exposure.

SIMPLIFIED PHOTOMICROGRAPHY

Method

A very simple method of photomicrography, available to anyone who is engaged in work with a compound microscope, utilizes principles already discussed in this chapter. No darkroom is needed, the lack of which is often otherwise a valid excuse that prevents many people from making photomicrographs. By using daylight-loading roll film and a commercial photofinisher, available at most drugstores, the microscopist need be concerned only with exposing the film as in amateur photography. Any hand camera that is around the house or laboratory will do, although for wide applicability it should be possible to hold the shutter open, on "Bulb" or "Time," to allow variation of exposure. Unfortunately, very simple cameras of the Brownie* and Star type no longer provide more than a single snap exposure. This limits their application, which is

---

*It is recognized that cameras that are no longer sold are both mentioned in the text and included in the illustrations. This emphasizes the important point that almost any camera can be used but that some of the older types of camera with lenses of longer focal length and simpler construction are actually advantageous for this application.

unfortunate, since they are otherwise most acceptable. Some individuals may have an older type of Brownie around the house; they are most useful. There is a definite advantage also in having a camera with a lens of relatively long focal length and reasonably large frame size. The old Tourist type of folding camera can well be taken off the shelf. Polaroid cameras are also very adaptable for this purpose. However, 35 mm cameras are not only useful but are the ones most easily available.

As already discussed, the simple lens of an inexpensive camera has an actual advantage for this purpose over a lens with many components. The optical efficiency of the method is high, and the professional microscopist can achieve micrographs of high quality.

The general method consists of first independently focusing the microscope visually and then placing above (or behind) the ocular a roll-film or film-pack camera which has already been focused at infinity or some other predetermined distance.

There are two apparatus arrangements for carrying out this method, analogous to Methods 1 and 2 in the first part of this chapter. **The first is the very simple carrying out of the method as just stated. In the second a matte or transparent surface and an alternate lens are temporarily substituted for the camera and its film** but the final focusing is usually done as before, although film-plane focusing can now be employed. This second method, although somewhat more elaborate to prepare, is recommended as worth the extra trouble whenever the microscopist is prepared to determine his exposures with a photometer. Actually, the addition of a photometer (or exposure meter) really is a final factor to keep the whole method one of simplified photomicrography.

The directions for photomicrography by the first method (I) are given on p. 481 and for the second (II), on p. 486.

Focusing the Image and Camera

Unaided Eye. As discussed under "Principles," if the microscope is focused while looking into it as usual, a photomicrograph made by placing a hand camera over it with the lens focused at 25 ft will usually be well focused. Also, as already discussed, the focus setting is most correct and least critical when the front surface of the lens is placed at the eyepoint of the microscope. Some individuals, especially younger people, do accommodate with their eyes so that they should focus the camera differently. Spectacles, if worn, especially for distance, should always be used.

Use of Telescope. A fundamentally better method that ensures reliable results is the use of a telescope to focus the microscope image.

It must contain a reticle that is also in focus through the eyepiece. The cross hair, or other reticle pattern, is brought into focus by the eyepiece adjustment to suit the individual's eye. The whole telescope is then set for a long distance by focusing out of a window or down a long hall. The camera lens should be set for that same distance, for example, 100 or 25 ft. Then, if the microscope image is focused through the telescope, the photomicrograph will be found in perfect focus. This was true in test situations with a large number of people, irrespective of ordinary visual focusing habits. It makes no difference if the telescope image from the microscope is overmagnified and seems blurred; the picture is not taken through it and focusing is critical with it.

Almost any standard telescope that has a reticle and focusing eyelens will do. A telescope good enough for this purpose can be improvised, as frequently done by the author. The telescopes furnished for phase microscopy are of convenient focal length but usually not directly usable. They contain no reticles or cross hairs and sometimes will not focus at sufficient distance. Moreover, for their original purpose an optically good objective lens is not needed, and therefore it is usually not a very good lens.

Inexpensive, relatively small telescopes can be obtained from some of the optical supply houses that sell at a discount. Usually a reticle must be added; it is essential. The following improvisation proved quite satisfactory.

A 48 mm microscope objective was used for the telescope but with its normal back facing the field. It was threaded onto a petrographic objective changer, then pushed into a cork cylinder which could, in turn, be pushed into the brass tube that formed the telescope body. The cork cylinder is easily cut from a stopper and has a larger hole at its other end to clear the image rays. The microscope objective mount is needed as a handle in focusing the telescope by pushing the cork to and fro. At the other end a hand magnifying lens was pushed into the tube by a similar cork cylinder with a homemade reticle on a cork lying between the two in front of the eyepiece. A microscope ocular with reticle and focusing eyelens was sometimes used.

Two variations of this method, which allow simultaneous viewing, focusing, and photography, as in commercial apparatus, can be used. The first consists in the use of a binocular microscope with a focusing ocular in one tube which is carefully adjusted for equality of focus of both sides. In the inclined binocular this convenience is offset by the greater inconvenience of holding and manipulating the camera. This method does not give the independent check of the angular distance that a telescope does. Of course, a telescope can still be used with the binocular method. The second method consists in making a split-beam device and viewing the horizontally reflected beam with the telescope. A microscope cover slip, held at 45° above the ocular, will reflect about 4% of the beam horizontally.

If brightfield microscopy is done with relatively short camera exposures, this will be ample illumination for focusing. The clean cover slip can be held in a small box that sits on the microscope. In this case the camera-lens surface cannot lie in the plane of the eyepoint. However, it does not do so in some commercial apparatus. The telescope is now held horizontally in the emergent image beam by a separate vertical support clamp that allows quick and simple vertical alignment.

Centration

It is assumed that the microscope has been in use visually before finding the field to be recorded and that the instrument, with its illumination and image, is well centered before photography is undertaken. The process of setting up the microscope is discussed in Chapter 10.

Further alignment consists only in seeing that the middle of the visual field in the microscope is in the middle of the photograph and that the edges of the photographic field are symmetrically placed. Lopsided pictures are obtained if the microscope and camera axes are not aligned. The camera lens diaphragm is best kept wide open. The centration of the camera is exceedingly sensitive to the tilt of the camera, just as it is when photographing a subject at a great distance. It is therefore wise to place a spirit level on the back of the camera to make sure that it is parallel with the stage. **If a dummy control camera is used, as described in** Method II, **(p. 482) insurance that the photograph is well centered is easy.** If the camera is used alone above the microscope, it is wise to align it first with the camera back open and a matte film or tissue paper in the film plane. Since the camera is usually used with a vertical support, it is normally possible to fix a stop that will ensure that the camera will return to central alignment.

Magnification

As discussed in the previous chapter, the magnification in the negative or direct photograph will be less than that apparent from the direct observation through the microscope by the ratio of the focal length of the camera lens in millimeters to two hundred and fifty (F/250).

METHOD I

Apparatus and Its Setup

Method I merely consists in holding the camera, focused for 25 ft or more, behind or above the ocular of the microscope. The simplest equipment is involved when the microscope is horizontal, as illustrated in Figure 11-2.

A simpler procedure involves slightly more apparatus to hold the camera over the microscope. In this way a simple and quick decision can be made to take a picture of a field while using the microscope visually, with very little to do but swing the camera into position and snap the picture, assuming that the exposure is known. Yet the total cost of the equipment can be quite low.

A separate camera support, which should be relatively sturdy, is recommended. The support rod *should not be less than $\frac{3}{4}$ in. (19 mm) in diameter.*

The following apparatus, illustrated in Figure 11-3, is merely a suggestion but has proved practical, especially for larger cameras. It is assumed here that the microscope is mounted on a board also carrying its light source. For microscopes with built-in light sources of adequate brightness a board may still be the simplest way of ensuring that the camera is supported inflexibly with respect to the microscope.

In addition, two boards, for example, of pine, are required, one for the base of the apparatus, unless the support is directly fastened to a table top, and one for the camera support. The optimum dimensions of both can vary with the situation; with this listed support ring, the camera board can be 7 in. square.

The whole device is illustrated in Figure 11-3. The base plate is screwed to the board or the table top at a suitable distance from the microscope

Figure 11.2 Simplified photomicrography without stand.

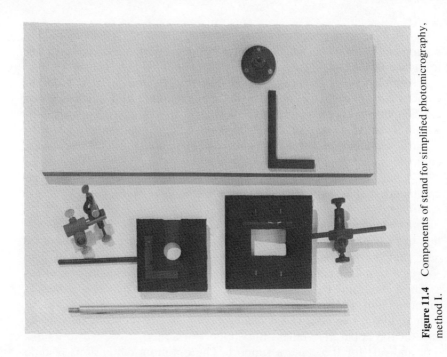

Figure 11.4 Components of stand for simplified photomicrography, method I.

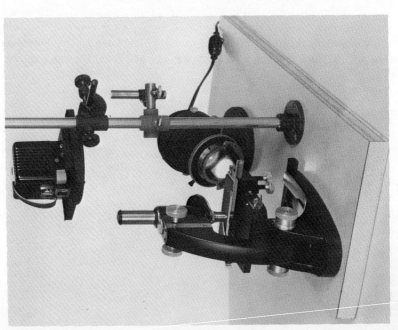

Figure 11.3 Simplified photomicrography, method I.

Number	Gaertner Catalog Number	Description
1	S622	Base plate for 19 mm rod
1	S306	Round support rod, 19 mm diameter, 60 cm long
1	S903	Right-angle clamp with round hole and V-groove
2	S1002	Right-angle clamp with two V-grooves
2	S215A	Round rod, unthreaded, 13 mm diameter, 15 cm long
2	S1019	Collars for 19 mm rod.

Number	Central Scientific Catalog Number	Description
1	18000	Extension ring, 7-in. diameter

Total cost (1962) less than $10.00

axis. The camera board is the wooden board stapled to the top of the laboratory ring support. The nature of the holes that are cut in the board will depend on the camera to be used. Protuberances in the face of the camera require only countersunk depressions in the board. It is important that two strips of wood be placed so that the camera will fit snugly and reproducibly into a corner; sometimes one strip plus one edge of the hole is more convenient for folding cameras. At any rate, it must be possible to place the camera in identically the same location by shoving the end and one side of the camera against two stops. Thus the camera can be removed for change of film, etc., without affecting centration above the microscope. Two pairs of right-angle screw hooks should be placed on each side of the hole, opposite each other. Then, if rubber bands are snapped on, always with the hook on the braced side fastened last, the camera is automatically brought to the identical location each time.

Variations of the support board will certainly have to be made for the contour of the individual camera. In Figure 11-4 two support boards are shown, one for the box camera of Figure 11-3 and one for a folding camera, Kodak Tourist-type, that can be seen in Figure 11-5. In one camera, the Kodak Starflex, which had flat metal bars on the side, it was simpler to make a hole in a board with parallel sides just wide enough for the camera to be inserted from below with the side bars as stop.

The camera was held snugly against the board by friction, plus the rubber bands over the hooks, and was easily withdrawn and replaced.

In the case of this camera and all double-lens reflex cameras the board is U-shaped around the hole, open at one end to allow the image to be viewed in the reflex finder when that lens is swung over the microscope. However, in this case it is necessary to fasten a piece of matte film, such as matte acetate sheet, over the lens of the reflex finder; the camera should not be dismantled to insert it underneath the lens. This viewing of the image has some advantage but not a great one. With the twin lens of wider aperture than the camera lens, its appearance is no guarantee that the camera image is well centered and it is not useful for determining exposure as is the dummy camera of Method II. For best results focusing should still be done directly over the microscope with the unaided eye or a telescope and not through the reflex finder.

A very simple way of supporting a small camera, such as a 35mm, is by means of a base plate (in a vertical plane) at the end of a horizontal support rod. The camera is supported by the ordinary tripod screw, that is, a $\frac{1}{4}$-20 bolt. The only problem is that of removing the camera quickly and easily and returning it to the identical position and level, also quickly, easily, and stably. This can be done, too, but it seems preferable to use Method II rather than otherwise elaborate this equipment which might do for routine pictures.

By placing a collar or a clamp below the right-angle clamp supporting the camera rod the camera can be swung out of the way when not in use. Moreover, by placing a short vertical rod in the manner shown in Figure 11-3 the camera can be instantly realigned over the microscope because its supporting clamp can be made to hit the vertical rod at the correct location. The proper level of the camera is, of course, with the front surface of the lens at the height at which the bright disk of light formed on a piece of paper held over the eyepiece is smallest.

The Shutter and Exposure Technique

Although the camera will be held firmly above the microscope, it is easy to allow a vibration to occur during the exposure of the film. The acceptable tolerance for the resulting reduction of definition is usually very small in a photomicrograph. Focal-plane (curtain) shutters usually cannot be tolerated for this purpose. A cable release should be used when possible and very carefully. Unfortunately, more and more of the simpler cameras, which are otherwise quite adequate, have no provision for this device. The trigger must then be squeezed very gently.

A between-lens shutter will be satisfactory only if the duration of its opening and closing action is short with respect to the total exposure time; the shutter, closing in like an iris diaphragm, will cause apparent nonuniformity of illumination (a vignetting effect) if the lens is not at the eyepoint and incorrect timing if it is in that plane.

A most satisfactory arrangement for avoiding all of these troubles from the camera shutter is to use a larger shutter in the light beam for exposure timing, opening the camera shutter independently before the exposure to the beam; its function, now, is merely to protect the film from light in general. An Ilex Universal Shutter No. 3 proved very useful as an outside shutter; others are available.

Another good way to control exposure is by the use of flash illumination. This is managed more easily by Method II, however.

As already noted, the simplest, least-expensive cameras now have no provision for varying exposure, it being assumed that they will be used for outdoor snapshots or indoor synchronized flash. Optically, they would be satisfactory. Such cameras are still useful in much photomicrography, since simple brightfield work with adequate illumination probably constitutes the majority of applications. Here the exposure technique consists in bringing the illumination to the specified level for the specific film at snapshot exposure time, usually about $\frac{1}{40}$ sec. This type of procedure is discussed in the chapter on exposure.

Some cameras are so simple that they do not possess a device for prevention of double exposure. This can be taken advantage of, although great care must be used. Using one click of the shutter of the Kodak Fiesta Camera as a unit of exposure, the author has built up to 16 units without appreciable degradation caused by vibration. The camera must be sturdily held with rods at least 19 mm in diameter.

The simplest 35 mm cameras with provision for more extensive exposure variation, hence much more practical for simplified photomicrography, are discussed under Method II.

Directions, Method I (for exposure determination, see Chapter 17, Vol. 2)

After the apparatus has been set up and a specimen is under the microscope, the procedure is as follows:

1. Determine the eyepoint of the microscope (illuminated disk of minimum diameter) with a piece of white paper.
2. Focus the camera at 25 ft or other predetermined distance.
3. Lower the camera to make the front surface of the lens coincide with the eyepoint or be slightly above it.
4. Bring the lower clamp up to the other one and clamp it.
5. Loosening the upper clamp, align the camera lens with the ocular.

This is best done the first time through the open back of the camera (containing no film) by looking through the lens. Tighten the upper clamp.

6. Rotate the lower clamp until the short vertical rod hits the upper clamp and again tighten it.

7. The camera can then be swung aside and a light-tight connector attached to the lens.

8. Focus the microscope visually, preferably through a telescope with a reticle.

9. Swing back the camera and take the picture.

For photomicrographs of other fields only Steps 8 and 9 need be repeated.

Method I is very satisfactory with a Polaroid Land Camera, if one is willing to use at least one frame to determine the proper exposure when conditions have changed. By using Method II, plus a photometer, even that use of extra film becomes unnecessary.

METHOD II WITH DUMMY CAMERA

Apparatus and Its Setup

Wide-Board Support. The same vertical support rod and support clamps are used, described for Method I and illustrated in Figure 11-4 except that the shelf bearing the camera must now be larger (Figure 11-5). It can be made from a piece of plywood $\frac{1}{2}$-in. thick. The shelf may be supported by a shelf support clamp sold for the purpose (Gaertner Cat. No. S1820). With this somewhat more sophisticated method it is also preferable to use right-angle clamps with two round supports instead of the V-groove. Sometimes aluminum tubes or rods were used to keep the extended weight down and thus reduce vibration. A double 19-mm round rod support is available (Gaertner S803). The same hole in the shelf should be made as for the camera, but it is now placed on one side of the middle of the board, as shown in Figure 11-5. At the other end of the board another hole is made for the view box. This box can be improvised from materials most easily available to hold a spectacle lens at the same relative height over the microscope as the camera lens and to hold a sheet of matte glass or, alternatively, a clear glass sheet in the same plane as the film in the camera. Fortunately, it is not important that it be exactly in the same plane.

The author's view box for a larger camera was made from a cylindrical waxed cardboard container of the type used for packaging ice cream; the inside was painted black and a 2.5-cm (1-in.) hole was cut in the cap.

Figure 11.5 Simplified photomicrography, method II, larger camera.

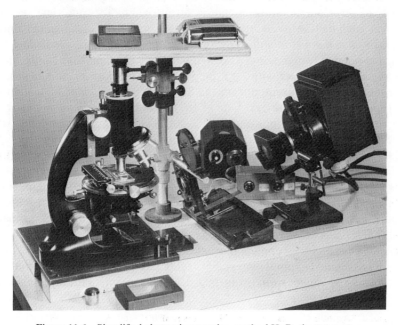

Figure 11.6 Simplified photomicrography, method II, Retinette camera.

This cap was held on with tape and could be slipped on and off, as desired. Thus the lens could be rested on this horizontal surface surrounding the hole and held in place by wax. The bottom of a 4- by 5-in. plate box with an appropriate hole cut in it was placed on top of the shelf, surrounding the cylinder. Different tops, each containing a piece of clear or ground glass, can be fitted on the view box for measuring the light or viewing the image.

The lens of the viewing box should have the same focal length as that of the camera. Since neither the focusing nor the photography will be done with it, a spectacle lens may be used. There is some advantage in substituting, for a single lens with strong curvatures, a stack of several lower power lenses whose diopter power adds up to that of the camera lens. If there is a 100 mm lens in the camera, for example, it is preferable to use two 5-diopter lenses instead of one 10-diopter lens. There is also some advantage in using a meniscus lens with the concave side toward the microscope. Because the camera lens and the viewing-box lens have the same focal length, the lens is focused best in the box by making the magnification on the glass of the view box the same as that in the camera, for which purpose the camera should be temporarily opened and a piece of matte film placed inside.

A simple, inexpensive, yet adequate camera for photomicrography must have an adequate shutter, that is, a reasonably accurate time series that will include the availability of "Bulb" or "Time" exposures and perferably a release that will allow more than one exposure per frame. It does not need a rangefinder. The writer knows of none that fulfills the first two conditions, yet without a rangefinder. Two 35 mm cameras that otherwise fill the requirements are the Sears-Roebuck Tower 65 Camera and the Kodak Retinette 1A. One of these cameras can be obtained for this purpose and incorporated into a relatively compact wide-board device with dummy camera, as shown in Figure 11-6. A box for $2\frac{1}{2}$ $\times 3\frac{1}{2}$ in. sheet film was used with the clear- and matte-glass covers. In this case a 45 mm achromatic lens was obtained quite inexpensively to match that of the camera from an optical-parts dealer,* but a composite of three spectacle lenses would do. Separate achromatic camera lenses are still listed, but an acceptable alternative is an unmounted achromat that *nearly* matches the camera lens, plus a spectacle-lens correcting element. This merely supplies the difference in power $(1/f)$ between the camera lens and the simple achromat (see Chapter 1). Both cameras can use almost the same supporting board and the same dummy camera,

*Edmund Scientific Co., Barrington, New Jersey 08007, and A. Jaegers, 691 Merrick Road, Lynbrook, New York, are two sources of these lenses.

very slightly refocused. A cable release should be used to make the exposure; one with a clamp lock should be obtained to hold the shutter open, when first setting up the equipment, or for checking alignment. Use of the lever to change frames between exposures is a little awkward with the Tower but not seriously so, since, with the rubber bands, it is so simple to lift the camera and to return it into alignment.

The arrangement, with dummy and real cameras on a wide board, is used like a twin-lens reflex camera in ordinary photography, except that the dummy is swung aside and the real camera substituted, using the prearranged mechanical stop to insure alignment. During this shift normally the whole board is temporarily shoved aside and the microscope focused with the eye or preferably through a telescope. However, with this dummy camera available, it can be set up so that its image can be used to focus the main camera image by ordinary view-camera focusing technique. For this purpose it is not even necessary that the two camera lenses be an exact match. It is necessary, however, that they be set once and for all with both in simultaneous focus. Use of a telescope is the fast way of setting up the photographic camera exactly, assuming that its distance scale is correct, but it can also be set up photographically. A test object of very critical detail should be chosen. The magnification should be as low or lower than will ever be used; there is too much depth of focus in the image area at high magnification. Then the dummy camera lens is focused, probably by pushing it in or out with a tube that will slip inside the supporting tube for the lens. As far as the author is concerned, the focusing technique with a telescope is so simple and reliable that the ultracalibration for focusing has never been utilized.

When it is known that the two lenses are closely the same, adjustment of the dummy lens until the two cameras, dummy and real, have the same magnification is the simplest and fastest procedure and quite sufficient with the independent focusing technique.

When a photoelectric photometer with a large probe (e.g., 2 in. in diameter) was used for exposure determination, as discussed in Chapter 17, Vol. 2, a block of wood was made to fit into the lower fixed half of the dummy camera box to extend the bellows length temporarily and provide adequate illuminated area. The block had a hole through it $2\frac{1}{8}$ in. in diameter and extended the distance from the camera lens by the factor $\sqrt{10}$, that is, 3.162 times. The photometer reading was then multiplied by 10.

Use of Kodak Instamatic Cameras for Photomicrography

The Kodak Instamatic Cameras may be used for photomicrography but are not quite so efficient with 35 mm film for this purpose as the Retinette

1A. The Instamatic 314 or 714 is quite useful, except that the coupled exposure meter, so helpful to general photography, is of no direct use here; the added expense of the camera is a waste *unless* one is available anyway. Blocking the window with a mask ensures that the lens diaphragm will remain wide open to make centration easier.

The inexpensive Kodak Instamatic 124, which can also be used, has neither exposure meter nor rangefinder to represent wasted money. The photographic definition obtained should be quite acceptable. The limitations are the small linear aperture of the lens diaphragm (about $\frac{5}{32}$ in.) and the shutter which has only two exposure times ($\frac{1}{40}$ sec for flash and $\frac{1}{90}$ sec for daylight snapshots) and neither "Bulb" nor "Time" exposure. Therefore the usual advice to place the front surface of the camera lens at the eyepoint should be ignored; it will cause vignetting of the field. An ocular that allows the eyepoint to lie within the lens near the diaphragm must be used. The Kodak Instamatic Reflex Camera with its microscope adapter consitutes, in practice, a system of simplified photomicrography. It is discussed earlier in this chapter.

The intensity of the illumination must be sufficient to utilize one of the two available exposure times. The longer one can be employed by inserting a *used* flash cube in the socket. Unless the illumination intensity happens to be that required for optimum exposure or lies within the acceptable exposure latitude of the film, the correct exposure must be obtained by reducing excessive illumination by the proper degree. This is most easily done with neutral density filters; the discussion of these filters in Chapter 9 should be considered. Most of the built-in illuminators have sufficient intensity for the available two exposure times for simple brightfield illumination, at least at medium powers. Because color film for daylight is the only color film available in the cassettes for this camera and because the best exposure modulation technique is by use of neutral densities anyway, flashbulbs or electronic flash to simplify the procedure are especially applicable to Instamatic cameras. This procedure is described in Chapter 13. It does require the use of a substage mirror. After the correct exposure has been determined by trial for one specimen, all subsequents exposures can be calculated by employing records of the variations that are produced by the factors of numerical aperture, magnification and specimen type, as discussed in Chapter 17, Vol. 2. If flash exposures are used, the following directions would apply to the use of a control lamp (see Chapter 13).

Directions Method 2

Most of the directions (p. 481) for Method I apply. It is assumed that the apparatus has been set up with the shelf supported by a sturdy rod and at a level so that the front of the camera lens is at the height of the eyepoint or can be quickly brought to it if the optics have been changed. The procedure will differ somewhat according to whether the dummy camera has been calibrated for focusing (as commonly done at the ground glass of a camera) or whether focusing is independently done by the unaided eye or, preferably, with a telescope. The camera is focused at infinity.

Most of the time the camera shelf will be swung aside while the microscopist is engaged in visual observations. Then a field is found for which a photomicrograph is wanted.

1. Focus the microscope field; use the telescope set for infinity, if that focusing method is used.

2. If the exposure is known from experience, the camera is swung directly over the microscope and the front of its lens is adjusted to the eyepoint. (The level and proper centration of the camera should have been ensured by original precautions and setting the lower stop clamp as indicated for Method I.)

3. Take the picture.

4. Advance the film. If the hook and rubber-band system (p. 480) is used, removal and replacement of the camera may be the simplest way.

Alternatively, if the exposure is unknown or there is a question about the field; the following directions apply:

1a. Swing the dummy camera into place. Its centration is easily done visually.

2a. If a photometer is available, put a clear glass over the dummy camera and photometer the image field (see Chapter 17 for photometry of the clear field).

An accessory block, to raise the photometer to be inside the beam, may be used (see p. 485).

Calculate the exposure by the simple E_c-method (see Chapter 17).

3a. Swing the camera into location over the microscope, as in Step 2.

4a. Take the picture.

When color photomicrography is being done and conditions are at all constant, use of flashbulbs or electronic flash for illumination (see Chapter 13) lends itself to simplified photomicrography. The biggest problem in this method would occur when there is an appreciable variation in exposure for different photomicrographs.

Appendix 1

Depth of Field
in Photomicrography

A

WITH THE SIMPLE MICROSCOPE

The discussion of resolution and definition, given in Chapter 2, assumed that the object field was a flat plane perpendicular to the optical axis and that the image was contained similarly. Also discussed were the points of object space and the corresponding antipoints of image space, which are really disks, hopefully smaller than anything the observer's eyes can distinguish from true points.

In reality, object space has three dimensions and so has image space. It is true that the image space is cut by the plane of a ground glass or a photographic film, which corresponds, critically, to only one plane of the object space. Before considering this problem, some definitions are in order.

Definitions

Depth of field is the term for the total distance, measured in the field, between the nearest point of acceptable focus, when the image or its picture is viewed, and the farthest point of acceptable focus along the optical axis or parallel to it.

Depth of focus has been used in two ways: (a) it is the depth along the axis of acceptable definition in image space; (b) it is the axial distance the lens can be moved without noticeable degradation of definition in a single plane in image space. This last factor is of great practical importance in photography at magnifications near 1X. Depth of focus has sometimes been used mistakenly instead of the term depth of field: (a) seems the preferable choice of definition, although it could be replaced by the term *depth of image*; (b) then, should be termed *latitude of focusing*.

Both points and antipoints are distributed throughout object space and image space, respectively. In Figure A.1-1 three object points are represented (with some exaggeration) at distances l_1, l_2, and l_3 from the lens; therefore their images (antipoints) will be at the distances l_1', l_2', and l_3', respectively. If a photographic film is placed at l_2' from the lens, l_2 becomes the object distance of the lens at this focal setting. The other two points are then represented on the film by disks, caused by the interception of the two cones of image rays, one coming from the point imaged at l', the other imaged at l_3'. Since the disks are represented as having the same size, $b\text{-}c$, the distances of the two antipoints will normally be unequal, $l_3'-l_2'$ being larger than $l_2'-l_1'$. In this figure we can assume that the disk size, $b\text{-}c$, represents the largest size of image disk acceptable in the picture as within limits of focus. Therefore $l_3'-l_1'$ is the image depth and $l_1\text{-}l_3$ $(= \Delta)$ is the depth of field. The diameter of the disk represented by $b\text{-}c$ has been called the **circle of confusion.** This will do. Its size may be affected by a variety of effects but most optical defects in a good camera and photographic system are usually of a different order of size. The corresponding disk in the object plane at l_2 will be smaller by the factor m of magnification when m is greater than 1.0; it is larger when m is a fraction, as in most photography.

Where c' = diameter of the circle of confusion in the image plane,
  c = diameter of the corresponding disk in the object plane,
  $c' = mc$.

The value of c will vary with projection distance and other factors affecting magnification, whereas c' is the directly observed and determinable quantity. Rather a number of formulations of depth of field use c as the limiting parameter but the directly determinable constant c' seems preferable to the author.

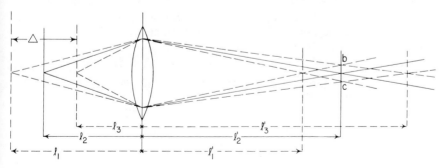

Figure A.1-1 The geometrical optics of depth of field.

Another criterion that expresses the visual limitation is the *limiting angle for the separation of two points in a plane being viewed*, that is, the limit of visual acuity. This is most frequently taken as one minute of arc $(0°1')$. It is a good method, of course, but the one using the circle of confusion is more adaptable to our study of the application to photomacrography.

Another criterion considers the value of c in the object field to be a constant fraction of the object distance of the camera lens, such as $f/2000$.

Since any criterion chosen will be somewhat rounded off numerically, the values of depth of field obtained will differ somewhat as found by the various formulas, even though each is consistently correct. The greatest difference is whether c or c' is taken as the fundamental parameter, usually a constant parameter.

Depth of field will be infinite in extent under two conditions:

1. When all antipoints and their circles of confusion projected in the picture plane are infinitely small points as far as the observer's eye is concerned. This condition is approached in some field photography.

2. When all circles of confusion are disks of approximately equal size, almost independent of the size. This is a practical but usually unsatisfactory method by the expedient of increasing the diameter of the antipoints with a smaller aperture or even poor lens quality up to a minimum that collects nearly all the diameters of the disks in the image field. The extra depth of field of a very inexpensive camera over that of a fine camera lens is well known.

Imaging of Three-Dimensional Space

The three-dimensional space in front of the lens is represented in an image space of three dimensions, that is, the antipoints are arranged in the same pattern. However, **longitudinal magnification** (along the axis) is **proportional to the square of the lateral magnification** m. This means that three-dimensional object space, including all objects in it, is distorted as image space except at the magnification $m = 1X$. It also means that the image of an object such as a sphere, is flattened at the fractional magnifications of field and studio photography when m^2 rapidly becomes larger with increasing projection distance. This is illustrated in Figure A. 1-2 and is also discussed by Dade (5. 1952). When a film is inserted into this image space and makes an acceptable photograph, it can be considered equivalent to a section of image space and of the distorted image of the sphere; the thickness of the section is determined by the

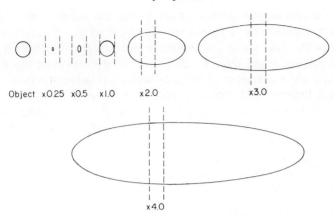

Object x0.25 x0.5 x1.0 x2.0 x3.0

x4.0

Figure A.1-2 Effect of magnification on image size and shape and apparent depth of field. The object is assumed to be a sphere of such size as to just occupy depth of field tolerance at 1X magnification. The distance between the limits of near and far focus tolerance (c') in image space is considered a constant and shown as parallel dotted lines.

acceptable size of the circles of confusion on both the near and far side of sharp focus on the film. Such an image space section is still very thin. It can be projected back into the object space by dividing this thickness by the magnification; it then becomes the depth of field.

Where m is a small fraction, as it may be in field photography, the image of the sphere and other objects may be so flattened that all of it is included in the section of acceptable focus. With the elongated three-dimensional image, obtained when m is appreciably larger than 1.0, very little of the whole image may lie within the optical section.

The concepts discussed above are important, possibly all that are needed in most cases involving the practical compromises that must be made in photomacrography between the desired picture definition and depth of field.

Depth of Field, Formulation

The formulas for calculation of the acceptable depth of field in photography have been derived by very simple geometry and algebra in nearly all texts of geometrical optics. The formula can be derived quite simply from the similar triangles of image space in Figure A. 1-1. However, the various authors, even when using the same criterion for c, may leave the formulas so that their essential identity is not easily recognizable; the formulas are often left in terms of object distance l, the focal length of the camera lens f, and its clear diameter a. A form that is common, but seems objectionable to this writer, leaves the value of the depth of field Δ as the difference between two numbers so nearly equal that the first digit of the final value may be determined by the fourth digit of the original numbers made large by the term f^2. By replacing the term of the absolute clear diameter of the lens with its relative aperture $\alpha = f/a$ and optical lengths with magnification, it is possible *to reduce all* the *apparent discrepant formulas* to the same one, in terms of magnification. An exception is some depth formulas for field photography which are simplified by assuming that magnification $m = l/f$, an assumption unacceptable for photomacrography.

The following symbols are used, some in only the formulas considered here to be intermediate:

f = focal length of camera lens,

l = object distance, l' = image distance,

a = clear diameter of lens

α = relative aperture or f-number,

$$\alpha = \frac{f}{a}$$

m = magnification of the image over the whole range (e.g., $1/1000\text{X}$– 1000X),

c' = the limiting circle of confusion,

c = the limiting circle of confusion projected to object plane,

$c' = m \cdot c$,

Δ = depth of field = near depth + far depth.

Two formulas* are

$$\Delta = \frac{cl}{a-c} + \frac{cl}{a+c} = \frac{0.00058l^2}{a+0.00058l} + \frac{0.00058l^2}{a-0.00058l}.$$

These formulations do emphasize that if the object distance *and* the *effective* linear diameter of the lens are constant the depth of field is constant. Obviously a lens of shorter focal length will have the same depth of field that requires a smaller f-number, α, for a lens of longer focal length.

Most of the equations found in the literature can be transformed

$$\Delta = \frac{2\alpha c' \cdot f^2 (m+1)}{f^2 m^2 - \alpha^2 c'^2}$$

*The first is by Hardy and Perrin (2. 1932); the other is in the Leica Manual (1. 1961), in which the circle of confusion is taken as $f/2000$ in inches; the original formula utilized an angular criterion.

but $\alpha^2 c'^2$ is negligible† with respect to $f^2 m^2$. Therefore the working formula for depth of field is

$$\Delta = \frac{2\alpha c' (m+1)}{m^2}. \qquad (A.1\text{-}1)$$

This formulation shows that if the relative aperture α *and* the magnification are constant the depth of field is constant for all focal lengths of lens. This is more typical of photomacrography than a constant object distance.

Applicability of Formula

The usual and certainly the apparently normal procedure has been to assume that c' has the value of calculated resolution for the picture. Some (5. Gibson, 1960) have assumed that the diameter of c' is increased because of lens defects and the limitations of the photographic materials. Both factors are certainly capable of doing just that, but with an excellent camera lens, adequate photographic materials, and proper processing these degradations should be small. They will tend to increase depth of field as they decrease the minimum size of the antipoints that give noticeable degradation of central definition when this size is above the visual limit.

The author therefore decided to study the phenomena and to determine the parameter c' empirically.

Experimental Study of Depth of Field. To prepare a suitable object field for a series of photomacrographic magnifications a set of photographs on paper was made from a single component of an engraved halftone screen bearing parallel lines at a series of fractional magnifications such that on the final photomicrographs the pattern had the same size; that is, there were 23 white plus black pairs per 10 mm. The width of the black lines was about 1.8 mm, since the photographic processing decreased their width. However, the intermediate photographic materials and optics had adequate resolving power. The test object consisted of a ruler perpendicular to the camera axis for magnification and field, whereas the test pattern, together with a ruler lying in the same plane, was at an angle θ of $25°50'$ from the optic axis. One of the photomacrographs of the test object, originally made at 2X, is reproduced as Figure A.1–3. The photomacrographs were made as a series of magnifications from $\frac{1}{2}$ to 20X, one set at $f/4.5$, another at $f/22$. The lens was a Kodak Enlarging Ektar of 100 or 50 mm focal length. The Kodak Panatomic-X Sheet Film used was developed for 5 min in Kodak Developer DK-50. Two sets of contrast

---

†Under some conditions this may not be strictly true; see discussion of case with compound microscope.

Figure A.1-3 Depth of field test object. Photographed at X2 magnification and f/22 with Enlarging Ektar Kodak Panatomic-X at gamma of 1.2.

prints were made, one of low contrast, the other of high.The observed depth of field could easily be obtained by multiplying the observed distance on the print by $(\cot \theta)/m = 2.07/m$.

The field depth is limited, in this case, by the narrowing of the black lines, which become points and disappear. The white lines, in effect, widen. The change in the width of the lines can be taken as a measure of c'. If the limitation is near the limit of visual acuity, a value of 0.1 to 0.2 mm can be assumed, say, 0.15 mm; this is a little more than 6 lines per millimeter (17/in.), counting both black and white lines of equal width. The test photomacrographs were examined with a traveling microscope that bore a micrometer eyepiece. This allowed accurate measurement of all dimensions with nearly all subjective factors removed. **Formula A1-1 was found to hold accurately in expressing the depth thus defined.** Naturally, if a different value of c' were taken and the length of the lines was measured and converted to depth, the formula also held.

The prints were examined with the unaided eye by three observers. The method used to observe the apparent depth was to lay a piece of fixed film over a print and mark the apparent boundaries of the near and far limits of focus. The results of the three independent observers were very similar and somewhat surprising, since the marked discrepancy with the first set of measurements must have been entirely due to subjective factors. Two sets of prints were made, one of quite high contrast, the other of low contrast. As might be expected of subjective judgement, this factor made some difference; those of each pair of high contrast showed less depth of field. Those of low contrast were considered more typical of most photomacrography. The depths of field were definitely greater than would be obtained by calculation from the formula, using a value of c', which seemed reasonable for acceptable definition. **The depths of field observed did seem more compatible with photomacrographs of natural objects such as flowers or insects,** although the methods of measurement with natural objects were crude.

The formula was then inverted to allow the calculation of c' for the observed depths of field by using the averaged data. The results were unbelievable under the original assumptions and without the empirical experience; c' seemed ridiculously large. The values obtained are given as a graph in Figure A.1-4. Note that the accepted diameter of the anti-point c' is independent of the lens aperture up to the magnifications at which diffraction disks are larger than the acuity limits of the viewer; that is, the lens becomes "diffraction limited." At 20X magnification and $f/22$ the minimum in-focus diffraction disks should be about 0.25 mm in diameter (see Table 5-1); out-of-focus disks will be larger. The poorer definition increases the accepted depth of field as usual, and obviously

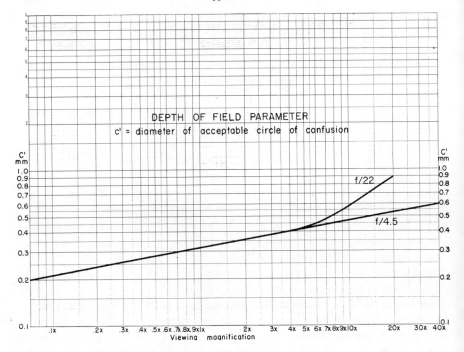

Figure A.1-4 Depth of field parameter, c', versus magnification.

out-of-focus disks of almost 1 mm diameter were the limit of acceptance for near and far focus! It does seem ridiculous.

By utilizing the values of c' shown as the acceptance limits in Figure A1-4, the depths of field for various magnifications were calculated at the two apertures used in the experiment and plotted against magnification on log-log paper (see Figure A.1-5). The curve for $f/8$ was added which required an estimation of c' above a magnification of 5X. The solid curves of the figure show the accepted depth of field by direct observation of the test photomacrographs. The curves for a value of $c' = 0.15$ mm have been added as dashed lines to show the limits of depth of field imposed by an accepted though generous value of c' and the *relaxation of definition actually accepted at the outer focus limits* in photomacrographs by comparison with the solid curves.

Depth of Field Theory

The author's conclusions from this experiment and other observations, including the important fact that **the standard formula applied excellently when c' could be accurately measured,** are as follows.

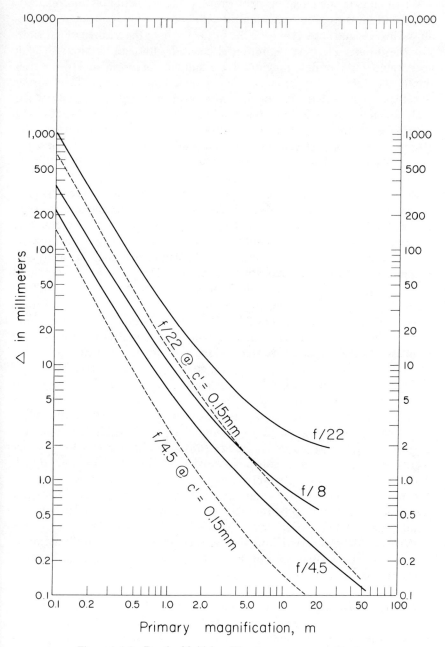

Figure A.1-5 Depth of field, in millimeters, versus magnification.

When an extreme close-up photograph is made, including photomacrographs, there is usually some central detail of special importance, especially of insects and flora. It may be the eyes of a fly or the stamen of a flower. For these parts the best definition is desired, although obtaining it with acceptable depth of field may involve a difficult compromise. However, if the facets of the eye of the insect are delineated, for instance, the observer is satisfied with relatively coarse delineation of the grosser parts of the body, including the legs, which are familiar. Not only is the **value of** c', the limiting circle of confusion, a variable according to the type of photograph, it **may be variable within the same photograph,** being larger near the edges as the gaze leaves the critical center of interest. Good definition in the center of interest, however, is still important.

Because the criterion is so largely subjective, it must differ, at least to some extent, among individuals and according to conditions. Obviously the best way is to determine c' for the individual situation and observer and to calculate the depth of field Δ. This is often impractical. Fortunately, the accurate value of Δ is usually unimportant; it is merely desired in planning a picture and need not be accurate. The values from the solid lines of Figure A.1-5 will usually be more realistic than those of the two dashed lines. In any case, in photomacrographs of three-dimensional objects rarely will the depth of field be as great as desired when the definition of the center of interest is adequate. **The lowest magnification and smallest camera lens aperture suitable are normally the best choice.** Therefore a photomacrographic lens that will go to very small apertures, such as $f/22$ or even $f/45$, is desirable. Incorporating a good iris diaphragm that will go to such small apertures is relatively expensive; many lens designers state that a lens is "diffraction limited" at them. It is difficult to persuade these people that for some subjects gaining appreciable depth of field may be more important than maintaining central definition below the level at which the eye can distinguish some degradation of very fine detail. This was discussed in Chapter 4 on p. 164.). Such lenses as the Kilfitt Macro-Kilars do offer very small apertures, according to the focal length; the same small linear aperture does become a smaller relative aperture (f-number) with longer focal length. Small disk stops would be satisfactory if they could be inserted into the correct plane in the lens but this is rarely possible. **A small stop in the wrong plane,** such as against the outside of the lens, **can be disastrous to image quality**.

Depth of Field with Secondary Enlargement.   The preceding discussion has involved the primary image from a camera lens only, as it might be recorded in one image plane by a photographic negative. In practice this may be a negative on 35 mm film used only to make an enlarged print.

When an image is recorded on a negative, the antipoints of one plane of the object are in focus (neglecting field curvature). There is a collection of image disks in image space of points throughout object space that have a distribution of sizes, from those of the in-focus antipoints, whose diameter is set by such limiting factors as resolution, to those of the nearest and farthest objects in the space before the lens, Most should be below the visual limit that recognizes them as other than image points. Enlargement of the negative now magnifies all of them, proportionally to the enlargement magnification m_2, where

$$M = m_1 \cdot m_2.$$

Therefore depth decreases as $1/m_2$. Hopefully most of the disks will still be below the limit of visual acuity.

There should be some gain in depth of field on enlargement to a specified magnification over that obtained by direct imaging to that magnification, but not by secondary *optical* magnification in which the primary pattern is not fixed by photography. The relations, including the advantage of utilizing an enlargement step, can be ascertained by inspection of Tables A.1-1 and 2, which show the depth of field, in millimeters, for various magnifications *if the values of c' taken from Figure A.1-4 are assumed applicable for the direct image.*

The depth of field for any final magnification M can be obtained by merely dividing the depth of field for the magnification of the negative m_1 by the magnification m_2, gained by the enlargement. *This assumes the same limiting value of c' originally assumed for the primary magnification.* **The depth of field for the primary magnification is given in the first column where $m_2 = 1$; this is equivalent to contact printing.** It is worthwhile to compare the depths of field for some double magnification sets $m_1 \times m_2$ with that for the same magnification obtained directly and listed in the first column. The biggest gains are naturally where the largest component magnification is by enlargement and where the primary magnification is fractional. For instance, if we should produce a final magnification of 20X by a primary magnification of 0.2X and a subsequent enlargement of 10X, an enormous gain in depth of field would be obtained. Unless the photographic materials are chosen for the negative that will keep graininess below objectionable size with 10X enlargement and unless the enlarger is excellent, the final quality may be poor. From Table 5-1 it can be seen that an aperture of $f/8$ is adequate for 20X magnification; with some specimens $f/11$ or even $f/16$ might be enough. These are the apertures that should be used for making the negative at the fractional magnification to be printed at 20X. If the aperture is $f/8$ and $c' = 0.24$, the depth

Table A.1-1

m_1	m_2 1	2	3	4	5	6	7	8	10	$\dfrac{m+1}{m^2}$
0.1	212		70.6							110
0.2	64.8		21.6		13.0	10.8			6.48	30
0.3	33.6		11.2	8.4						14.4
0.4	21.2		7.06		4.25				2.12	8.75
0.5	15.12	7.56	5.04	3.8		2.52		1.89	1.51	6.0
0.6	11.6		3.87							4.45
0.7	9.30		3.10							3.46
0.8	7.72		2.58		1.55					2.81
1.0	5.70	2.85	1.90	1.42	1.14	0.92	0.81	0.71	0.57	2.00
2.0	2.40	1.20	0.80	0.60	0.48	0.40	0.34	0.30	0.24	0.75
3.0	1.52		0.51	0.38	0.30					0.445
4.0	1.12		0.37		0.22					0.312
5.0	0.91			0.23						0.240
6.0	0.750	0.375	0.25							0.194
7.0	0.645									0.163
8.0	0.575									0.141
10.0	0.465									0.110
12.0	0.395									0.090
15.0	0.322									0.071
20.0	0.253									0.0525
25.0	0.214									0.0416
30.0	0.186									0.0344

Depth of field: $\Delta = 2\alpha c'\,(m+1)/m^2$ in millimeters.
$\alpha = 4.5 = f$-number.
c' is taken from the graph in Figure A.1-4.

of field at 0.2X is 115 mm. This is still 11.5 mm on enlargement to 20X. Direct imaging to 20X, *also with $c' = 0.24$ mm*, gives a depth of field of only 0.20 mm. To obtain this great gain in field depth the negative material must be able to image the antipoints so that they are acceptable after 10X enlargement, say, originally at 8 μ or 125 lines per millimeter. This requires a film of very high resolving power, especially if the subject contrast is not very high. Kodak Panatomic-X Film (135) may be satisfactory if well handled and for adequate subject contrast (see Chapter 16, Vol. 2). Color films might not quite do it for usual subject contrasts, but the picture might still seem quite good.

Now the phenomenon previously discussed becomes important. The actual observed depth of focus at 20X direct imaging is about 0.56 mm instead of 0.2 mm because of the greater subjective tolerance for coarse detail; therefore c' has not remained constant.

Table A.1-2

m_1	1	2	3	4	5	6	7	8	10
			m_2						
0.1	1031		344						
0.2	317		106		63.4				31.7
0.3	164		54.6						
0.4	104		34.6						10.4
0.5	74	37.0	24.6	18.5		12.3		9.1	7.4
0.6	56.6		18.9						
0.7	45.6		15.2						
0.8	37.5		12.5		7.5				
1.0	27.8	13.9	9.3	7.0	5.56	4.6	4.0	3.5	2.8
2.0	11.8		3.9			1.97			
3.0	7.45		2.48	1.86					
4.0	5.48		1.83		1.09				
5.0	4.44		1.48	1.11					
6.0	3.84	1.92	1.28						
7.0	3.44		1.15						
8.0	3.16		1.05						
10.0	2.80	1.40	0.93						
12.0	2.53								
15.0	2.31								
20.0	2.08								

Depth of field in millimeters.
$\alpha = 22 = f$-number.
c' is taken from the graph in Figure A.1-4.

It is very instructive to compare the values for depth of field obtained at various enlargements for the same final M value with that of the primary image in the first column. When the final magnification is low, for example, below 4X, the gain on enlargement is obvious. However, take some value for M, such as 12X, for which there are a number of simple primary and secondary magnification factors. For the combinations of 3X by 4X, 4X by 3X, or 6X by 2X, the depth of field by direct imaging seems actually greater than with an enlarger step included! This, of course, is because the values on a horizontal line are based on the same value of c' as that for the primary magnification of that line; but the increase of c' with increasing primary magnification more than offsets the normal gain by enlargement.

The author has done no quantitative work on enlargement of subjects capable of accurate measurement. From crude observation it would seem that an observer would accept the same value of c' for the same M, however arrived at, if some antipoints were below the limit of visual acuity to give some excellent definition of the important areas. Thus there still may be some gain in an enlargement step at these higher magnifications, since fewer of the circles of confusion will have grown as large by enlargement as direct imaging, but the gain is probably very small at appreciable magnification.

Summary Procedure for Photomacrography

Knowledge of the depth of field is principally desirable in planning a photomacrograph for a specific specimen. For this purpose Figure A.1-5 may be of some help in saving time. The values of c' assumed here may not be suitable for the observer and specimen. Some other value of c' can be used; if the values of the graph in Figure A.1-4 seem too generous, an estimate or a test may be made. The lowest value of c' likely to be applicable is 0.15 mm. To anyone handy with a slide rule a new value of Δ can be quickly calculated for Formula A.1-1, even the whole set represented by Column 1 of the tables, by multiplying the last column, $(m+1)/m^2$ by $2\alpha c'$.

If the negative is to be enlarged, divide the depth of field obtained for the direct magnification from Formula A.1-1, by the enlargement magnification.

B

DEPTH OF FIELD WITH THE COMPOUND MICROSCOPE

When considering the compound microscope, almost everyone has approached the problem the same way, that is, the limit of resolution is considered to be the limitation of the depth of field. This also involves conversion to the nomenclature for the aperture of the compound microscope. The limit of resolution is

$$z = \frac{\lambda}{2\,\mathrm{NA}}, \qquad (2\text{-}4)$$

where z is the distance between two just-resolved points. It should be pointed out that this assumption involves a constant circle of confusion in the object plane, $c = 2z$, whereas in photomacrography $c' = mc$ is the considered parameter. The constant value of c will give a calculated constant depth of field at different projection distances with the varying magnification, even though the depth of the total image will vary according to the longitudinal magnification. **Visual accomodation will affect the**

depth of field for visual work compared with photomicrographs. In a most interesting paper Brattgard (4. 1954) used atropine in the microscopist's eyes to confine the depth of focus to that at a single accomodation and claimed that it allowed him to verify and then use the depth-of-field equations. The author has done little experimental work in the depth of field with the compound microscope; but one would expect, by analogy with experience in photomacrography, that there would be somewhat greater depth of field than that calculated from the strictest criterion of the limit of resolution. Several people, however, including Beadle (4. 1940), state that the depth of field, as observed, is considerably less than than evaluated from the formulas. The fact that one often is not familiar with the gross appearance of the subject in microscopy at higher magnifications would keep the mind from being satisfied by good gross appearance, as in photomacrography. Also, the factors dropped in the derivation of the formulas may no longer be insignificant (see note in Part A). If

$$\alpha = \frac{nm}{2\,\mathrm{NA}\,(m+1)}, \tag{5-1}$$

it can easily be shown that Formula A.I.1 becomes

$$\Delta = \frac{\lambda\,\sqrt{n^2 - (\mathrm{NA})^2}}{(\mathrm{NA})^2} \tag{A.1-2}$$

A different criterion was selected by A. E. Conrady and used by L. C. Martin (2. 1930, p. 82) for the limitation to depth of focus, namely, where the difference in the optical path becomes a quarter of a wavelength. Then

$$\Delta = \frac{\lambda}{4n\,\mathrm{sine}^2\,(\tfrac{1}{2}\,\mathrm{sine}^{-1}(\mathrm{NA}/n))}. \tag{A.1-3}$$

Finally we can assume that the limitation is that of the minimum *angle* of visual resolution. This can lead to Berek's modification of Abbe's formula which was used by Brattgard:

$$\Delta = \frac{n\lambda}{2(\mathrm{NA})^2} + \frac{n \cdot 250{,}000\,w}{\mathrm{NA}\cdot m}, \tag{A.1-4}$$

where w is the tangent of this limiting angle. The number 250,000 is the standard projection distance, 250 mm, taken in microns; $w = 0.0003 =$ tangent 1 minute of arc.

The easiest way to compare these formulas is numerically. Two examples may be used.

	Objective	NA	Ocular	m	n
Case A	25X	0.63	12.5X	312.5X	1.0
Case B	100X	1.32	12.5X	1250.0X	1.54
$\lambda = 0.5\,\mu$					

Case A

$$\Delta = \frac{1-(0.63)^2}{2(0.63)^2} = 0.98\,\mu, \tag{A.1-2}$$

$$\Delta = \frac{0.5}{4\,\text{sine}^2(19°31\tfrac{1}{2}')} = 1.12\,\mu, \tag{A.1-3}$$

$$\Delta = \frac{1}{4\times0.397} + \frac{250,000\times0.0003}{0.63\times312.5} = 1.01\,\mu. \tag{A.1-4}$$

Case B

$$\Delta = \frac{\sqrt{(1.54)^2-(1.32)^2}}{2(1.54)^2} = 0.148\,\mu, \tag{A.1-2}$$

$$\Delta = \frac{0.5}{4(1.54)\,\sin^2(29°30')} = 0.335\,\mu, \tag{A.1-3}$$

$$\Delta = \frac{(1.54)(0.5)}{2(1.742)} + \frac{(1.54)(75)}{1.32(1250)} = 0.291\,\mu. \tag{A.1-4}$$

Appendix 2

The Determination of Equivalent Focal Length and Location of Gauss Points of a Lens

The accuracy required for the value of focal length can vary extensively. For a quick and relatively simple determination Method 1 is recommended. Although it is fundamentally the method used on a lens bench, without its nodal slide to determine the nodes, it is not so accurate as the other two methods. If the equivalent focal length, together with the principal planes from which it should be measured, is desired, in order to calculate magnification when the image plane is unavailable (as in roll film cameras), the focal length must be determined quite accurately; Method 3 is then recommended. Method 1 is usually sufficiently accurate for choosing an illumination condenser.

Method 1. Set up the lens to image a distant scene on a white card, ground glass, or clear glass which bears a mark for fixing focus. The mark will then be observed with a focusing magnifier and the coincidence of the image of the object (a scale or cross) and that of the camera back determined by parallax. Measure the distance from the middle of a lens to the image plane, unless the lens seems quite unsymmetrical. In a plano-convex lens, for instance, the principal planes are shifted toward the convex side, possibly one-third of the way. The lens-to-image distance is the focal length f. If it can be measured from the known back principal plane, it is the focal length.

Method 2. Set up a transparent test object, preferably a scale with a diffusing surface directly behind it; in fact, the scale can be directly on the matte surface. A scale on white reflecting material may be used with sufficient illumination. Image the scale at 1X magnification. Since the exact magnification is determined by identity of object and image size, a scale is not necessary. However, the test object, lens, and receiving surface for the image must be set up in a good line. At 1X a symmetrical lens is exactly halfway between the object and image planes. If the lens has appreciable thickness, there are two points on the optical axis (Gauss points), separated by a distance Δ from which all optical distances are measured. See Chapter 1, *focal scale*. This relation, from the focal scale, is

$$\text{object to image plane} = L = l + l' + \Delta = \left(m + \frac{1}{m} + 2\right)f + \Delta$$

$$= \left[\frac{(m+1)^2}{m}\right]f + \Delta. \tag{A.2-1}$$

At 1X, $L = 4f + \Delta$, where l and l' are the object and image distances, respectively, $f =$ focal length, and $m =$ magnification. As in Method 1, the extent of the inter-Gaussian distance Δ cannot be determined from the one setup. Often this distance is small, even negligible, and the lens reasonably symmetrical. In this case the middle of the total distance can be taken as the point of origin for measurements of l and l', thus assuming that $\Delta = 0$.

If the equivalent focal length is otherwise known, as in the case of a marked lens, the positions of the principal planes on the lens barrel and the distance Δ can be determined by measurement of $2f$ from the image plane. However, depth of focus is greatest at 1X and focal lengths are not always exactly as listed or marked. Use of Method 3 is much more reliable. A method that combines Methods 1 and 2, which probably is not quite so accurate as Method 3 but does not determine a magnification, hence requires no ruled scales, is described by D. H. Cronquist (2. 1963).

Method 3. With this method the equivalent focal length of a lens f and the location of its Gauss points (i.e., the axial intercepts of the principal planes) can be determined as accurately if not as conveniently as with a special optical bench with a nodal slide. Care and accuracy of setup and measurement are required, however. If the focal length f is to be used later to determine a magnification with any accuracy, *the positions of the support* for the photographic lens *should be determined to a fraction of a millimeter*. This is because the final values are obtained by subtraction of larger numbers. Moreover, the focal length is usually used, as in determination of magnification, by measuring distances that are large multiples of it. Certain accessories are required for reasonable facility of determination.

An optical bench and scale along it, preferably metric, should be used to *allow axial movement of the lens, stage, and receiving plane without misalignment*. However, this bench can be assembled for this job. An assembly from commercial components, like the two-rod bench supplied by the Gaertner Co. of Chicago, Ill., is most convenient. A meter stick with provisional pointers extending from the carriages will do for the accurate measurement of lens movement. A simple but perfectly practical optical bench consists of a flat board such as a plank 6 in. wide or a bench top with another board such as a 2×4 running along its edge and fastened down. The carriages bearing the bench components are simple rectangular blocks of wood that are pressed against this but can be slid along it. The inner bottom edge of each block should be beveled. The lens should be held at the top of a rod bearing a V, discussed in Chapter 3. A pair of *identical* scales is required for the object and image planes. The author used transparent scales made from an engraved rule, as supplied by Gaertner and Zeiss, by printing them onto Kodak 649 Spectroscopic Plates. Photographic rules can be purchased; they are used for measurement of magnification. A flat board or metal sheet can be erected on a block to hold the rules, which can be taped in place. Time is saved if one of them is held on the vertical stage with a mechanical stage so that it can be aligned. Otherwise, the rules should rest on a horizontal ledge on the vertical stage to allow lateral adjustment of the scales. Both the ledges and the scales can be taped to the stage.

The object scale is illuminated from behind by using the double condenser method. It will resemble Case 7 of the setup for low-power photomicrography, except that there is a diffusing film behind the scale on the object stage. Hence the filament is now imaged in the second condenser of the lamp and the first condenser is imaged on the scale by the second. The matte film should be removed while the assembly is being aligned. The test lens should be rotated while looking along the axis, first vertically then horizontally until the reflections of the filament lie in one line. This alignment is very important. The axial points of the object and image planes must be aligned also. This is most easily done with a vertical pointer that can be placed on the optical bench.

A magnifier is needed to view the image from behind to ensure that it is superimposed on the second scale. This magnifier must be supported by the same block or carriage as the image stage and its scale so that the focus of the magnifier will remain unchanged as the image stage is moved. The most efficient and convenient magnifier is a compound microscope tube bearing a 2 or 4X objective. It also must be aligned well, but this is an advantage for accuracy.

The whole concept of equivalent focus, etc., is a paraxial one. Therefore only the center of the field must be used. As J. S. Anderson (2. 1931) has pointed out, the chief source of error is due to curvature of field. The curved field has the optical axis of the lens as its important diameter. Two points widely separated are measured as on a chord of the arc and as a shorter distance from the lens than the diameter. This leads to an error of too short a computed focal length.

Procedure. The complete determination consists of two parts. The position of the object stage is unchanged throughout the procedure in which only the lens and image stage are moved.

1. Set the distance to obtain a magnification of $\frac{1}{5}$X or a similar simple fraction appreciably smaller than $\frac{1}{2}$. The focusing of the image of the first scale into the plane of the second should be done by the method of parallax through the observing microscope, wagging the eye to and fro to ensure that there is no motion of the image of the original with respect to the second scale. Note that the apparent motion of the image of the first scale reverses *with respect to the direction of movement of the eye* as it passes through the coincidence plane of the second scale. The magnification m should be noted. Assume that m should be $\frac{1}{5}$X. The image is focused by moving the image plane, resetting the lens forward a bit if the magnification is too small and conversely. The position s_1 of the lens, or some fiducial mark on its carrier, is now read very accurately.

With the object and image planes well fastened in place move the lens forward to its other conjugate point. In this case the focus should be best with $m_2 = 5$X. Again read the position s_2 of the lens, or the same fiducial mark on the lens support, very accurately. Actually the position of the fiducial mark can be anywhere on the block, since $s_1 - s_2 = l - l'$, where l and l' are object and image distances and the scale zero is eliminated.

Equivalent focus $= f = (l - l')/(m_1 - m_2)$. If $m_1 = \frac{1}{5}$X, $m_1 - m_2 = 4.8$.

Remove the lens in its support and measure the distance L between object and image planes, preferably with a good steel tape. This whole determination at $\frac{1}{5}$ and 5X should be repeated at least twice.

2. Then set up as before but at $\frac{1}{2}$ and 2X, leaving the setup at 2X for location of the principal planes. The rear principal plane is marked on the lens barrel (or its position noted in a pattern) by measuring a distance of three focal lengths (i.e., $m_2 + 1$) from the image plane. The front principal plane is measured as $1\frac{1}{2}$ focal lengths from the object plane.

The following data sheet was made in measuring a camera lens:

M	Bench Position of Lens, s	$l-l'$	f	L	
$\frac{1}{5}$X	267.8				$f = \dfrac{l_1-l_2}{m_1-m_2}$
5X	45.5	222.3	46.3	332.4	
$\frac{1}{5}$X	266.6				
5X	45.6	221.0	46.04	332.0	$\Delta = L - f(m + 1/m + 2)$
$\frac{1}{5}$X	266.3				$\Delta = L - 7.2f = 0.06$ mm
5X	45.5	220.8	46.00	332.0	
$\frac{1}{5}$X	266.8				
5X	45.5	221.3	46.10	331.5	
$\frac{1}{2}$X	129.0				
2X	59.8	69.2	46.13	283.0	

Gauss point at $3f = 138.33$ mm from object plane.

Note. Although the lens appeared appreciably long, Δ was so small it could be neglected, although the principal plane was not quite central.

Appendix 3

Low-Power Photomicrography by Transmitted Illumination

A

EQUATIONS FOR CASE 2a OF CHAPTER 5*

The nomenclature is that consistent with the designations in Figure A.3-1 and in Chapter 5. It is assumed that the condenser lies in the plane of the field F. This constitutes an error, sometimes not quite negligible because of the thickness of the condenser. Moreover, the image of the light source is behind the objective diaphragm, being preferably in the exit pupil.

Lens I = condenser of focal length f_1,
Lens II = camera objective of focal length f_2,
$\quad \alpha_0$ = f-number of camera lens directly illuminated,
$\quad Q_2$ = diameter of image of light source in lens II,
$\quad s$ = diameter of light source,
$\quad m$ = magnification of camera image,
$\quad m_s$ = magnification of light source in objective,
$\quad l_1$ = distance between the light source and condenser,
$\quad l_2$ = distance of condenser, lens I, from objective, lens II.

The development of the useful relations consists in taking the same distance, such as l_1, and considering it first as the object distance of lens I for the specimen and then as the image distance of lens I for the light source. The two expressions, being for an identity, can be equated.

$$l_2 = f_2 \frac{(m+1)}{m} = f_1(m_s + 1) \qquad \text{(from focal rule)},$$

but $\qquad m_s = Q_2/s \text{ and } \alpha_0 = f_2/Q_2;$

*(Most satisfactory with objectives of fairly long focal length).

510

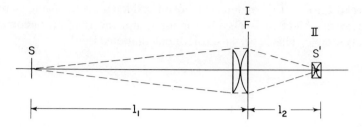

Figure A.3-1 Optics of illumination system of Case 2a with simple microscope.

therefore

$$m_s = \frac{f_2}{\alpha_0 s};$$

$$f_1 = \frac{\alpha_0 \cdot s \cdot f_2(m+1)}{(f_2 + \alpha_0 \cdot s)m}, \tag{A.3-1}$$

$$f_2 = \frac{m\alpha_0 s f_1}{\alpha_0 s(m+1) - mf_1}, \tag{A.3-2}$$

$$\alpha_0 s = \frac{mf_1 f_2}{f_2(m+1) - mf_1}. \tag{A.3-3}$$

obviously $\alpha_0 s$ will be negative if mf_2 is larger than $(m+1)f_2$. Therefore lens II cannot be kept directly behind the field, if f_1 is larger than $[f_2(m+1)]/m$.

Example 1. Let $m = 5\times$, $f_2 = 72$ mm, $\alpha_0 = 25$ (see Table 5-1). Then $[f_2(m+1)]/m = 86.4$ mm, which is the maximum focal length for f_1 if it is to be directly behind the field. Assume that a lens $f_1 = 60$ mm is available, either as a single lens or preferably a composite of simpler elements, and of sufficient diameter:

$$s = \frac{1}{25} \cdot \frac{5 \cdot 72 \cdot 60}{(6 \cdot 72) - (5 \cdot 60)} = 6.5 \text{ mm} \qquad \text{(diameter of light source)}.$$

If the photomicrograph at $5\times$ is to be enlarged or projected for viewing, the aperture of the objective should, of course, be larger than $f/25$. Moreover, the objective aperture should be wider open than that, if all measurements are as given in Example 1, because the image of the source will undoubtedly be larger than the linear dimension of Q_2 at $f/25$, which is about 3 mm, and the image plane will be unevenly lighted unless all of the light from the relatively poor image of the light source is admitted.

Special Case. If a condenser whose focal length is the same as that of the camera objective is chosen, or made up, the relation becomes especially simple. This is very useful for calculations. Let

$$f_1 = f_2.$$

Then,

$$mf = \alpha_0 \cdot s \qquad\qquad (A.3\text{-}4)$$

Here f can represent the focal length of either lens and

$$s = \frac{mf_1}{\alpha_0} = \frac{mf_2}{\alpha_0}.$$

Also

$$l_1 = ml_2.$$

Example 2.

$$m = 5\times, \qquad f_1 = f_2 = 72 \text{ mm}.$$

Then

$$s = 14.4 \text{ mm}.$$
$$l_1 = 5l_2,$$

where

$$l_2 = (6 \cdot 72)/5 = 86.4 \text{ mm}.$$

Example 3.

$$m = 3\times, \quad f_1 = f_2 = 100 \text{ mm}, \quad \alpha_0 = 36.$$
$$s = 8.3 \text{ mm}, \quad l_2 = (4 \cdot 100)/3 = 133.3.$$

Example 3a. If it is assumed that the photomicrographs will be enlarged about twice, thus requiring $f/16$, then $s = 19$ mm diameter of light source.

Example 3b. If $\alpha_0 = 36$, but a 50mm condenser is available that will cover the field, then by (A.3-2)

$$s = 1.4 \text{ mm}.$$

A ribbon-filament lamp would be sufficiently big.

Case 2b does not lend itself to helpful formulation. However, by making some quick calculations for Case 2a and endeavoring to meet them it should not be necessary to back the condenser far from the field, with consequent gain in the size of the uniformly illuminated field.

B

EQUATIONS FOR CASE 7 OF CHAPTER 5*

Method of Derivation

The equations are simply derived, by thin lens formulas, by considering the two imaging systems involved, that with the filament and that with the lamp condenser as object. The same length (distance) can be expressed by both systems and then equated. The considerable but valid algebriac juggling required is not reproduced here.

Nomenclature (see Figure A.3.B-1)

α_L = aperture of lamp condenser in terms of f/α, $\alpha_L = 1/A_L$,

A_L = aperture of lamp condenser (*proportional to* used *diameter*),

α_0 = relative aperture of camera objective actually filled by illumination beam, that is, the f- number used; $\alpha_0 = 1/A_0$,

A_0 = relative aperture of camera objective (*proportional to* used *diameter* and to NA),

Q_1 = diameter of lamp condenser, actually used (lens I),

Q_3 = diameter of illuminated field = Q_1',

$Q_3' = mQ_3$ = diameter of illuminated image field in camera,

f_1, f_2 = focal lengths of lenses I and II,

s = diameter (or width) of light source,

m = magnification of camera.

$$A_L = \frac{1}{\alpha_L} = \frac{Q_1}{f_1} = Q_3\left(\frac{m}{\alpha_0 \cdot s \cdot (m+1)} + \frac{1}{f_2}\right) \qquad \text{(B.3-1)}$$

*Recommended method at 5 × and above. See p. 232

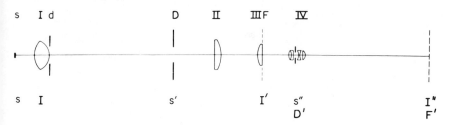

Figure A.3.B-1 Optics of Case 7 with simple microscope.

Comments. The focal length of lens II has a slight effect on the field or aperture size, but usually in low-power work $1/f_2$ is a small quantity compared with the others. The ratio $m/(m+1)$ can be considered a constant (almost equal to 1.0) for a given set-up. With a limited number of factors, only so many are available for choice. If these two factors are negligible, then

$$A_L \cdot s = Q_3 \cdot A_0 = \frac{Q_3}{\alpha_0} \qquad \text{(B. 3-2)}$$

If the objective aperture filled is set by resolution requirements or otherwise, **the size of the field Q_3 is proportional to the product of the size of the light source and the aperture of the lamp condenser.** The optical system can be refocused to make almost any field *or* aperture size available but at the direct expense of the other factor. **To increase one *without* sacrificing the other requires a larger light source *or* lamp condenser.**

The following equations are sometimes helpful in setting up the system. They are intermediate equations (in setting up B.3-1.)

$\Delta =$ distance of diaphragm D from lens I,
$L =$ distance of field F from lens I,

$m_s =$ magnification of the filament on the diaphragm D whose diameter $= s'$.

$$\Delta = \frac{Q_1 \cdot f_2}{Q_3} = f_1 \left[\frac{m(f_2 + 1)}{(m+1) \cdot s \cdot \alpha_0} \right], \qquad \text{(B. 3-3)}$$

$$L = f_2 \left[\frac{(Q_1 + Q_3)^2}{Q_1 \cdot Q_3} \right], \qquad \text{(B. 3-4)}$$

$$Q_3 = \frac{f_2}{\alpha_L \cdot (m_s + 1)}. \qquad \text{(B. 3-5)}$$

The values of the formula are limiting values; for instance, with a given arrangement of optics, the aperture of the illumination α_0 can be reduced by closing the diaphragm at D (or the objective iris) without affecting the size of the illuminated field at all. When α_0 is to be reduced, however, the elements *could* be refocused to give a smaller filament size, hence smaller aperture, but a larger field Q_3. This can be a great advantage.

Figure I-28 Eggs of Fasciola hepatica (Sheep Liver Fluke) X100. Taken from whole mount of trematode, stained with hematein.

Ultropak with 250-watt Siemen's Solid Source with +20 mired photometric filter to 3200°K. Kodak Color Compensating Filters after removal of ultraviolet with Kodak Wratten Filter No. 2B. Film: Kodak Ektachrome, Type B. Exposure time: 5 seconds.

Figure I-30 Calf Hide (Cross Section) X50.

Stain: Mallory's. Optics: Simple microscope, arrangement as for Case 7, Chapter 5. Objective: Cine Ektar, 25mm, $f/1.4$ (reversed in mount). Illumination: Ribbon filament + Kodak Wratten Filter, No. 78C. Filters: eliminator (Kodak Wratten Filter No. 2A) + Kodak Color Compensating Filters + neodymium acetate liquid filter. Film: Kodachrome, Type B.

Figure IV-7 South American Indian Relic X15. (Gold Mechanically Alloyed with Platinum)

Illumination: (see Index) G–E Xenon Flashtube FT-429 at 200 W-sec; two xenon flash tubes with foil reflectors, 125 and 300 watt-seconds, opposite each other and at very low vertical angle. Illumination Quality: (see Index) Kodak Wratten Filter, No. 2B plus dilute compensating filters. Optics: Enlarging Ektar Lens, 2-inch, reversed in mount, at $f/11$. Film: Kodak Ektachrome Film, Daylight Type (sheet).

From American Museum of Natural History and Courtesy of H. E. Searle.

Figure IV-10 Larva of Tropaea luna (Luna Moth) X1.0. (4th Instar)

(a). Retina Reflex III. Focused at infinity with R 1:2 supplementary lens at $f/22$. Original magnification $\frac{1}{2}$X. M-2 flash lamp (bare) at 6 inches, 45°. Kodacolor-X. Picture taken by H. L. Gibson.

(b). Miranda Camera. Soligor 50 mm lens focused directly at $\frac{1}{2}$X magnification at $f/16$. 1 Bare AG-1 flash lamp at 45°, 15 inches at left. 1 Diffused AG-1 flash lamp at 20°, 24 inches at right, photoelectric slave unit. Kodacolour-X. Guide Number 31, $\frac{1}{30}$ sec.

Figure IV-19 Radioactive Particle X3000

Illumination: (a). Transmitted; (b). Reflected; (c). Transmitted + reflected. Illumination oblique at high aperture, conical; carbon arc, dc 10-A automatic for vertical illuminator; AO Universal Lamp + inconel densities for transmitted light. For details, see Index. Color filters: none. Objective: B & L metallographic apochromat, 3 mm, 1.40 NA. Film: Kodacolor C135 (35 mm), enlarged 2X in printing.

Figure X-2 Sambucus (Elder) Stem (Cross Section) X150

Stains: Safranin and fast green FCF
Filters: Red, Kodak Wratten No. 25. Green, Kodak Wratten No. 9 + No. 65
Optics: Objective, Zeiss Planapochromat 10X 0.32 NA. Ocular, Komplans 8X
Material: Kodak Separation Negative Plate, Type 1 DK-50 (1:1) $3\frac{1}{2}$ min. at 68°F.

Figure X-3 Human (White) Scalp (Horizontal Section) X75

Stains: Hematoxylin and Eosin
Optics: Apochromat 16 mm 0.30 NA. Homal II
Filters: Red, Kodak Wratten No. 25. Blue-green, Kodak Wratten No. 45
Material: Kodak Panchromatic Plate

Figure XIX-1 Bronze Silver Soldered on Steel X100. Ammonia-Peroxide Etch

Illuminant: Carbon arc, 4½ amperes.
Coated vertical illuminator (B. & L.)
Objective: 16 mm (10X) Apochromat. NA 0.30 (0.9 filled)
Ocular: Ampliplan, special compensating
Filters: Kodak Wratten No. 2A + blue and magenta liquid filters. Color balance point determined by trichroic ratio method
Film: Type B Sheet

Figure XIX-2 Puccinia graminis X100

(Rust on Wheat Showing Spores)
Conant's Quadruple Stain

Illuminant:	Ribbon filament at 2926°K
Substage Condenser:	Zeiss achromatic 1.0 NA with top lens removed
Objective:	16 mm (10X) apochromat
Ocular:	Ampliplan for low power
Filters:	Kodak Wratten No. 78C (to bring lamp to 3200°K)
	Kodak Wratten No. 2A
	Blue and magenta liquid filters

Figure I-28. Eggs of *Fasciola hepatica* (Sheep Liver Fluke)   X100
Taken from whole mount of trematode, stained with hematein.

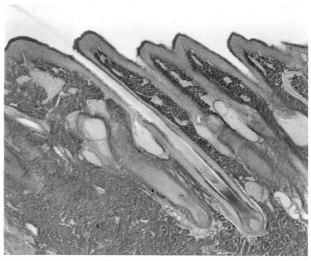

Figure I-30.   Calf Hide (Cross Section)   X50

Figure IV-7. South American Indian Relic   X15
(Gold Mechanically Alloyed with Platinum)
From American Museum of Natural History and Courtesy of H. E. Searle

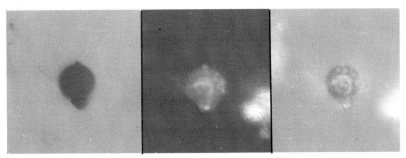

Figure IV-19. Radioactive Particle   X3000

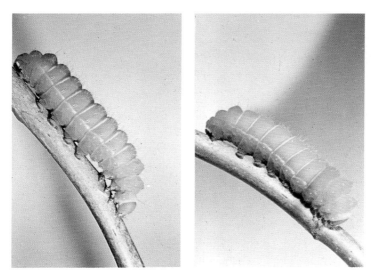

Figure IV-10.   Larva of *Tropaea luna* (Luna Moth) X 1.0

Figure XIX-1.   Bronze Silver Soldered on Steel   X100
(Ammonia-Peroxide Etch)

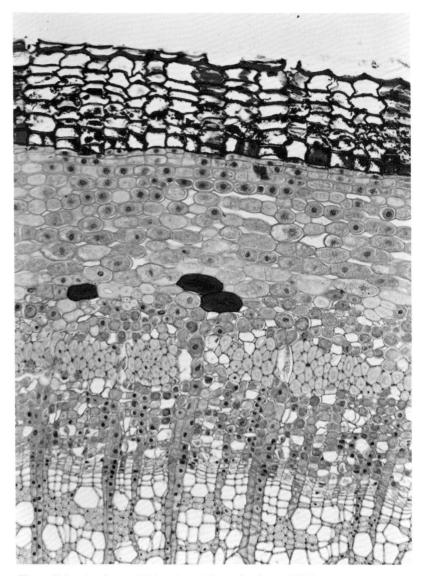

Figure X-2.   Sambucus (Elder) Stem (Cross Section)   X150

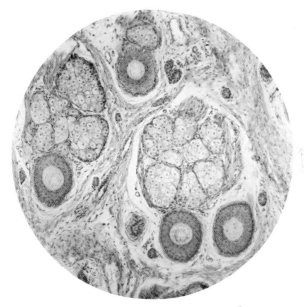

Figure X-3.   Human (White) Scalp (Horizontal Section)   X75

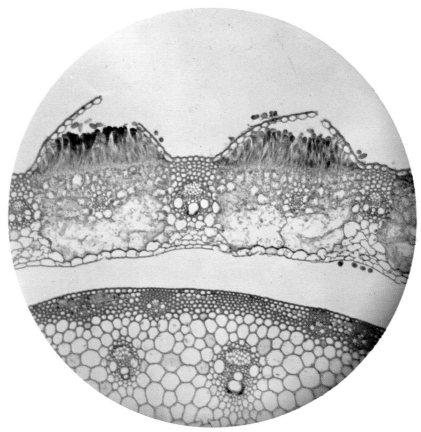

Figure XIX -2.   *Puccinia graminis*   X100 (Rust on Wheat Showing Spores)
Conant's Quadruple Stain

Example 1. Let m = 10X, $\alpha_0 = 16$ (close to maximum useful aperture; see Table 5-1): $f_1 = 60$ mm, $\alpha_L = 1.0$, $f_2 = 200$ mm, $s = 2.0$ mm (ribbon filament).

$$1 = Q_3 \left[\frac{10}{16 \cdot 2 \cdot 11} + \frac{1}{200} \right], \qquad \text{(from B. 3-1)}$$

$$Q_3 = 29.9 \text{ mm.}$$

Example 1a. If $f_2 = 100$ mm, $Q_3 = 26.2$ mm, $= Q'_3 = 26.2$ cm. All lengths could, of course, have been expressed in inches.

Example 2. Assume that a larger field is required in order to make an 8 × 10-in. photomicrograph with a little factor of safety for extreme edge defects of illumination: $m = 10X$, $\alpha_0 = ?$, $Q'_3 = 13$ in. (diameter $Q_3 = 1.3$ in.), $s = 2$ mm $= 0.0787$ in., $f_2 = 7.87$ in., lamp condenser $= f/1$,

$$1 = 1.3 \left(\frac{10}{\alpha_0 \cdot 0 \cdot 787 \cdot 11} + \frac{1}{7.87} \right), \qquad \text{(from B. 3-1)}$$

$$\alpha_0 = 18.$$

C

COMPOUND MICROSCOPE, KÖHLER ILLUMINATION

Optical relations (see chapter 8)

Nomenclature. For consistency with Cases A and B, the objective is still called lens IV; lenses II and III of Case 7 are combined as lens II.

Lens	E. Focus	Illuminated Diameter	Function
I	f_1	Q_1	Lamp condenser
II	f_2	Q_2	Substage condenser
		Q_3	Illuminated field
IV	f_4	Q_4	Objective
		Q'_3	Illuminated field in camera back

$\alpha_L = \dfrac{f_1}{Q_1} =$ aperture of lamp condenser in terms of f/α,

$A_L = \dfrac{Q_1}{f_1} =$ aperture of lamp condenser (relative to used diameter),

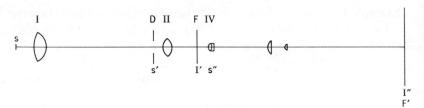

Figure A.3.C-1   Optics of Köhler illumination with compound microscope.

$D = s'$ = illuminated aperture, at substage diaphragm,
Δ = distance of substage diaphragm D from lens I.
$m_s = \dfrac{D}{s}$ = magnification, on diaphragm, of lamp filament.

(a) Derivation of equation, with the light source in rear focal plane of lens II:

$$\Delta = f_1(m_s + 1) \qquad \text{in terms of imaging } s \text{ at } D \text{ by lens I,} \qquad \text{(C.3-1)}$$

$$\Delta = \frac{Q_1 f_2}{Q_3} \qquad \text{in terms of imaging I at } F \text{ by lens II,} \qquad \text{(C.3-2)}$$

$$Q_3 = \left(\frac{Q_1}{f_1}\right)\frac{f_2}{m_s + 1} = \frac{f_2}{\alpha_L(m_s + 1)} = \frac{A_L f_2}{m_s + 1}, \qquad \text{(C.3-3)}$$

$$Q_1 = \frac{Q_3(m_s + 1)f_1}{f_2}. \qquad \text{(C.3-4)}$$

(b) If the light source is imaged *in* the substage condenser, rather than in its rear focal plane, which is not Köhler illumination but sometimes tends to be done, the illuminated field becomes larger.
Equation C.3-1 becomes

$$\Delta = f_1(m_s + 1) - f_2. \qquad \text{(C.3-2')}$$

Then

$$Q_3 = \frac{Q_1 \cdot f_2}{f_1(m_s + 1) - f_2}, \qquad \text{(C.3-3')}$$

With low magnifications, the last arrangement is the double condenser method discussed in Chapters 4 and 5. With a compound microscope and a substage condenser whose focal length is short with respect to that of the lamp condenser, *when multiplied by the magnification of the light source*, the difference between the two sub methods is not casually distinguishable. Moreover, with high aperture substage condensers, the focal plane is most often inside the condenser. Therefore it is not surprising that the equations of section (a) hold even when the source seems to be imaged within the usual substage condensers.

Aperture Relations

The aperture relations for Köhler illumination are expressed by the famous sine theorem (see Index) if the lamp condenser is considered the object, the substage condenser is the imaging lens, and the *illuminated field* of the microscope is the image. This has already been briefly discussed in Chapter 8. The nomenclature is consistent.

$$s(NA)_L = Q_3(NA)_o; \tag{C.3-5}$$

$(NA)_L$ and $(NA)_o$ are the apertures of the lamp condenser and the objective, respectively, on the image side of the lamp and object side of the objective, since these constitute the same NA's as the two sides of the substage condenser.

It is inconvenient to consider the lamp condenser aperture in terms of numerical aperture, and this is the way in which that of the objective is given. To convert the expression of the aperture of the lamp condenser to the more familiar ratio of diameter to focal length we can utilize Formula 5-2, sometimes called the formula for the effective aperture. In this case the magnification involved is that of the lamp filament m_s on the diaphragm of the substage condenser.

$$(NA)_L = \frac{m_s}{2(m+1)\alpha_L} = \frac{m_s A_L}{2(m_s+1)} = \frac{D \cdot s \cdot A_L}{2(D+s)} \tag{5-2}$$

Then (C.3-5) becomes

$$\frac{m_s \cdot s}{2(m_s+1)\alpha_L} = \frac{D \cdot s}{2(D+s)\alpha_L} = Q_3(NA)_o \tag{C.3-5'}$$

These two equations [(C.3-3) and (C.3-5) or (C.3-5')] are most valuable for making evident **the inexorable relations between the size of the lamp and the size of the object field** (or image field at a given magnification). **The lamp size includes both light source and lamp condenser size in equivalent relation.** Unlike the equations for the simple microscope, they probably will not be much used to help speed the setting up of a system by actual calculation of values. These calculations, however, may prove useful here as an explicit and quantitative demonstration of practical principles.

Example 1.

$m = 100\times$ magnification of image,

$(NA)_o = 0.25$ aperture of objective (about 0.9 the NA 0.30 of $10\times$ objectives),

$f_1 = 34$ mm equivalent focal length of lamp condenser,

$f_2 = 37$ mm equivalent focal length of substage condenser (top element off),

$s = 2.0$-mm diameter of light source (width of ribbon filament),

$D = 20$ mm-size of filament image on substage condenser.

This size just fills the 0.25 NA of the objective with this substage condenser of e.f. $= 37$ mm. It must be assumed that the distance of the source gives exactly this size of filament image with the source size and condenser focal length. Therefore $m_s = 10\times$ magnification of filament.

The largest *useful* field may be assumed to be limited to 100 mm (4 in.) in the image plane due to curvature of the field of 1.0 mm in the object plane. This, however, may not be the value of the field size Q_3, since the field size is defined as the image of the lamp condenser Q_1 in the object plane.

(a) Let us assume no field diaphragm to control the effective size of the lamp condenser Q_1

$Q_1 = 30$ mm actual diameter of lamp condenser as a diaphragm,

$$Q_3 = \frac{30}{34} \cdot \frac{37}{20+1} = 1.55 \text{ mm} - \text{diameter of field from (C.3-3).}$$

The extra diameter of 1.55 mm over the used diameter of 1.0 mm can only produce some flare. The field is 55% too large.

(b) Let us reduce the diameter of the lamp condenser to make the field size $Q_3 = 1.0$ the diameter of the useful field.

$$Q_1 = \frac{1.0 \times 21 \times 34}{37} = 19.3\text{-mm} \quad \text{diameter of lamp condenser [(from}$$

(C.3-4)].

Therefore $\alpha_L = 1.76 \left(= \dfrac{34}{193} \right)$ is the f-number of lamp condenser, which is quite efficient.

(c) Note that **with change of the focal length of the condenser**, *but* imaging the light source to maintain the prescribed and just filled aperture at the substage condenser iris, the relative aperture f/α remains the same when the lamp condenser is diaphragmed to make the field size the same. Let

$f_1 = 45$ mm, object condenser,

$Q_1 = 1.0$ mm, $Q_1 = \dfrac{1.0 \times 21 \times 45}{37} = 25.5$-mm diameter of lamp condenser,

$\alpha_L = 1.76.$

However, the distance from lamp to condenser is, of course, different in these two cases.

$\Delta = f_1(m_s + 1)$ distance, lamp condenser to substage diaphragm,
$\Delta = 374$ mm with 34-mm condenser,
$\Delta = 494$ mm with 45-mm condenser.

Example 2

$m = 1000\times$ magnification of image,
$(NA)_o = 1.20$ (about 0.9 of NA 1.3),
$f_1 = 34$-mm equivalent focal length of lamp condenser,
$f_2 = 10.5$ mm,
$Q_3 = 0.1$ mm ($\frac{1}{1000}$ of 4 in.) is still the largest useful field,
$s = 2.0$ mm — limiting size of ribbon filament.

Consider the Zeiss Aplanatic Condenser 1.4 NA to be used:

$f_2 = 10.5$ mm equivalent focal length,
$D = 22$-mm size of aperture of diaphragm at NA 1.20; hence the size of the filament image to just fill it,
$\therefore m_s = 11\times$ magnification of the filament image.

The size of the useful field in the image plane is still about 100-mm diameter (4 in.); hence the size of the usefully illuminated object field is 0.10 mm.

(a) Assume no diaphragm against the lamp condenser:

$$Q_3 = \frac{30}{34} \cdot \frac{10.5}{12} = 0.77 \text{ mm.}$$

The illuminated field is almost 8× too large, which cannot but produce considerable flare: $\alpha_L = f/1.1$.

(b) With $Q_3 = 0.10$ mm, that is, with the condenser diaphragmed down to illuminate only the useful field,

$$Q_1 = \frac{0.10 \times 12 \times 34}{10.5} = 4.1 \text{ mm diameter of lamp condenser,}$$

$\alpha_L = 8.3$ aperture of lamp condenser.

(c) Let us reduce the size of the source and use a 40-W zirconium arc. The size of the aperture at NA 1.20 remains the same:

$$s = 0.7 \text{ mm, } m_s = \frac{22}{0.7} = 31\times,$$

$$Q_1 = \frac{0.1 \times 32 \times 34}{10.5} = 10.4 \text{ mm}$$

$$\alpha_L = 2.9.$$

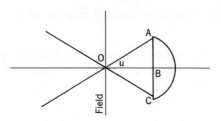

Figure A.3.C-2   Illumination aperture as an angle.

Note that the increase in the aperture from $f/8$ to $f/3$ shows why, at high magnifications with the small object fields to be illuminated, the small and lower wattage light sources are adequate for image brightness and photographic exposure and give better contrast because of less flare.

Sometimes, instead of using m_s or D, it is desirable to use the source size, s and the NA of the objective that is filled, since these may be known. In this case the optical medium is air ($n = 1.0$) and $NA = $ sine u = sine AOB (Figure A.3.C-2). In our formula tangent u is used. To obtain tan u merely refer to a table of natural trigonometric functions and note the value of the tangent corresponding to the sine that is equal to the objective NA. The equations can be brought to the form

$$\frac{f_2 Q_1 s}{f_1} = \frac{f_2 \cdot s}{\alpha_L} = 2f_2 \cdot Q_3 \cdot \tan u + \frac{s Q_1 Q_3}{Q_1 + Q_3}. \qquad \text{(C.3-6)}$$

Here f_2, f_1, and either Q_1 or Q_3 should be known.

This equation would normally be used to determine Q_3, the illuminated field, as in a predetermination of the flare involved with excessive field size. With this equation the value of Q_3 may be in error by as much as 10%.

Equations

The nomenclature will be found on the page of reference.

Simple Microscope: Case 2a

$$f_1 = \frac{\alpha_0 \cdot s \cdot f_2(m+1)}{(f_2 + \alpha_0 \cdot s)m}.$$ (A.3-1) 511

$$f_2 = \frac{m\alpha_0 s f_1}{\alpha_0 s(m+1) - mf_1}.$$ (A.3-2) 511

$$\alpha_0 s = \frac{mf_1 \cdot f_2}{f_2(m+1) - mf_1}.$$ (A.3-3) 511

$$mf = \alpha_0 \cdot s.$$ (A.3-4) 512

Simple Microscope: Case 7

$$A_L = \frac{1}{\alpha_L} = \frac{Q_1}{f_1} = Q_3\left[\frac{m}{\alpha_0 \cdot s \cdot (m+1)} + \frac{1}{f_2}\right].$$ (A.3B-1) 513

$$A_L \cdot s = Q_3 \cdot A_0 = \frac{Q_3}{\alpha_0}$$ (A.3B-2) 514

$$\Delta = \frac{Q_1 f_2}{Q_3} = f_1\left[\frac{m(f_2+1)}{(m+1) \cdot s \cdot \alpha_0}\right].$$ (A.3B-3) 514

$$L = f_2\left[\frac{(Q_1 + Q_3)^2}{Q_1 \cdot Q_3}\right].$$ (A.3B-4) 514

$$Q_3 = \frac{f_2}{\alpha_L \cdot (m_s + 1)}.$$ (A.3B-5) 514

Compound microscope: Kohler illumination

$$\Delta = f_1(m_s + 1).$$ (A.3C-1) 516

$$\Delta = \frac{Q_1 \cdot f_2}{Q_3}.$$ (A.3C-2) 516

$$Q_3 = \frac{Q_1}{f_1}\frac{f_2}{m_s + 1} = \frac{f_2}{\alpha_L(m_s + 1)} = \frac{A_L f_2}{m_s + 1}.$$ (A.3C-3) 516

$$Q_1 = \frac{Q_3(m_3 + 1)f_1}{f_2}.$$ (A.3C-4) 516

Compound Microscope: light source imaged *in* condenser

$$\Delta = f_1(m_s + 1) - f_2.$$ (A.3C-2') 516

$$Q_3 = \frac{Q_1 \cdot f_2}{f_1(m_s + 1) - f_2}.$$ (A.3C-3') 516

$$s(NA)_L = Q_3(NA)_0.$$ (A.3C-5) 517

Index